Lecture Notes in Artificial Intelli

T0237849

Edited by R. Goebel, J. Siekmann, and W. Wahlster

Subseries of Lecture Notes in Computer Science

Yong Gao   Nathalie Japkowicz (Eds.)

Advances in Artificial Intelligence

22nd Canadian Conference
on Artificial Intelligence, Canadian AI 2009
Kelowna, Canada, May 25-27, 2009
Proceedings

 Springer

Series Editors

Randy Goebel, University of Alberta, Edmonton, Canada
Jörg Siekmann, University of Saarland, Saarbrücken, Germany
Wolfgang Wahlster, DFKI and University of Saarland, Saarbrücken, Germany

Volume Editors

Yong Gao
University of British Columbia Okanagan
Irving K. Barber School of Arts and Sciences
Department of Computer Science
3333 University Way, Kelowna, BC V1V 1V5, Canada
E-mail: yong.gao@ubc.ca

Nathalie Japkowicz
University of Ottawa
School of Information Technology & Engineering
800 King Edward Avenue, Ottawa, ON K1N 6N5, Canada
E-mail: nat@site.uottawa.ca

Library of Congress Control Number: Applied for

CR Subject Classification (1998): I.2, I.5, J.3, H.3.1, J.5

LNCS Sublibrary: SL 7 – Artificial Intelligence

ISSN   0302-9743
ISBN-10   3-642-01817-3 Springer Berlin Heidelberg New York
ISBN-13   978-3-642-01817-6 Springer Berlin Heidelberg New York

springer.com

© Springer-Verlag Berlin Heidelberg 2009
Printed in Germany

Typesetting: Camera-ready by author, data conversion by Scientific Publishing Services, Chennai, India
Printed on acid-free paper   SPIN: 12683819   06/3180   5 4 3 2 1 0

Preface

This volume contains the papers presented at the 22nd Canadian Conference on Artificial Intelligence (AI 2009). The conference was held in Kelowna, British Columbia, during May 25-27, 2009, and was collocated with the 35th Canadian Graphical Interface Conference and the 6th Canadian Conference on Computer and Robot Vision.

The Program Committee received 63 submissions from across Canada and around the world. Each submission was assigned to three reviewers. The Program Committee selected 15 regular papers each of which was allocated 12 pages in the proceedings and a 30-minute presentation at the conference. We accepted 21 short papers among which two were withdrawn by the authors. Each short paper was allocated 4 pages in the proceedings, and was presented as a poster plus a 15-minute brief talk. Also included in the proceedings are eight student abstracts presented at the Graduate Student Symposium.

The conference program featured three keynote presentations by Evgeniy Gabrilovich (Yahoo Research), Alan Mackworth (University of British Columbia), and Jonathan Schaeffer (University of Alberta). The one-page abstracts of their talks are also included in the proceedings.

Two pre-conference symposiums, each with their own proceedings, were held on May 24, 2009. The Second Canadian Semantic Web Working Symposium was organized by Weichang Du and Harold Boley. The International Symposium on Teaching AI in Computing and Information Technology was organized by Sajid Hussain and Danny Silver.

This conference would not have been possible without the hard work of many people. We would like to thank all Program Committee members and external referees for their effort in providing high-quality reviews on time. We thank all authors who submitted their work to this conference. Many thanks to Maria Fernanda Caropreso and Svetlana Kiritchenko for their effort in putting together an excellent program for the Graduate Student Symposium.

We are in debt to Andrei Voronkov for developing the EasyChair conference management system and making it freely available to the academic world.

The conference was sponsored by the Canadian Artificial Intelligence Association (CAIAC), and we thank the CAIAC Executive Committee for the constant support. We would like to express our gratitude to Yves Lucet, the AI/GI/CRV General Chair, and the local Organizing Chairs for their excellent work that made the three joint events AI 2009, GI 2009, and CVR 2009 an enjoyable experience. We also thank Jan Paseska for helping us collect the copyright forms.

March 2009

Yong Gao
Nathalie Japkowicz

Organization

Program Committee Chairs

Yong Gao   University of British Columbia Okanagan,
                              Canada
Nathalie Japkowicz   University of Ottawa, Canada

Graduate Student Symposium Chairs

Maria Fernanda Caropreso   University of Ottawa
Svetlana Kiritchenko   National Research Council of Canada

Program Committee

Esma Aimeur   University of Montreal
Ebrahim Bagheri   University of New Brunswick
Sabine Bergler   Concordia University
Cory Butz   University of Regina
Brahim Chaib-draa   Laval University
Yllias Chali   University of Lethbridge
Joseph Culberson   University of Alberta
Lyne Da Sylva   Université de Montréal
Christopher Drummond   National Research Council of Canada
Larbi Esmahi   Athabasca University
Atefeh Farzindar   NLP Technologies
Michael Fleming   University of New Brunswick
Robert Hilderman   University of Regina
Graeme Hirst   University of Toronto
Diana Inkpen   University of Ottawa
Igor Jurisica   University of Toronto
Ziad Kobti   University of Windsor
Grzegorz Kondrak   University of Alberta
Vlado Keselj   Dalhousie University
Philippe Langlais   University of Montreal
Guy Lapalme   Université de Montréal
Kate Larson   University of Waterloo
Guohui Lin   University of Alberta
Alejandro Lopez-Ortiz   University of Waterloo
Choh Man Teng   Institute for Human and Machine Cognition,
                              USA
Joel Martin   National Research Council of Canada

Evangelos Milios	Dalhousie University
David Nadeau	National Research Council of Canada
Gerald Penn	University of Toronto
Fred Popowich	Simon Fraser University
Doina Precup	McGill University
Mohak Shah	McGill University
Michael Shepherd	Dalhousie University
Daniel L. Silver	Acadia University
Marina Sokolova	Université de Montréal
Stan Szpakowicz	University of Ottawa
Thomas Tran	University of Ottawa
Andr Trudel	Acadia University
Marcel Turcotte	University of Ottawa
Peter van Beek	University of Waterloo
Herna Viktor	University of Ottawa
Shaojun Wang	Wright State University, USA
Ren Witte	Concordia University
Yang Xiang	University of Guelph
Ke Xu	Beihang University, PRC
Yiyu Yao	University of Regina
Jia-Huai You	University of Alberta
Xiaokun Zhang	Athabasca University
Nur Zincir-Heywood	Dalhousie University

External Referees

William Bares	Marek Lipczak
Yongxi Cheng	Guohua Liu
Sadrul Chowdhury	Emilio Neri
Fida Kamal Dankar	Maxim Roy
Qing Dou	Mahdi Shafiei
Oana Frunza	Kaile Su
Ashley George	Yuefei Sui
Yeming Hu	Milan Tofiloski
Sittichai Jiampojamarn	Yisong Wang
Fazel Keshtkar	Ozge Yeloglu
William Klement	Reza Zafarani

AI/GI/CRV General Chair

Yves Lucet	University of British Columbia Okanagan

Local Organization Chairs

Patricia Lasserre	University of British Columbia Okanagan
Ramon Lawrence	University of British Columbia Okanagan

Sponsoring Institutions

Canadian Artificial Intelligence Association/Association pour l'intelligence artificielle au Canada

Table of Contents

Invited Talks

Regular Papers

Short Papers

Graduate Student Symposium

AI in Web Advertising: Picking the Right Ad Ten Thousand Times a Second

Evgeniy Gabrilovich

Yahoo! Research
701 First Ave, Sunnyvale, CA 94089, USA
gabr@yahoo-inc.com
http://research.yahoo.com/Evgeniy Gabrilovich

Abstract. Online advertising is the primary economic force behind many Internet services ranging from major Web search engines to obscure blogs. A successful advertising campaign should be integral to the user experience and relevant to their information needs as well as economically worthwhile to the advertiser and the publisher. This talk will cover some of the methods and challenges of computational advertising, a new scientific discipline that studies advertising on the Internet. At first approximation, and ignoring the economic factors above, finding user-relevant ads can be reduced to conventional information retrieval. However, since both queries and ads are quite short, it is essential to augment the matching process with external knowledge. We demonstrate how to enrich query representation using Web search results, and thus use the Web as a repository of relevant query-specific knowledge. We will discuss how computational advertising benefits from research in many AI areas such as machine learning, machine translation, and text summarization, and also survey some of the new problems it poses in natural language generation, named entity recognition, and user modeling.

Y. Gao and N. Japkowicz (Eds.): Canadian AI 2009, LNAI 5549, p. 1, 2009.

Living with Constraints

Alan K. Mackworth*

Department of Computer Science
University of British Columbia
Vancouver, B.C., Canada V6T 1Z4

Abstract. In order to thrive, an agent must satisfy dynamic constraints deriving from four sources: its internal structure, its goals and preferences, its external environment and the coupling between its internal and external worlds. The life of any agent who does not respect those constraints will be out of balance. Based on this framing of the problem of agent design, I shall give four perspectives on the theme of living with constraints, beginning with a theory of constraint-based agent design and a corresponding experiment in robot architecture. Second, I shall touch briefly on a personal historical note, having lived with the evolving concept of the pivotal role of constraints throughout my research life. Third, I shall outline our work on the design of two assistive technology prototypes for people with physical and mental disabilities, who are living with significant additional constraints. Finally, I shall suggest our collective failure to recognize, satisfy and live with various constraints could explain why many of the worlds we live in seem to be out of kilter. This approach hints at ways to restore the balance. Some of the work discussed is joint with Jim Little, Alex Mihailidis, Pinar Muyan-Ozçelik, Robert St-Aubin, Pooja Viswanathan, Suling Yang, and Ying Zhang.

* Canada Research Chair in Artificial Intelligence.

Y. Gao and N. Japkowicz (Eds.): Canadian AI 2009, LNAI 5549, p. 2, 2009.

Computer (and Human) Perfection at Checkers

Jonathan Schaeffer

Department of Computing Science
University of Alberta

Abstract. In 1989 the Chinook project began with the goal of winning the human World Checkers Championship. There was an imposing obstacle to success ?the human champion, Marion Tinsley. Tinsley was as close to perfection at the game as was humanly possible. To be better than Tinsley meant that the computer had to be perfect. In effect, one had to solve checkers. Little did we know that our quest would take 18 years to complete. What started out as a research project quickly became a personal quest and an emotional roller coaster. In this talk, the creator of Chinook tells the story of the quest for computer perfection at the game of checkers.

Decision Tree Learning Using a Bayesian Approach at Each Node*

Mirela Andronescu[1] and Mark Brodie[2]

[1] University of Washington, Seattle WA 98195 USA
andrones@u.washington.edu
[2], Simpson College, Indianola IA 50125 USA
mark.brodie@simpson.edu

Abstract. We explore the problem of learning decision trees using a Bayesian approach, called TREBBLE (TRE building by Bayesian LEarning), in which a population of decision trees is generated by constructing trees using probability distributions at each node. Predictions are made either by using Bayesian Model Averaging to combine information from all the trees (TREBBLE-BMA) or by using the single most likely tree (TREBBLE-MAP), depending on what is appropriate for the particular application domain. We show on benchmark data sets that this method is more accurate than the traditional decision tree learning algorithm C4.5 and is as accurate as the Bayesian method SimTree while being much simpler to understand and implement.

In many application domains, such as help-desks and medical diagnoses, a decision tree needs to be learned from a prior tree (provided by an expert) and some (usually small) amount of training data. We show how TREBBLE-MAP can be used to learn a single tree that performs better than using either the prior tree or the training data alone.

1 Introduction

Decision tree learning is a well studied problem in machine learning, with application to various domains, such as help desk systems and medical diagnoses. In traditional decision tree learning, a training set is used to construct a decision tree [11]. However, in many applications it may be prohibitively expensive, or even impossible, to obtain sufficient training data to learn a good tree, but relatively easy for a human expert to provide an initial tree, based on their knowledge and experience. In such situations the challenge is to learn a decision tree using both the expert's prior tree and the training data that is available, since the expert may possess knowledge not reflected in the data (particularly if the data set is small). In this work, we describe and test a new approach which combines the prior tree with training data using a probabilistic (Bayesian) approach at each node to build decision trees that are more accurate than using either the prior tree or the training data alone. If no prior tree exists, the method

---

* This paper is based on work at IBM T.J. Watson Research Center, Hawthorne NY 10532 USA.

Y. Gao and N. Japkowicz (Eds.): Canadian AI 2009, LNAI 5549, pp. 4–15, 2009.

can be used to build decision trees from training data alone; it outperforms traditional algorithms such as C4.5 [11] and is as accurate as, and much simpler than, Bayesian methods like SimTree [15,16].

Problem Statement

We define the problem of "decision tree learning with a prior tree" as follows.

- Given:
    1. A set of n attributes $A = \{a_1, a_2, \ldots, a_n\}$, each attribute a_i having n_i possible values (or outcomes);
    2. A set of m classes (or labels) $C = \{c_1, c_2, \ldots, c_m\}$;
    3. A set of N data points $D = \{d_1, \ldots, d_N\}$, where each data point consists of the values for all attributes in A and the class in C;
    4. A decision tree T^0 provided by an expert, and an "expert level" parameter λ representing our level of confidence in this tree.
       Each internal node of T^0 is an attribute in A, though all the attributes need not appear in T^0. Each attribute node a_i has n_i child nodes, and each leaf node contains a probability distribution over the classes in C. (This distribution may be a single class with probability 1.) The expert may also provide the branch probabilities at each internal node.
- Output either:
    1. A single best decision tree, to be used for prediction (classification);
    2. Multiple decision trees, each with an associated probability. Predictions are made by averaging the predictions of the different trees.

If no prior tree is available, we obtain the traditional problem of learning a decision tree from training data. This can be regarded as a degenerate case of the above problem. By assuming a uniform prior the method we develop is also applicable to this case.

2   Algorithm

The algorithm we propose starts from the empty tree and adds one node at a time, deciding what to insert at each node by using a probability distribution over the attributes and classes which could be inserted. We may either pick the most probable attribute or class or sample from the distribution.

For each node L being added to the tree, we need a discrete probability distribution $\{p_1, \ldots, p_{n+m}\}$, where p_i, $i = 1, \ldots, n$ denotes the probability of attribute a_i being questioned at L and p_i, $i = n + 1, \ldots, n + m$ denotes the probability of assigning label c_i at L; thus $p_i \geq 0 \ \forall i$ and $\sum_{i=1}^{n+m} p_i = 1$.

Prior: We use the prior tree T^0 and the partial tree T constructed so far to generate a prior distribution $P(L|T, T^0)$ for the attribute or class to be inserted at node L. The details of how this is done are described in Section 2.2.

Likelihood: The training data D provides us with another 'opinion' about what should be inserted at node L. We use the information gain of the attributes

Procedure **TREBBLE** (**TRE**e Building by **B**ayesian **LE**arning)

> **Input:** Data set D, prior tree T^0, expert level λ;
> **Output:** Posterior tree T;
> **Algorithm:**
>> T = Tree with one 'undefined' node;
>> For each 'undefined' node L of T:
>>> Set $\mathcal{A} = \{$all attributes not appearing in the parent path in T, all classes$\}$;
>>> Step 1: Prior = $P(L|T,T^0)=$**Compute Prior**$(L,T^0,T,\mathcal{A},\lambda)$;
>>> Step 2: Likelihood = $P(D|L,T)=$**Compute Likelihood**$(L,D,T,\mathcal{A})$;
>>> Step 3: Posterior = $P(L|D,T,T^0) \propto P(D|L,T)P(L|T,T^0)$;
>>> Step 4: 'Define' node L: $L \sim$ Posterior (or pick MAP), where $L \in \mathcal{A}$;
>>> If L is an attribute node:
>>>> For each outcome of L, add branches and 'undefined' leaf nodes to T;
>>> Else: add a leaf node, containing posterior probabilities for each class;
>> Return T.

Procedure 1: Pseudocode for the TREBBLE algorithm. At each node, an attribute or a probability distribution over classes is inserted in the tree. If Step 4 is performed by picking the most likely element of $\mathcal{A}$, the procedure generates a single tree and is called TREBBLE-MAP. If Step 4 samples from the Posterior, the procedure generates a family of trees whose predictions are averaged at classification time, and is called TREBBLE-BMA.

and the frequencies of the classes in the data to define a likelihood distribution $P(D|L,T)$ that reflects how likely we would be to see this data given the partial tree T constructed so far and a choice for the attribute or class at L. The details of how this is done are described in Section 2.3.

Posterior: Given these prior and likelihood distributions, we define the posterior distribution for which attribute or class to insert at the current node in the usual way, by Bayes rule: Posterior $\propto$ Likelihood $\times$ Prior, or $P(L|D,T,T^0) \propto P(D|L,T)P(L|T,T^0)$ (we assume the data D is independent of the prior tree T^0). This posterior distribution is the desired probability distribution $\{p_1,\ldots,p_{n+m}\}$ described above.

It is convenient to use a Dirichlet distribution $Dirichlet(\mathbf{p}|\alpha) \propto \prod_{i=1}^{n+m} p_i^{\alpha_i-1}$ for the prior and a multinomial distribution $Multinomial(\mathbf{k}|\mathbf{p}) \propto \prod_{i=1}^{n+m} p_i^{k_i}$ for the likelihood, since then the posterior can be obtained in closed form as a Dirichlet distribution $Dirichlet(\mathbf{p}|\alpha + \mathbf{k}) \propto \prod_{i=1}^{n+m} p_i^{\alpha_i+k_i-1}$.

We call the approach TREBBLE (**TRE**e **B**uilding by **B**ayesian **LE**arning, see Procedure 1). If the desired output is a single tree, we choose at each node the attribute with the maximum-a-posteriori (MAP) value. If a class is chosen, the leaf node is assigned probabilities for all classes. We call this algorithm TREBBLE-MAP.

If a single tree is not necessary, a population of trees can be generated by repeatedly sampling from this distribution. To classify a new example, we use Bayesian Model Averaging, as follows: the probability of a tree T is given by: $P(T|D,T^0) = P(L_0|D,T^0) \prod_{i=2}^{|T|} P(L_i|parent(L_i),D,T^0)$, where L_0 is the root node of T, $|T|$ is the number of nodes in T, and $parent(L_i)$ is the parent of node L_i in T.

For classification of a new data point x, we sample S independent trees in order to approximate the posterior class probability $P(y|x, D) \approx \frac{1}{S} \sum_{i=1}^{S} P(y|T_i, x, D)$ and we choose the most likely class - we call this TREBBLE-BMA.

2.1   Example of Decision Tree Learning with a Prior Tree

As an example, we apply TREBBLE-MAP to the task of classifying contact-lens type, based on Tear-rate, Sightedness, and Eye-shape. We are given a prior tree and a training set of labelled data, some of which is shown in Figure 1.

Consider first which node should be inserted at the root (Iteration 1). The expert's prior tree has Tear-rate at the root, so the prior distribution should have a strong preference for this choice. The likelihood distribution is computed using the information gain of each attribute in the training data. In this example the data set yields a strong preference for Eye-shape at the root; the expert's choice of Tear-rate is only weakly supported by the data. The relative weight of the prior and likelihood depends on our confidence in the expert's tree versus how much data we have. In this example the strong prior preference for Tear-rate outweighs the strong data preference for Eye-shape, and so Tear-rate is chosen at the root. Branches for each outcome of Tear-rate are added to the tree, and the process continues.

At Iteration 2, Eye-shape (suggested by the data) is inserted. As a consequence, since the prior tree had Sightedness for that node, we need to restructure the prior tree at Iteration 3 (otherwise we would not know what the prior tree preference is for the next nodes). Note that the restructuring step increases the prior tree size. For this example, it takes seven iterations to obtain the final posterior tree with TREBBLE-MAP.

2.2   The Prior Probability Distribution

We now explain in detail how the prior probability distribution $P(L|T, T^0)$ for the attribute or class which should be inserted at the current node L of the tree is defined, as a function of the prior tree T^0 and the partial tree T constructed so far (see Procedure 2). Recall that p_i, $i = 1, ..., n + m$, denotes $P(L = \{a_1, ..., a_n, c_1, ..., c_m\})$.

The prior distribution is defined using a Dirichlet distribution, given by $Dirichlet(\mathbf{p}|\alpha) \propto \prod_{i=1}^{n+m} p_i^{\alpha_i - 1}$. To define the prior distribution we need to estimate α_i for each attribute and class that could be chosen at L; α_i represents the 'weight' of attribute or class i, according to the prior tree, and therefore depends on our level of confidence λ. Although we do not assume that the prior tree was generated from training data, if this were the case we can think of λ as representing the amount of such data.

If there is no prior tree, we assume a uniform prior; i.e. $\alpha_i = 1\ \forall i$. If there is a prior tree T^0, then computing the prior distribution at L requires that the partial **posterior** tree T is the same as the **prior** tree T^0 up to the level of node L. If

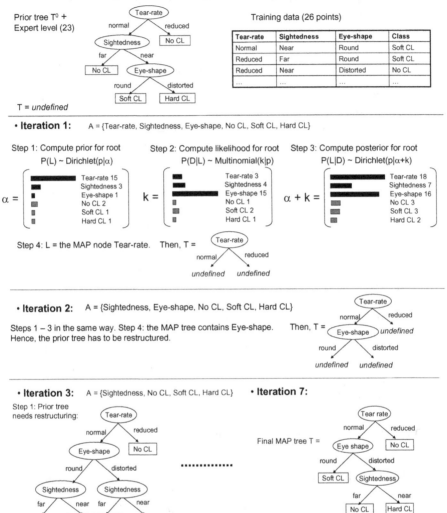

Fig. 1. Example of applying TREBBLE-MAP to a contact lenses (CL) application where a prior tree and training data are given. The possible attributes are Tear-rate, Sightedness and Eye-shape. The possible classes are No Contact Lenses (No CL), Soft Contact Lenses (Soft CL) and Hard Contact Lenses (Hard CL). The example shows in detail our probabilistic approach at the root level. It also shows how the prior tree gets restructured if a different node is selected.

Procedure **Compute Prior**

> **Input:** Current node L, prior tree T^0, partial posterior tree T,
> attributes and classes $\mathcal{A}$, strength λ;
> **Output:** Prior distribution for next node L to be added to posterior tree T;
> **Algorithm:**
> Compute the expert level (prior strength) $\lambda(L)$ at node L;
> Restructure prior subtree if needed, and obtain T_s^0;
> For each attribute or class k in set $\mathcal{A}$, compute $\alpha(k)$;
> Normalize all $\alpha(k)$ values to sum up to $\lambda(L)$;
> Return Dirichlet(vector α).

Procedure 2: The prior distribution of the attributes and classes for the current node is a Dirichlet distribution, with hyperparameters α. This procedure describes how we obtain the parameters α. The λ value at node L is the given expert level if L is the root node; otherwise it is the parent's λ value multipled by the probability of the branch. If the previous node was not the node suggested by the prior tree, then the prior tree needs to be restructured.

this is not the case, T^0 has to be restructured so that the parent of the node L is the same in T^0 as in T, as illustrated by the example in Figure 1. We adopt the simple restructuring algorithm used by Utgoff [14]. A more efficient restructuring algorithm could be implemented, as described by Kalles and Morris [10].

After restructuring, we have a subtree T_s^0 of T^0 whose root node represents the prior tree's 'opinion' about which attribute or class should be inserted at node L. Recall that λ represents the expert's 'level of expertise' or experience. Our confidence in the prior tree's 'opinion' at node L, denoted by $\lambda(L)$, reflects the amount of "virtual" data at that level, and depends on the likelihood of following the path from the root to L. Thus, $\lambda(root) = \lambda$ and $\lambda(L) = \lambda(p(L))$ $\times$ P(branch from p(L) to L), where p(L) is the parent node of L. If the branch probabilities are not provided by the expert, they can be computed from the data used to generate the prior tree (if available), estimated from the training data D, or assumed uniform.

2.3 The Likelihood Distribution

We now explain in detail how the likelihood distribution $P(D|L, T)$ is defined, using the multinomial distribution $Multinomial(\mathbf{k}|\mathbf{p}) \propto \prod_{i=1}^{n+m} p_i^{k_i}$ (see Procedure 3). Recall that p_i, $i = 1, ..., n + m$, denotes $P(L = \{a_1, ..., a_n, c_1, ..., c_m\})$. We can think of k_i as meaning the 'weight' of attribute or class i, according to the training data D.

The partial tree T constructed so far defines a path from the root to L; this path specifies particular values for some of the attributes. Let $D(L)$ be the subset of D whose attribute values match the values along the path from the root to L.

If $D(L)$ is empty, we set $k_i = 0$ $\forall i$. Otherwise, compute the information gain $G(a_i)$ of each attribute, for $i = 1$ to n, and the frequency $F(c_{i-n})$ of each class, for $i = n + 1$ to $n + m$. The weights for each attribute and class are:

Procedure **Compute Likelihood**

> **Input:** Current node L, data D, partial posterior tree T, attributes and classes $\mathcal{A}$;
> **Output:** Likelihood for the next node to be inserted into T;
> **Algorithm:**
>> Select the data subset $D(L)$, to satisfy the attribute values of the parent path;
>> Compute weight β of attributes versus classes, depending on class entropy;
>> For each attribute item a_k from $\mathcal{A}$:
>>> Compute information gain $G(a_k)$;
>>> Compute 'attribute number of occurences' $k(a_k) = \beta \times |D(L)| \times \frac{G(a_k)}{\sum G(a_k)}$;
>> For each class item c_k from $\mathcal{A}$:
>>> Compute class frequency $F(c_k)$;
>>> Compute 'class number of occurences' $k(c_k) = (1 - \beta) \times |D(L)| \times \frac{F(c_k)}{\sum F(c_k)}$;
>> Return Multinomial (vector k);

Procedure 3: The likelihood is a multinomial distribution whose parameters 'simulate' the number of times each attribute or class is seen. This value is proportional to the information gain for attributes, and to class frequency for each particular class. The weight between attributes and classes is represented by a value β, which depends on the class entropy (see text).

$$k_i = \frac{\beta |D(L)| G(a_i)}{\sum_{j=1}^{n} G(a_j)}, i = \{1, \ldots, n\},$$

$$k_i = \frac{(1-\beta)|D(L)| F(c_{i-n})}{\sum_{j=1}^{m} F(c_j)}, i = \{n+1, \ldots, n+m\}.$$

The k_i values sum up to $|D(L)|$, the number of data points in $D(L)$.

The parameter $\beta \in [0, 1]$ controls the relative importance of attributes versus classes. One choice for β is the class-entropy $\sum_{j=1}^{m} F(c_j) \log F(c_j)$, normalized to be between 0 and 1. If $\beta = 1$ all the probability mass is assigned to the attributes, in proportion to their information gain. If $\beta = 0$, then there is only one class with non-zero frequency, so this class should be chosen. Note that β in effect performs pruning, because even if some attribute has non-zero information gain, it will not necessarily be chosen for the current node if its information gain is outweighed by the relative frequency of a class.

3   Related Work

Traditional decision tree algorithms, such as C4.5 [11], use training data as input and greedily construct a tree, which is then post-pruned to avoid overfitting and to decrease the tree size. The greedy construction involves adding the most informative attribute to the tree, such that the number of attributes tested is minimized, while the classification accuracy is maximized. Such algorithms do not deal with a prior tree given as input.

Incremental decision tree learning algorithms [14] use the training data sequentially. Given some training data, a tree is constructed in the traditional way. Statistics about this first batch of training data are kept, although the training data itself is not used again. When new training data becomes available, the current tree and the statistics are updated. In our case we do not have any statistics about the data from which the prior tree was constructed (other than, perhaps, branch and class probabilities).

Random forest algorithms [5,9] are similar to our TREBBLE-BMA algorithm in that they generate many decision trees and pick the class that is voted by the most number of trees. However, there are several key differences: every tree generated by a random forest uses a traning set obtained by bootstrapping the original data set (therefore is not applicable to small training sets), whereas our algorithm samples nodes according to distributions given by a fixed training data set. Again, random forests do not deal with prior trees as input.

Bayesian decision tree learning methods [6,7,15,16] usually start from some prior tree, chosen from some distribution. Random walks or sophisticated Markov Chain Monte Carlo methods are used to obtain a family of trees that fits the training data. Our approach starts from a fixed prior tree and, by adopting a Bayesian approach at the node-level instead of at the tree level, avoids the necessity of performing long Markov Chain Monte Carlo simulations.

4   Experimental Results

In this section, we present experimental results of our TREBBLE algorithms (implemented in Java). We compare our results against results obtained with the C4.5 program version 8.3 [11] and SimTree [15,16] (both with default options).

First, we present a detailed analysis of the Wisconsin Breast Cancer data set [3] from the UCI machine learning repository [2]. At the end of this section, we test our algorithms on five additional data sets (see Table 1. We report the *misclassification rate* (i.e. the percentage of misclassified test data points), averaged over 10 random partitionings of the data into the specified prior, training and test set sizes.

4.1   Detailed Analysis of the Wisconsin Breast Cancer Data Set

Results when no prior tree is given. First, we assume no prior tree is given. We ran TREBBLE-MAP and TREBBLE-BMA (averaging over 500 trees) with uniform priors, C4.5 (the pruned version), and SimTree with default prior (averaging over 500 trees, with 500 "burn-in" trees, i.e., trees before the SimTree's Markov Chain Monte Carlo algorithm converges). Figure 2 shows the results for various training set sizes (the test set size is 250).

On these sets, TREBBLE-MAP obtains similar accuracy as C4.5, within 3% either better or worse for different training set sizes. This result is expected, since both algorithms construct the tree one node at a time by greedily picking the attribute with the highest information gain. The difference consists in the "pruning" method.

It has been previously shown [6,7,15,16] that Bayesian approaches give better misclassification rate than "point estimate" approaches. Our Bayesian version

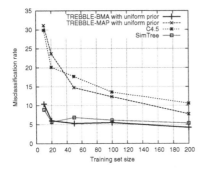
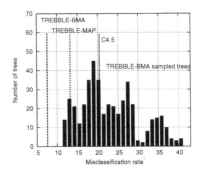

Fig. 2. Misclassification rate measured on test sets of 250 points. **Left:** For TREBBLE-BMA, we sampled 500 trees. For SimTree, we sampled 500 trees, and we considered 500 trees were "burn-in" trees. **Right:** A histogram of the misclassification rate for the 500 TREBBLE-BMA trees, for one split with training set size 100.

TREBBLE-BMA (with uniform prior) also performs much better than C4.5 and TREBBLE-MAP (see the left plot of Figure 2). TREBBLE-BMA performs comparably with SimTree (even slightly better), while being conceptually much simpler, since there is no need for Markov Chain mixing and tree moves. The right plot of Figure 2 shows that each sampled tree has high misclassification rate, however the average prediction is very accurate, because the errors the trees make are somewhat independent.

In addition, TREBBLE-BMA runs faster than SimTree: it took TREBBLE-BMA 7.6 CPU minutes to sample 500 trees on a training set of 100 data points, whereas SimTree took 17.6 minutes (we used an Intel Xeon CPU with 2.4GHz, 512 KB cache and 1GB RAM, running Linux 2.6.16).

Results when a prior tree is given, and the desired output is one single tree. Next, we assume we have a prior decision tree, given by an expert, and the desired output is one single tree. For example in help-desk applications, the questions asked should follow one single tree.

In order to evaluate our algorithms when a prior tree is given by an expert, we use C4.5 (with pruning) to create the prior tree from the prior data. This allows us to set the expert level λ in accordance with the amount of prior data. We also set the class and branch probabilities according to the data.

Figure 3 shows the results. TREBBLE-MAP with the expert prior is more accurate than the prior tree alone and the C4.5 tree obtained from the training data alone.

In addition, TREBBLE-MAP with the expert prior is slightly more accurate than the naive approach where we artificially generate data from the prior tree, we add the training data, and we use C4.5. This is expected, since sampling artificial data from the prior tree adds a fair amount of noise to the data.

Using the naive approach may have other disadvantages as well. First, we expect a large variation of accuracy for different samplings, especially when the prior tree strength is small and the training set size is small. Second, the resulting

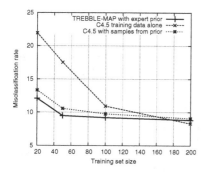 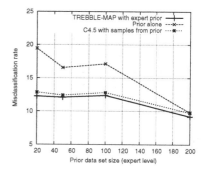

Fig. 3. Performance of TREBBLE-MAP when a prior tree is given by an expert. **Left**: Prior tree strength (expert level) is fixed to 200, while the training set size varies from 20 to 200 data points. The misclassification rate of the prior tree varies from 12% to 9%. **Right**: Training set size is fixed to 100 data points, while the prior tree strength varies from 20 to 200. The misclassification rate of the training data, as given by the C4.5 tree, varies between 12% and 8%.

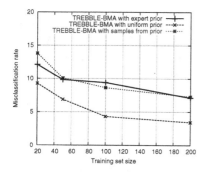 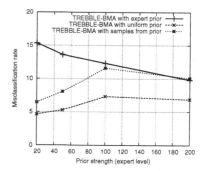

Fig. 4. Misclassification rates of the BMA approaches when a prior tree is given. **Left**: Prior data set size is fixed at 200. **Right**: Training set size is fixed at 50.

trees may differ substantially from the prior tree, because the prior tree structure is not used at all in this approach. In contrast, TREBBLE-MAP keeps consistent with the prior tree unless there is strong evidence from the training data to use a different attribute or class for a specific node.

These results show that TREBBLE-MAP improves accuracy as compared to the prior tree alone or the training data alone, while being more robust than a naive approach, and more consistent with the prior tree given.

Results when prior tree is given, and the desired output is many trees. If the application domain permits averaging the prediction results given by many trees, then it is interesting to see whether Bayesian Model Averaging also gives better results when a prior tree is given.

Table 1. Misclassification rates of our TREBBLE algorithms and C4.5 on six data sets from the UC Irvine repository. The total number of instances (shown in column 2) was randomly split in three equal sizes, for the prior tree, training and testing. We have performed ten splits, and averaged the results. Columns 3 and 4 give the number of attributes and the average number of outcomes for each attribute. All data sets have two classes. For TREBBLE-BMA we sampled 100 trees.

Data set	#inst N	#attr n	avg #outc $avg(n_i)$	Training only TREBBLE-BMA	C4.5	Training + prior TREBBLE-MAP	Prior only
Balance scale [13]	576	4	5	**16.20**	27.97	28.23	29.90
Breast cancer [3]	681	10	10	**3.35**	7.89	*6.48*	7.88
Car evaluation [4]	1593	6	3.5	**6.74**	8.83	*7.66*	8.36
Congressional votes [12]	231	16	2	5.32	4.68	*3.25*	4.03
Hayes-Roth [1]	129	4	3.75	**27.67**	34.88	*31.63*	33.95
Mushroom [8]	5643	22	5.68	**0.46**	0.58	1.89	0.97

We compare TREBBLE-BMA with the given prior tree, TREBBLE-BMA with uniform prior and TREBBLE-BMA with uniform prior and where the "training set" is composed of the original training set plus artificial data sampled from the prior tree (alternatively, one can use SimTree to sample trees from this data). Figure 4 shows the results. In the left plot, we keep the prior tree strength fixed to 200, and vary the training set size. Surprisingly, TREBBLE-BMA with the expert prior performs worse than when the prior tree is not used. The reason is that, since the given prior is only one tree, it restricts the posterior sampled trees to be similar to the prior tree. Thus, most of the posterior trees make the same mistakes, while when no prior tree is used, posterior trees are wrong on different test points (see the right plot of Figure 2).

What is interesting however is that, when the number of sampled trees is small (e.g. five), TREBBLE-BMA with the expert prior has much better accuracy than when a uniform prior is used (e.g. 7-8% versus 15-20% on a prior size of 200 and training set size of 100).

TREBBLE-BMA with the naive approach always performs worse than when no prior is used. In the left plot it performs similarly with TREBBLE-BMA with the expert tree, because the prior strength is high, and so the induced noise is high. The right plot also shows the misclassification rate of the naive approach goes up when the prior strength increases, since the noise is higher.

4.2   Results on Other Data Sets

Table 1 shows the results of C4.5 and TREBBLE on six data sets (we could not sucessfully run SimTree on any data sets other than the Wisconsin breast cancer data set). In five out of six cases, TREBBLE-BMA performs better than C4.5 (see the bold values), and in four out of six cases, TREBBLE-MAP with prior performs better than both C4.5 trained on the training data alone, and the prior tree alone (see the italic values). These results demonstrate that our TREBBLE algorithms are successful on data sets with various characteristics.

5   Conclusions

In this work, we propose a probabilistic algorithm to solve the problem of decision tree learning when a prior tree (or uniform prior) and training data are given. Our strategy is to iteratively construct a tree by inserting one attribute or class node at a time, picked from a probability distribution which is a combination of two derived probability distributions: one coming from the prior tree, and one coming from the training data. When no prior tree is available, our Bayesian averaging approach TREBBLE-BMA performs slightly better than the state-of-the-art Bayesian CART approach, while our approach is much simpler and faster to run. When a prior expert tree is available, our algorithms typically perform better than the prior tree alone and the training data alone.

References

1. Anderson, J., Kline, P.: A learning system and its psychological implications. In: Proc. of the Sixth International Joint Conference on Artificial Intelligence, pp. 16–21 (1979)
2. Asuncion, A., Newman, D.: UCI machine learning repository (2007)
3. Bennett, K.P., Mangasarian, O.L.: Robust linear programming discrimination of two linearly inseparable sets. Optimization Methods and Software 1, 23–34 (1992)
4. Bohanec, M., Rajkovic, V.: Knowledge acquisition and explanation for multi-attribute decision making. In: Proc. of the 8th Intl. Workshop on Expert Systems and their Applications, pp. 59–78 (1988)
5. Breiman, L.: Random Forests. Machine Learning 45(1), 5–32 (2001)
6. Chipman, H.A., George, E.I., McCulloch, R.E.: Bayesian CART Model Search. Journal of the American Statistical Association 93(443), 935–947 (1998)
7. Denison, D.G., Mallick, B.K., Smith, A.F.: Bayesian CART. Biometrika 85(2), 363–377 (1998)
8. Duch, W., Adamczak, R., Grabczewski, K., Ishikawa, M., Ueda, H.: Extraction of crisp logical rules using constrained backpropagation networks - comparison of two new approaches. In: Proc. of the European Symposium on Artificial Neural Networks, pp. 109–114 (1997)
9. Ho, T.K.: Random Decision Forest. In: Proc. of the 3rd Int'l. Conf. on Document Analysis and Recognition, pp. 278–282 (1995)
10. Kalles, D., Morris, T.: Efficient Incremental Induction of Decision Trees. Machine Learning 24(3), 231–242 (1996)
11. Quinlan, J.R.: C4.5: Programs for Machine Learning. Morgan Kaufmann Publishers, San Francisco (1993)
12. Schlimmer, J.C.: Concept acquisition through representational adjustment. PhD thesis (1987)
13. Shultz, T., Mareschal, D., Schmidt, W.: Modeling cognitive development on balance scale phenomena. Machine Learning 16, 59–88 (1994)
14. Utgoff, P.E.: Incremental Induction of Decision Trees. Machine Learning 4, 161–186 (1989)
15. Wu, Y.: Bayesian Tree Models. PhD thesis (2006)
16. Wu, Y., Tjelmeland, H., West, M.: Bayesian CART: Prior Specification and Posterior Simulation. Journal of Computational and Graphical Statistics 16(1), 44–66 (2007)

Generating Satisfiable SAT Instances Using Random Subgraph Isomorphism

Călin Anton[1,2] and Lane Olson[2]

[1] Grant MacEwan College, Edmonton, Alberta, Canada
[2] Augustana Faculty, University of Alberta, Camrose, Alberta, Canada

Abstract. We report preliminary empirical results on Generating Satisfiable SAT instances using a variation of the Random Subgraph Isomorphism model. The experiments show that the model exhibits an easy-hard-easy pattern of empirical hardness. For both complete and incomplete solvers the hardness of the instances at the peak seems to increase exponentially with the instance size. The hardness of the instances generated by the model appears to be comparable with that of Quasigroup with Holes instances, known to be hard for Satisfiability solvers. A handful of state of the art SAT solvers we tested have different performances with respect to each other, when applied to these instances.

1 Introduction

Satisfiability(SAT) is the prototypical problem for Artificial Intelligence. It has practical applications such as formal verification of software and hardware - which reduces to checking if a Boolean formula is satisfiable. Incomplete Satisfiability solvers, which can not prove unsatisfiability, can be very useful for these applications, as they can find a solution if one exists. There is a permanent competition between finding new, challenging Satisfiability instances and designing efficient Satisfiability solvers able to tackle the challenges. This competition has resulted in many improvements of the Satisfiability solvers in the last decade.

While there are several models for generating hard SAT instances in general, there are only a few such models for generating Satisfiable SAT instances - instances guaranteed to have solutions. Generating hard satisfiable instances has theoretical importance - related to problems in cryptography - as well as practical importance - evaluating the performance of incomplete SAT solvers.

A natural strategy is to generate a random total assignment of the variables and then add random clauses which do no contradict the initial assignment. This approach, known as solution hiding, has resulted in easy satisfiable random k-SAT instances, due to the clustering of many solutions around the initial assignment [1,2]. In order to disperse the solutions, a variation has been proposed [3] in which the clauses are rejected if they contradict the original assignment or its complement. It has been shown that this variation generates harder satisfiable instances.

The RB and RD models for random Constraint Satisfaction Problem are used in [4] to generate satisfiable instances of SAT. The solution is hidden in the

Y. Gao and N. Japkowicz (Eds.): Canadian AI 2009, LNAI 5549, pp. 16–26, 2009.

natural way, but the properties of the models insure that the expected number of solutions of the generated instances is the same as the expected number of solutions in the random case. Therefore the instances must be hard.

Quasigroup with Holes (QGWH)[1] generates satisfiable SAT instances by converting a satisfiable Quasigroup Completion instance to SAT. Quasigroup Completion asks to fill a partial Latin square. In QGWH the partial Latin square is generated by deleting some colors ("punching holes") in a Latin square. An easy-hard-easy pattern (depending on the fraction of holes) has been noticed [1] for QGWH. This is surprising as the instances are guaranteed to be satisfiable. On the other hand this offers the possibility of varying the hardness of the satisfiable instances.

Given a pair of graphs, the Subgraph Isomorphism Problem(SGI) asks if one graph is isomorphic to a subgraph of the other graph. It is an NP-complete problem with many applications in areas like pattern recognition, computer aided design and bioinformatics. Random Subgraph Isomorphism instances converted to SAT are very hard (in terms of running time) for state of the art SAT solvers. For example, none of the solvers which participated at the 2003 SAT competition could solve the hardest SAT-translated SGI instances[5]. After four years and three more competitions, the last two of these instances were eventually solved at the 2007 SAT competition, by 4, respectively 2 solvers.

In this paper we report empirical results of complete and incomplete SAT solvers on SAT translated random SGI instances generated such that the instances are satisfiable. The main findings are:

- The variation of the empirical hardness of these instances, exhibits an easy-hard-easy pattern.
- The hardness of the instances seems to increase exponentially with one of the parameters of the model and with the instance size.
- When the comparison is made in terms of number of variables, the hardest instances of the model are empirically more difficult than the hardest QGWH instances. If the number of literals is used as comparison criteria, the empirical hardness of the former is comparable to that of the latter.
- The complete and incomplete solvers have different performances on these instances. Some state of the art complete solvers which we tested on these instances had different performance with respect to each other.

2   Definitions

We consider finite, undirected graphs without multiple edges. A graph G is defined as a pair $G = (V_G, E_G)$, with V_G being the set of vertices of G, and $E_G \subset V_G X V_G$ being the set of edges of G. The number of vertices of a graph G is referred to as the order of the graph G and we will use n_G to denote it. The number of edges of a graph G is referred to as the size of the graph, and we will use m_G to denote it. A subgraph G of a graph $H = (V_H, E_H)$, is a graph (V_G, E_G) such that $V_G \subseteq V_H$, and $E_G \subseteq E_H$. If $E_G = E_H \cap V_G X V_G$ then G is an induced subgraph of H. Two graphs $G = (V_G, E_G)$ and $H = (V_H, E_H)$ are isomorphic, if and only if there exists a bijective mapping $f : V_G \to V_H$ such that for each $v_1, v_2 \in V_G$, $v_1 \neq v_2$, $(v_1, v_2) \in E_G$ if and only if $(f(v_1), f(v_2)) \in E_H$. Given two graphs $G = (V_G, E_G)$, $H = (V_H, E_H)$ the subgraph isomorphism problem (SGI) asks if G is isomorphic to a subgraph of H. If the answer is yes, then a subgraph(H_1) of H to which G is isomorphic, as well as a bijective mapping of G to H_1 which defines the isomorphism, should be provided. The induced subgraph isomorphism problem is similar to SGI but it asks for G to be an induced subgraph of H. The subgraph isomorphism problem is empirically harder than the induced subgraph isomorphism problem, as the extra constraints of the latter - forcing non-edges of G to be matched into non-eges of H - helps a search algorithm to find a solution in fewer steps.

For $n \in \mathbb{N}^*$, $m \in \{0 \ldots \binom{n}{2}\}$, the **G(n,m)** random graph model generates graphs of order n, and size m. The m edges are randomly chosen without replacement from the set of all possible $\binom{n}{2}$ edges.

For $n \leq m \in \mathbb{N}$, $q \in \{0 \ldots \binom{m}{2}\}$, $p \in (0, 1)$ a satisfiable random (n, m, p, q) SGI[1] consists of a graph H and a subgraph G. H is a **G(n,m)** random graph of order m and size q. G is obtained by the following two steps: (1)select at random an order n induced subgraph G' of H; (2) randomly remove $\lfloor pm_{G'} \rfloor$ distinct edges of G'. As SRSGI has 4 parameters, it is difficult to analyze it when all parameters are allowed to vary. In most cases three parameters are fixed and only one is allowed to vary. Usually n, m and q are fixed and p is varied, from 0 to 1. If p=0 no edge is removed from G' during the second phase of the process, and thus G is an induced subgraph of H. If $p = 1$, G has no edges and therefore it is isomorphic to any subgraph of H with order n and size 0. Hence it is conceivable that the hardness of SRSGI varies with p, as for p=0, instances are similar to induced subgraph isomorphism ones which are empirically easier, and for p=1, the instances are trivial. Based on these assumptions it is expected that the hardest instances of SRSGI will be obtained for values of p strictly between 0 and 1.

The direct encoding is used to convert SRSGI instances to SAT. For each $v_i \in V_G$ and each $w_j \in V_H$ the direct encoding defines a Boolean variable b_{ij}. The variable b_{ij} is assigned TRUE, if and only if v_i will be mapped into w_j. The direct encoding will produce a SAT instance with $n_G n_H$ variables. For each $v_i \in V_G$, a clause of the form $b_{i1} \vee b_{i2} \cdots \vee b_{in_H}$ is added, to make sure that each vertex of G is mapped into some vertices of H. There are n_G such clauses. For each $v_i \in V_G$, and each pair $(j < k) \in \{1, \ldots, n_H\}^2$ a clause of the form $\neg b_{ij} \vee \neg b_{ik}$ is added to insure that each vertex of G is mapped into at most one vertex of H. For each $i \in \{1, \ldots, n_H\}$, and each pair $(j < k) \in \{1, \ldots, n_G\}^2$ a clause of the form $\neg b_{ji} \vee \neg b_{ki}$, is added to guarantee that at most one vertex of G is mapped into any vertex of H. For each pair $(i < j) \in \{1, \ldots, n_G\}^2$ such that $(v_i, v_j) \in E_G$ and each pair $(k < l) \in \{1, \ldots, n_H\}^2$ such that $(w_k, w_l) \notin E_H$ the clauses $\neg b_{ik} \vee \neg b_{jl}$ and $\neg b_{il} \vee \neg b_{jk}$, are added to make sure that no edge of the subgraph is mapped into a non-edge of the graph.

---

[1] When no confusion arises we will use SRSGI to refer to satisfiable random (n, m, p, q) SGI.

Therefore, the direct conversion of an SGI creates a SAT instance which has $n_G n_H$ variables, n_G clauses of size n_H, containing only positive literals and $n_G n_H (n_G + n_H - 2)/2 + m_G(n_H(n_H - 1) - 2m_H)$ 2-clauses with negative literals.

A model similar to SRSGI was used in [6] for testing variations of the Survey Propagation algorithm for the Constraint Satisfaction Problem. The main differences between the two models reside in the way edges are added (removed) to (from) $H(G)$. In SRSGI a certain number of edges are randomly added/removed while in the other model each edge is added/removed with a fix probability.

3 Empirical Results

We intended to investigate the main characteristics of SAT encoded SRSGI and especially their empirical hardness. We chose 20, 21, 22, 23, 24, 26 and 28 as values for m. For each value of m we varied n from $m - 6$ to $m - 2$. Previous experiments with SRSGI indicated that for these values of m and n the hardest instances are generated when q varies between $\binom{m}{2}/2$ and $\binom{m}{2}$ and p varies between 0 and 0.4. Based on these observations we varied q between $0.65\binom{m}{2}$ and $0.80\binom{m}{2}$ in increments of $0.05\binom{m}{2}$ [2], and p between 0% and 40% in increments of 5%. For each combination of values for m, n, q and p we generated 10 SAT encoded SRSGI instances.

We used five complete solvers: BerkMin[7], minisat[8], picosat[9], Rsat[10] and march-ks[11], and two incomplete solvers adaptg2wsat+ [12] and gnovelty+ [13]. Minisat, picosat, Rsat and march-ks, are winners of different categories at the last Satisfiability Competition (2007). For our experiments we used their competition versions. Our previous experience indicates that BerkMin is very efficient for SGI instances so we decided to use it in the experiment. Adaptg2wsat+ and gnovelty+ are winners of the random satisfiable category at the 2007 SAT competition. For all solvers we set the cutoff time limit to 900 seconds. For the incomplete solvers we set the runs parameter to 10.

3.1 Empirical Hardness; Easy-Hard-Easy Pattern

In this subsection we consider m, n and q fixed and let p vary. March-ks could only solve a few difficult instances, and so, we decided not to report the results of this solver. For minisat, RSat and picosat (in this order) some $m = 24$ instances were very difficult (the solvers timed out). In several cases these solvers solved none of the hardest instances. For this reason we could not use them for instances with m larger than 24. Thus, we used only BerkMin, adaptg2wsat+ and gnovelty+ to complete the experiment as designed.

For all solvers and for all combinations of m, n and q with $n \geq m - 4$, we noticed an easy-hard-easy pattern in the variation of the empirical hardness (estimated by the running time) with p. The same pattern occurred when the number of visited nodes - for complete solvers, or steps - for incomplete solvers were plotted against p. The pattern appears clearly for the complete solvers,

---

[2] In some cases we used finer increments such as $0.01\binom{m}{2}$.

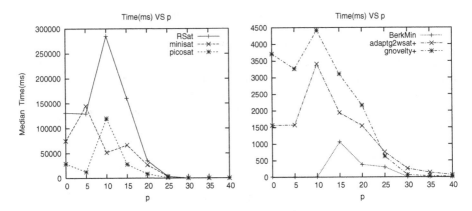

Fig. 1. Evolution of empirical hardness with p (m=22, n=18, q=161). Notice the different time scales.

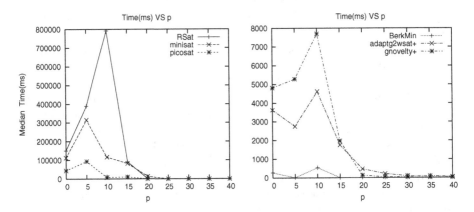

Fig. 2. Evolution of empirical hardness with p (m=24, n=18, q=193). Notice the different time scales.

while for the incomplete solvers in some cases it may appear as hard-harder-easy. For small values of n, such as $m - 6$, and large values of q, such as $0.80\binom{m}{2}$, RSat, minisat and picosat found the instances with $p = 0$ to be the hardest. Hence, in these cases they did not exhibit the easy-hard-easy pattern. We believe these "irregularities" are consequences of some form of preprocessing done by the solvers. For all values of m and large $n(n \geq m - 4)$ the location of the peak in empirical hardness is consistent among complete and incomplete solvers, see Figs. 1 and 2. With a few exceptions, the location of the hardest instances is the same for the three solvers used in the extended experiment (BerkMin, adaptg2wsat+ and gnovelty+) - see Figs. 3 and 4. We consider this to be an indication of the fact that the peak in hardness is a characteristic of the model and not an artifact of the solvers.

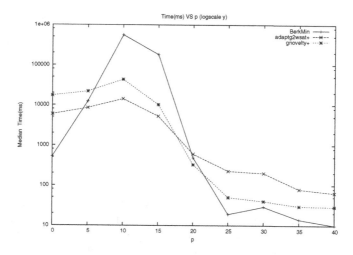

Fig. 3. Location of the hardness peak ($m=26$, $n=20$, $q=243$). Notice the logarithmic time scale.

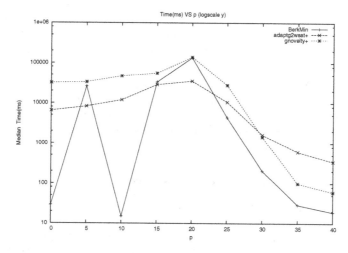

Fig. 4. Location of the hardness peak ($m=28$, $n=22$, $q=264$). Notice the logarithmic time scale.

3.2 Location of the Hardness Peak

For m and n fixed we investigated how the location of the hardness peak varies with q. The experiment shows that the location of the hardness peak is not the same for complete and incomplete solvers; adaptg2wsat+ and gnovelty+ found the same instances to be the hardest. RSat, picosat and minisat also agreed on the location of the hardest instances, but it is different than the location indicated by the incomplete solvers. For BerkMin, the hardest instances were located between the ones found by incomplete solvers and the ones indicated by

Table 1. Combinations of parameters for the hardest SRSGI instances

m	q	n	p
20	152	18	5
21	168	19	10
22	173	20	20
23	189	21	25
24	193	22	30
26	243	24	30
28	302	26	25

RSat, minisat and picosat. This variation shows that the model can be used to differentiate between SAT solvers. Despite this variation, for all pairs m and n and for all solvers, the value of p at which the hardness peaks, decreases as q increases. As q increases, H becomes denser; hence less edges need to be removed from G for H to contain several copies of G, and this can be the reason for the occurrence of the hardness peak.

A natural question is which combination of parameters produces the hardest instances - the highest hardness peak. To answer this question we fixed m and investigated the relationship between n and the value of q which produced the hardest instances. For all values of m we noticed an inverse correlation between n and the value of q which resulted in the highest peak. The correlation is almost linear but this may be a consequence of the sample size. The positive correlation between n and p is an immediate consequence of the two inverse correlations presented above. In table 1 we present, for each value of m, the values of n, q and p which produced the hardest instances[3]. It appears that the hardest instances are generated when $n=m-2$.

3.3   Growth Rate of the Hardest Instances

The Subgraph Isomorphism Problem is NP-complete, and thus it is assumed that it is exponentially hard for deterministic algorithms. This statement does not immediately transfer to SRSGI, but the analogy with random models of other NP-complete problems, such as, random satisfiable k-SAT or QGWH suggests that the SAT encoded SRSGI instances may grow exponentially. To check this hypothesis we fixed m and plotted the hardness of the instances from the peak against n. The resulting curves look exponential - see Fig. 5 for both complete and incomplete solvers. Due to the linear variation of the number of variables with n, the previous observation implies that the empirical hardness of the instances grows exponentially with their number of variables. For the incomplete solvers we noticed a similar exponential growth when we plotted the hardness of instances against their size (number of literals) - see Fig 6. In the case of complete solvers the shape is essentially exponential, but it has some irregularities. The

---

[3] The hardest instances are those for which most of the solvers had the largest running time.

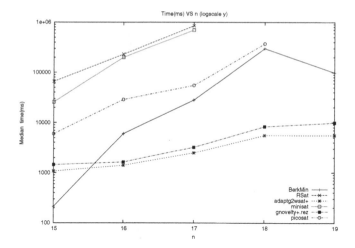

Fig. 5. Evolution of empirical hardness at peak with n ($m=21$). Some points are missing due to time out. Notice the logarithmic time scale.

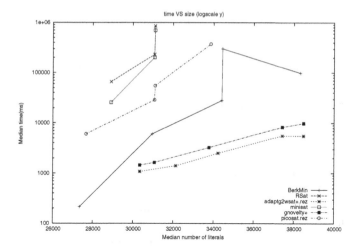

Fig. 6. Evolution of empirical hardness at peak with the size of the instances ($m=21$). Some points are missing due to time out. Notice the logarithmic time scale.

difference between the performances of complete and incomplete solvers may be a consequence of the difference between the perceived hardest instances noticed in the previous subsection.

Due to the encoding scheme used, the SAT instances generated by SRSGI are large. It is therefore possible that the empirical hardness of the instances is solely a consequence of their size. To check this hypothesis we compared SRSGI instances with QGWH[4] instances having a comparable number of variables.

---

[4] It is considered that QGWH generates hard satisfiable instances[1].

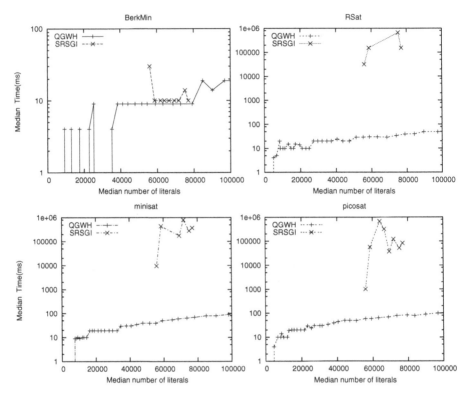

Fig. 7. Empirical hardness of SRSGI (m=24, q=193, n=20) and QGWH(order 20) for complete solvers. Size of the instances is used as comparison criteria. Some points are missing from the SRSGI plots due to time out. Notice the logarithmic time scale.

At comparable number of variables, the hardest SRSGI instances are harder than their QGWH counterparts; all solvers but gnovelty+ found the SRSGI instances significantly harder. When the comparison is done in terms of number of literals the results are different for complete and incomplete solvers. The complete solvers still found the SRSGI instances harder then the QGWH ones - see Fig. 7. Gnovelty+ timed out on almost all large QGWH instances and therefore we must conclude that they are harder than the SRSGI ones. For adaptg2wsat+ , the SRSGI instances are harder than the QGWH ones, but they have significantly more literals than the latter. If we assume that the hardness of the QGWH instances increases exponentially it follows that at comparable size QGWH instances would be harder than the SRSGI ones - see Fig. 8.

The difference in size between the two models is caused be their encoding schemes. QGWH uses a more compact SAT encoding than SRSGI. This encoding scheme can not be immediately transferred to SRSGI. To compensate for this difference, we compared the empirical hardness of SRSGI instances with that of QGWH instances encoded to SAT by a method similar to the direct encoding used for SRSGI. Under these circumstances, at comparable size (number

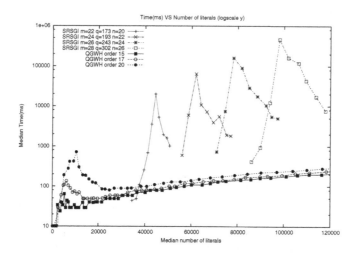

Fig. 8. Empirical hardness of SRSGI and QGWH for adaptg2wsat+ . Size of the instances is used as comparison criteria. Notice the logarithmic time scale.

of literals), the SRSGI instances were more difficult than the QGWH ones for all solvers. The comparison indicates that the hardness of SRSGI is comparable with that of QGWH instances, which are considered to be "hard". Therefore we can say that the SRSGI instances are also hard.

4   Conclusion

We investigated the empirical hardness of a generator of satisfiable SAT instances, based on subgraph isomorphism. We noticed an easy-hard-easy pattern of the evolution of the empirical hardness of the instances, which allows for a fine tuning of the hardness. The location of the hardest instances is consistent among the set of complete and incomplete solvers used in the experiments. We identified several correlations between the location of the hardness peak and the parameters of the model. The experiments indicate an exponential growth of the empirical hardness, and this is consistent for both complete and incomplete solvers. We noticed a difference in the performance of the complete and incomplete solvers on SAT instances generated from SRSGI. Even within the class of complete solvers, the performance of some state of the art solvers on these instances varies significantly with respect to each other. All these features recommend this model as an alternative for generating satisfiable SAT instances and as a generator of Pseudo Boolean instances.

Acknowledgments. We want to thank to the members of the Intelligent Information Systems Institute at Cornell University and especially to Carla Gomes for making the QGWH generator available to us.

References

1. Achlioptas, D., Gomes, C., Kautz, H., Selman, B.: Generating satisfiable problem instances. In: Proceedings of AAAI 2000, pp. 256–261 (2000)
2. Clark, D.A., Frank, J., Gent, I.P., MacIntyre, E., Tomov, N., Walsh, T.: Local search and the number of solutions. In: Proceedings of CP 1996, pp. 119–133 (1996)
3. Achlioptas, D., Jia, H., Moore, C.: Hiding satisfying assignments: Two are better than one. In: Proceedings of AAAI 2004, pp. 131–136 (2004)
4. Xu, K., Boussemart, F., Hemery, F., Lecoutre, C.: A simple model to generate hard satisfiable instances. In: Proceedings of IJCAI 2005, pp. 337–342 (2005)
5. The international SAT competitions web page, http://www.satcompetition.org
6. Chieu, H.: Finite energy survey propagation for constraint satisfaction problems (2007); Singapore MIT Alliance Symposium 2007
7. Goldberg, E., Novikov, Y.: Berkmin: A fast and robust SAT-solver. In: Proceedings of DATE 2002, pp. 142–149 (2002)
8. Eén, N., Sörensson, N.: Minisat a sat solver with conflict-clause minimization. In: Proceedings of SAT 2005 (2005), http://minisat.se/Papers.html
9. Biere, A.: Picosat essentials. Journal on Satisfiability, Boolean Modeling and Computation 4, 75–97 (2008)
10. Pipatsrisawat, K., Darwiche, A.: Rsat 2.0: Sat solver description. Technical Report D–153, Automated Reasoning Group, Computer Science Department, UCLA (2007)
11. Heule, M.J.H., van Maaren, H.: March-dl: Adding adaptive heuristics and a new branching strategy. Journal on Satisfiability, Boolean Modeling and Computation 2, 47–59 (2006)
12. Wei, W., Li, C.M., Zhang, H.: Switching among non-weighting, clause weighting, and variable weighting in local search for sat. In: Stuckey, P.J. (ed.) CP 2008. LNCS, vol. 5202, pp. 313–326. Springer, Heidelberg (2008)
13. Pham, D.N., Thornton, J., Gretton, C., Sattar, A.: Combining adaptive and dynamic local search for satisfiability. Journal on Satisfiability, Boolean Modeling and Computation 4, 149–172 (2008)

Enhancing the Bilingual Concordancer TransSearch with Word-Level Alignment

Julien Bourdaillet, Stéphane Huet, Fabrizio Gotti,
Guy Lapalme, and Philippe Langlais

Département d'Informatique et de Recherche Opérationnelle
Université de Montréal
C.P. 6128, succursale Centre-ville
H3C 3J7, Montréal, Québec, Canada
http://rali.iro.umontreal.ca

Abstract. Despite the impressive amount of recent studies devoted to improving the state of the art of Machine Translation (MT), Computer Assisted Translation (CAT) tools remain the preferred solution of human translators when publication quality is of concern. In this paper, we present our perspectives on improving the commercial bilingual concordancer TransSearch, a Web-based service whose core technology mainly relies on sentence-level alignment. We report on experiments which show that it can greatly benefit from statistical word-level alignment.

1 Introduction

Although the last decade has witnessed an impressive amount of effort devoted to improving the current state of Machine Translation (MT), professional translators still prefer Computer Assisted Translation (CAT) tools, particularly *translation memory* (TM) systems. A TM is composed of a *bitext*, a set of pairs of units that are in translation relation, plus a search engine. Given a new text to translate, a TM system automatically segments the text into units that are systematically searched for in the memory. If a match is found, the associated target material is retrieved and output with possible modifications, in order to account for small divergences between the unit to be translated and the one retrieved. Thus, such systems avoid the need to re-translate previously translated units. Commercial solutions such as SDL Trados[1], Deja Vu[2], LogiTerm[3] or MultiTrans[4] are available; they mainly operate at the level of sentences, which narrows down their usefulness to repetitive translation tasks.

Whereas a TM system is a translation device, a *bilingual concordancer* (BC) is conceptually simpler, since its main purpose is to retrieve from a bitext, the pairs of units that contain a *query* (typically a phrase) that a user manually submits.

---

[1] http://www.trados.com
[2] http://www.atril.com
[3] http://www.terminotix.com
[4] http://www.multicorpora.ca

Y. Gao and N. Japkowicz (Eds.): Canadian AI 2009, LNAI 5549, pp. 27–38, 2009.

It is then left to the user to locate the relevant material in the retrieved target units. As simple as it may appear, a bilingual concordancer is nevertheless a very popular CAT tool. In [1], the authors report that `TransSearch`,[5] the commercial concordancer we focus on in this study, received an average of 177 000 queries a month over a one-year period (2006–2007).

This study aims at improving the current `TransSearch` system by providing it with robust word-level alignment technology. It was conducted within the `TS3` project, a partnership between the RALI and the Ottawa-based company Terminotix.[6] One important objective of the project is to automatically identify (highlight) in the retrieved material the different translations of a user query, as discussed initially in [2]. The authors of that paper also suggested that grouping variants of the same "prototypical" translation would enhance the usability of a bilingual concordancer. These are precisely the two problems we are addressing in this study.

The remainder of this paper is organized as follows. We first describe in Section 2 the translation spotting techniques we implemented and compared. Since translation spotting is a notoriously difficult problem, we discuss in Section 3 two novel issues that we think are essential to the success of a concordancer such as `TransSearch`: the identification of erroneous alignments (Section 3.1) and the grouping of translation variants (Section 3.2). We report on experiments in Section 4 and conclude our discussion and propose further research avenues in Section 5.

2   Transpotting

Translation spotting, or *transpotting*, is the task of identifying the word-tokens in a target-language (TL) translation that correspond to the word-tokens of a query in a source language (SL) [3]. It is therefore an essential part of the `TS3` project. We call *transpot* the target word-tokens automatically associated with a query in a given pair of units (sentences). The following example[7] illustrates the output of one of the transpotting algorithms we implemented. Both `conformes à` and `fidèles à` are French transpots of the English query `in keeping with`.

S_1 = These are important measures in keeping with our international obligations.
T_1 = Il s'agit d'importantes mesures conformes à nos obligations internationales.

S_2 = In keeping with their tradition, liberals did exactly the opposite.
T_2 = Fidèles à leur tradition, les libéraux ont fait exactement l'inverse.

---

[5] `www.tsrali.com`
[6] `www.terminotix.com`
[7] The data used in this study is described in Section 4.1.

As mentioned in [4], translation spotting can be seen as a by-product of word-level alignment. Since the seminal work of [5], statistical word-based models are still the core technology of today's Statistical MT. This is therefore the alignment technique we consider in this study.

Formally, given a SL sentence $S = s_1...s_n$ and a TL sentence $T = t_1...t_m$ in translation relation, an IBM-style alignment $a = a_1...a_m$ connects each target token to a source one ($a_j \in \{1, ..., n\}$) or to the so-called NULL token which accounts for untranslated target tokens, and which is arbitrarily set to the source position 0 ($a_j = 0$). This defines a word-level alignment space between S and T whose size is in $O(m^{n+1})$.

Several word-alignment models are introduced and discussed in [5]. They differ by the expression of the joint probability of a target sentence and its alignment, given the source sentence. We focus here on the simplest form, which corresponds to IBM models 1 & 2:

$$p(t_1^m, a_1^m) = \prod_{j=1}^{m} \sum_{i \in [0,n]} p(t_j|s_i) \times p(i|j, m, n)$$

where the first term inside the summation is the so-called transfer distribution and the second one is the alignment distribution.

A transpotting algorithm comprises two stages: an alignment stage and a decision stage. We describe in the following sections the algorithms we implemented and tested, with the convention that $s_{i_1}^{i_2}$ stands for the source query (we only considered contiguous queries in this study, since they are by far the most frequent according to our user logfile).

2.1   Viterbi Transpotting

One straightforward strategy is to compute the so-called Viterbi alignment ($\hat{a}$). By applying Bayes' rule and removing terms that do not affect the maximization, we have:

$$\hat{a} = \underset{a_1^m}{\arg\max}\, p(a_1^m|t_1^m, s_1^n) = \underset{a_1^m}{\arg\max} \frac{p(s_1^n) \times p(t_1^m, a_1^m|s_1^n)}{p(s_1^n) \times p(t_1^m|s_1^n)} = \underset{a_1^m}{\arg\max}\, p(t_1^m, a_1^m|s_1^n).$$

In the case of IBM models 1 & 2, this alignment is computed efficiently in $O(n \times m)$. Our first transpotting implementation, called VITERBI-SPOTTER, simply gathers together the words aligned to each token of the query. Note that with this strategy, nothing forces the transpot of the query to be a contiguous phrase. In our first example above, the transpot of the query produced by this formula for in keeping with is the French word conformes.

2.2   Contiguous Transpotting

This method, which was introduced by M. Simard in [4], forces the transpot to be a contiguous sequence. This is accomplished by computing for each pair $\langle j_1, j_2 \rangle \in [1, m]^2$, two Viterbi alignments: one between the phrase $t_{j_1}^{j_2}$ and the query $s_{i_1}^{i_2}$, and one between the remaining material in those sentences, $\bar{s}_{i_1}^{i_2} \equiv s_1^{i_1-1}s_{i_2+1}^n$ and $\bar{t}_{j_1}^{j_2} \equiv t_1^{j_1-1}t_{j_2+1}^m$. The complexity of this algorithm is $O(nm^3)$:

$$\hat{t}_{j_1}^{j_2} = \underset{(j_1,j_2)}{\mathrm{argmax}} \left\{ \max_{a_{j_1}^{j_2}} p(a_{j_1}^{j_2}|s_{i_1}^{i_2},t_{j_1}^{j_2}) \times \max_{\bar{a}_{j_1}^{j_2}} p(\bar{a}_{j_1}^{j_2}|\bar{s}_{i_1}^{i_2},\bar{r}_{j_1}^{j_2}) \right\}.$$

This method, called CONTIGVIT-SPOTTER hereafter, returns the transpot conformes à for the query in keeping with in the first example above.

2.3   Maximum Contiguous Subsequence Transpotting

In this transpotting strategy, called MCSS-SPOTTER, each token t_j is associated
with a score computed such as:

$$\mathrm{score}(t_j) = \mathrm{score}_0(t_j) - \tilde{t}$$

where $\mathrm{score}_0(t_j) = \sum_{i=i_1}^{i_2} p(t_j|s_i)p(i|j,m,n)$ and $\tilde{t} = \frac{1}{m}\sum_{j=1}^m \mathrm{score}_0(t_j)$.

The score corresponds to the word alignment score of IBM model 2 minus the
average computed for each token t_j. Because of $\tilde{t}$, the score associated with a
token t_j is either positive or negative.

The Maximum Contiguous Subsequence Sum (MCSS) algorithm is then applied. Given this sequence of scores, it finds the contiguous subsequence whose
sum of scores is maximum over all subsequences. When processing the sequence,
the trick is that if a contiguous subsequence with a negative sum is encountered,
it cannot be a MCSS; therefore, either the MCSS occurred before this negative
sum subsequence or it will occur after. Our implementation runs in $O(m)$. Finally, the MCSS corresponds to the transpot of the given query $s_{i_1}^{i_2}$. In the first
example above, the unsatisfying transpot à is returned by this method.

2.4   Baseline

To challenge the transpotting algorithms we implemented, we also considered
a strategy which does not embed any statistical alignment. It consists in pro-
jecting the positions of the query $s_{i_1}^{i_2}$ by means of the length ratio between the
two sentences. The transpot is determined by $t_{j_1} \ldots t_{j_2}$ where $j_1 = \lfloor \frac{m}{n} i_1 \rfloor$ and
$j_2 = \lceil \frac{m}{n} i_2 \rceil$. In our example, this method, called BASELINE-SPOTTER hereafter,
returns the transpot importantes mesures conformes à.

3   Post-Processing

Frequent queries in the translation memory receive numerous translations by the
previously described transpotting process. Figure 1 illustrates the many transpots returned by CONTIGVIT-SPOTTER for the query in keeping with. As can
be observed, some transpots (those marked by a star) are clearly wrong (*e.g.*
à), while many others (in italics) are only partially correct (*e.g.* conformément).

conforme à (45)   conforme aux (18)   conformes à (11)
conformément à (29) *conforme* (13)   *conformément* (9)
à* (21)   conformément aux (13) , conformément à (3)
dans* (20)   conforme au (12)   correspondant à (3)

Fig. 1. Subset of the transpots retrieved by CONTIGVIT-SPOTTER for the query **in keeping with** with their frequency shown in parentheses

Also, it appears that many transpots are indeed very similar (*e.g.* conforme à and conformes à).

Since in TS3 we want to offer the user a list of retrieved translations for a query, strategies must be devised for overcoming alignment errors and delivering the most salient information to the user. We investigated two avenues in this study: detecting erroneous transpots (Section 3.1) and merging variants of the same prototypical translation (Section 3.2).

3.1   Refining Transpotting

As detailed in Section 4.1 below, we analyzed 531 queries and their transpots, as computed by CONTIGVIT-SPOTTER, and manually annotated the erroneous transpots. This corpus served to train a classifier designed to distinguish good transpots from bad ones. To this end, we applied the *voted-perceptron* algorithm described in [6]. Online voted-perceptrons have been reported to work well in a number of NLP tasks [7,8]. In a nutshell, a weighted pool of perceptrons is incrementally acquired during a batch training procedure, where each perceptron is characterized by a real-valued vector (one component per feature on which we train the classifier) and its associated weight, computed as the number of successive training examples it could correctly classify before it fails. When the current perceptron misclassifies a training example, a new one is added to the pool, the coefficients of which are initialized from the current perceptron according to a simple delta-rule and kept fixed over the training procedure.

We computed three groups of features for each example of the annotated corpus, that is, each query/transpot (q, t) pair. The first group is made up of features related to the size (counted in words) of q and t, with the intuition that they should be related. The second group gathers various alignment scores computed with word-alignment models (min and max likelihood values, etc.). The last group gathers clues that are more linguistically flavored, among them the ratio of grammatical words in q and t, or the number of prepositions and articles. In total, each example is represented by at most 40 numerical features.

3.2   Merging Variants

Once erroneous transpots have been filtered out, there usually remain many translation variants for a given query. Some of them are very similar and are therefore redundant for the user. For instance, returning the inflected forms of nouns or verbs is often useless and may prevent more dissimilar and potentially more interesting variants from being shown to the user when the number of displayed translations for a query is limited. This phenomenon is more acute for the French language with its numerous verb conjugation forms. Another problem that often shows up is that many transpots differ only by punctuation marks or by a few grammatical words, *e.g.* conformément aux and , conformément à in Figure 1.

Merging variants according to their closeness raises several difficulties. First, the various transpots must be compared, which represents a costly process. Second, we need to identify clusters of similar variants. Lastly, a prototype of the selected clusters must be selected and output to the user. We now describe our solution to these problems.

Comparing the transpots pairwise is an instance of multiple sequence alignment, a well studied problem in bioinformatics [9]. We adopt the approach of progressive alignment construction. This method first computes the distance between each pair of transpots to align and progressively builds a tree that aims at guiding the alignment of all pairs. At each step, the most similar pair is merged and added to the tree, until no transpot remains unaligned. In order to build this tree, we use a bottom up clustering method, called *neighbor-joining* [10].

The main interest of this approach is its computational efficiency, since pairwise aligning the transpots is carried out in polynomial time, which allows us to use it even when a large set of transpots is returned. This property is obtained thanks to the greedy nature of the algorithm. Indeed, it is based on a metrics that can be straightforwardly computed between a new node —associated with a joined pair of sequences— and the other sequences from the metrics previously computed for the sequences just joined. Although this clustering method is greedy and may not build the optimal tree, it has been extensively tested and usually finds a tree that is quite close to the optimal one.

The neighbor-joining algorithm requires computing a distance matrix between each pair of transpots to align. A word-level specific edit-distance was empirically developed to meet the constraints of our application. Different substitution, deletion or insertion costs are introduced according to the grammatical classes or possible inflections of the words; it is therefore language dependent. We used an in-house lexicon which lists for both French and English the lemmas of each wordform and its possible part-of-speech. A minimal substitution cost was empirically engineered between two inflected forms of the same lemma. An increasing edition cost was set empirically to account respectively for punctuation marks, articles, grammatical words (prepositions, conjunctions and pronouns), auxiliary verbs and finally all the remaining words (verbs, nouns, adjectives and adverbs).

Thus, we obtain a tree whose leaves are transpots. The closest leaves in the tree correspond to the closest variants, according to our edit-distance calculation. Therefore, clusters of similar variants can be formed by traversing the tree in a post-order manner. The transpots which are associated with two neighboring leaves and which differ only by grammatical words or by inflectional variants are considered as sufficiently similar to be merged into a cluster. This process is repeated until all the leaves have been compared with their nearest neighbor and

Fig. 2. Merging of close transpots by the process described in the text

no more similar variants are found. For each pair of merged leaves, a pattern is built from the alignment of the two transpots and regular expressions are used to represent the grouped variants.

Figure 2 illustrates this process with an extract from the output obtained for the examples in Figure 1. `conforme à` and `conformes à` are neighboring transpots in the tree which are grouped into the pattern `[conforme] à` and added to the tree. Similarly, the prototype `conforme [au]` is built from the two neighboring transpots `conforme au` and `conforme aux`. Once these merges are done, the two new prototypes become close in the tree; their comparison in turn leads to the decision to group them and to create the pattern `[conforme] [à|au]`.

4   Experiments

4.1   Corpora

Translation Memory. The largest collections in **TransSearch** come from the Canadian Hansards, that is, parallel texts in English and French drawn from official records of the proceedings of the Canadian Parliament. This material is aligned at the sentence-level by an in-house aligner. For our experiments, we indexed with Lucene[8] a translation memory comprising 3.3 million pairs of French-English sentences. This was the maximum amount of material we could train a statistical word-alignment model on, running the `giza++` [11] toolkit on a computer equipped with 16 gigabytes of memory.

Automatic Reference Corpus. In order to evaluate the quality of our approach, we developed a reference corpus called REF. In [4], the author manually identified the transpots in 4 100 pairs of sentences produced by 41 queries, a slow and difficult process. Indeed, the time spent analyzing one query with 30 pairs of sentences to annotate was in the order of 5-10 minutes.

We devised a way to get a much larger reference corpus without manual annotation. The 5 000 most frequent queries submitted by users to the system were extracted from the logs of **TransSearch**. Besides, we used an in-house bilingual phrase dictionary collected for the needs of various projects, which includes 59 057 English phrases with an average of 1.4 French translations each. Among the indexed pairs of sentences, only those that contain the phrases of the dictionary and their translation are kept.

---

[8] http://lucene.apache.org

According to this method, 4 526 of the 5 000 most frequent queries submitted by users to the system actually occured in our translation memory; of these 2 130 had a translation either in the bilingual phrase dictionary (713) or in a classical bilingual word dictionary (the remaining). From the memory, we retrieved a maximum of 5 000 pairs of sentences for each of those 2 130 queries, leading to a set of 1 102 357 pairs of sentences, with an average of 517 pairs of sentences per query. Altogether, these examples contain 7 472 unique pairs of query/transpot; each query received an average of 3.5 different transpots, and a maximum of 37.

Human Reference. In order to train the classifier described in Section 3.1, four human annotators (a subset of the authors) were asked to identify bad transpots among those proposed by the best of our transpotting algorithm. We decided to annotate the query/transpot pairs without their contexts. This allows a relatively fast annotation process, in the order of 40 seconds per query, but leaves some cases difficult to annotate. To go back to the query in keeping with, though some translations like conforme à are straightforward, other such as suivant, dans le sens de or even tenir compte de can be the correct transpots of this query according to its context of use.

We ended up with a set of 531 queries that have an average of 22.9 transpots each, for a total of 12 144 annotated examples. We computed the inter-annotator agreement and observed a 0.76 kappa score [12], which indicates a high degree of agreement among annotators.

4.2   Translation Spotting

To compare our transpotting algorithms, we conducted two series of evaluation: one at the sentence level, and one at the query level. In the former case, the ability of each algorithm to identify the reference translation $\hat{t}$ for a query q was measured according to precision and recall ratios, computed as follows:

$$\text{precision} = |t \cap \hat{t}|/|t| \quad \text{recall} = |t \cap \hat{t}|/|\hat{t}| \quad \text{F-measure} = 2\frac{|t \cap \hat{t}|}{|t|+|\hat{t}|}$$

where t is the transpot identified by the algorithm, and the intersection operation is to be understood as the portion of words shared by t and $\hat{t}$. A point of detail is in order here: since several pairs of sentences can contain a given query/reference transpot pair $(q, \hat{t})$, we averaged the aforementioned ratios measured per unique pairs $(q, \hat{t})$.[9] This avoids biasing our evaluation metrics toward frequent pairs in REF. Those average ratios are then averaged over the set of different pairs $(q, \hat{t})$ in REF.

Table 1 shows the results obtained by the different methods. We observe that MCSS-SPOTTER and CONTIGVIT-SPOTTER obtain the best results. MCSS-SPOTTER has a higher recall than CONTIGVIT-SPOTTER, meaning that its transpots t match more of the words of the references $\hat{t}$, but it has also a lower precision, meaning that its transpots are longer. A caricature of this strategy

---

[9] Without this normalization, results would be increased by a range of 0.2 to 0.4 points.

Table 1. Results of the different transpotting algorithms presented in Section 2 measured on the REF corpus

method	precision	recall	F-measure
BASELINE-SPOTTER	0.127	0.222	0.149
VITERBI-SPOTTER	0.139	0.349	0.190
MCSS-SPOTTER	0.198	**0.744**	0.295
CONTIGVIT-SPOTTER	**0.303**	0.597	**0.376**
CONTIGVIT-SPOTTER + best voted-perceptron	**0.372**	**0.757**	**0.470**

Table 2. Performance of different algorithms for identifying bad transpots using the test subset of the human reference

		CCI	Bad precision	recall	F-measure
Baselines:	all good	0.62	0.00	0.00	0.00
	grammatical ratio > 0.75	**0.81**	0.91	0.56	**0.69**
Features:	size	0.72	0.69	0.46	0.55
	IBM	**0.82**	0.75	0.80	**0.77**
	grammatical-ratio	0.80	0.90	0.55	0.68
	all	**0.84**	0.80	0.77	**0.78**

would be to propose the whole sentence as a transpot. This is very undesirable at the sentence level where transpotting must highlight with precision a transpot in a sentence. Finally, the results of CONTIGVIT-SPOTTER are more balanced: the behavior of this transpotter is more in keeping with what is expected. We comment on the last line of Table 1 in Section 4.3.

While at the sentence-level evaluation, each pair of sentences containing a query and a reference translation counts, at the query-level, we directly evaluate the set of different transpots found for each query. On average, the CONTIGVIT-SPOTTER transpotting algorithm identifies 40 different transpots per query, and at least one reference translation was proposed for 91% of the queries.

4.3   Filtering Spurious Transpots

As described in Section 3.1, we trained various classifiers to identify spurious transpots. For this, 90 % of the human reference presented in Section 4.1 was used as a training corpus and 10 % as a test corpus. The examples (query/transpot pairs) are represented by three kinds of feature sets. All the classifiers, plus a few challenging baselines are evaluated according to the ratio of Correctly Classified Instances (CCI). Since in our application, we are interested in filtering out bad transpots, we also report precision, recall and F-measure rates related to this class. These figures are shown in Table 2.

To begin with, the simplest baseline we built (line 1) classifies all instances as good. This results in a CCI ratio of 0.62. A more sensible baseline that we engineered after we investigated the usefulness of different feature sets consists in classifying as bad those transpots whose ratio of grammatical words is above 0.75. It receives a CCI ratio of 0.81.

Among all voted-perceptron configurations we tested, with the exception of the one with all feature sets, the one making use of word alignment scores based on an IBM model 1 obtains the best CCI ratio of 0.82. Although this is barely better than the performance of the best baseline, the voted-perceptron shows a much higher gain in F-measure for bad transpots: 0.77 compared to 0.69, which is the task we are trying to optimize. Finally, the voted-perceptron trained using all feature sets (last line) obtains a CCI ratio of 0.84 and a F-measure for bad transpots of 0.78, which clearly surpasses the baseline. It should be noticed that while the best baseline has a better precision than the best voted-perceptron, precision and recall are more balanced for the latter. Because it is not clear whether precision or recall should be favored for the task of bad transpot filtering, we optimized the F-measure.

The last line of Table 1 presents the results of the best transpotter CONTIGVIT-SPOTTER after filtering bad transpots with the best voted-perceptron. Significant gains can be observed: the F-measure increases from 0.376 to 0.470. It outperforms the MCSS-SPOTTER recall score and has a higher precision of nearly 0.17. This demonstrates the interest of filtering bad transpots.

4.4   Merging Variants

The second post-processing stage, which merges variants, was evaluated on the pairs of sentences collected from the same 5 000 queries as those of the REF corpus. Contrary to the evaluation of transpot filtering, which requires a dictionary to build the reference, the retrieved pairs of sentences were not discarded here if the translations did not occur in our in-house bilingual phrase dictionary. This allowed us to obtain a more important number of unique (q, t) pairs (389 989 after filtering spurious transpots).

The reduction of the variants to display for each query was quantitatively evaluated with two versions of our merge process, which differ only in the edit distance used to compare various forms. The first version merges only transpots that are inflected forms of the same lemmatized sequence or that only varies by punctuation marks; it leads to a significant decrease in the number of forms to display since we get 1.22 variants per pattern. The second version merges not only the transpots that would be clustered by the previous one, but also variants of word sequences that differ by the use of grammatical words; this method results in a higher number of variants per pattern (1.88 on average).

From these numbers, it can be seen that merging grammatical words dramatically decreases the number of outputs of our system, thereby allowing for the display of more different translations. It often leads to patterns that are easily understandable. For example, our system merges the sequences manière générale, de manière générale, une manière générale and d'une manière générale into the pattern [de]? (une)? manière générale where (.)? or [.]? notes optional words and [w] indicates that w is a word lemmatized from several merged words inflected from w. Our algorithm also builds patterns such as (avec|sur|durant) (des|les|plusieurs) années that succeed in grouping similar variants.

Sometimes, merging grammatical words generates patterns from expressions that are not synonymous, which requires a refinement of our method. For instance, (tout)? [faire] (tout)? (ce)? was built, whereas two different patterns would have been better: tout [faire] and [faire] tout (ce)?. In another example, the pattern (qui)? (s')? (en|à)? [venir] made from the variants à venir, qui vient, qui viennent or qui s'en viennent is difficult to understand. We are counting on the input of end-users to help us decide on the optimum manner of displaying variant translations.

5 Discussion

In this study, we have investigated the use of statistical word-alignment within TransSearch, a bilingual concordancer. Overall, we found that our best transpotting algorithm CONTIGVIT-SPOTTER, a Viterbi aligner with a contiguity constraint, combined with a filter to remove spurious transpots, significantly outperforms other transpotting methods, with a F-measure of 0.470. We have demonstrated that it is possible to detect erroneous transpots better than a fair baseline, and that merging variants of a prototypical translation can be done efficiently.

For the time being, it is difficult to compare our results to others in the community. This is principally due to the uniqueness of the TransSearch system, which archives a huge translation memory. To give a point of comparison, in [13] the authors report alignment results they obtained for 120 selected queries and a TM of 50 000 pairs of sentences. This is several orders of magnitude smaller than the experiments we conducted in this study.

There are several issues we are currently investigating. First, we only considered simple word-alignment models in this study. Higher-level IBM models can potentially improve the quality of the word alignments produced. At the very least, HMM models [14], for which Viterbi alignments can be computed efficiently, should be considered. The alignment method used in current phrase-based SMT is another alternative we are considering.

Acknowledgements

This research is being funded by an NSERC grant. The authors wish to thank Elliott Macklovitch for his contribution to this work.

References

1. Macklovitch, E., Lapalme, G., Gotti, F.: Transsearch: What are translators looking for? In: 18th Conference of the Association for Machine Translation in the Americas (AMTA), Waikiki, Hawai'i, USA, pp. 412–419 (2008)

2. Simard, M., Macklovitch, E.: Studying the human translation process through the TransSearch log-files. In: AAAI Symposium on Knowledge Collection from volunteer contributor, Stanford, CA, USA (2005)
3. Véronis, J., Langlais, P.: 19. In: Evaluation of Parallel text Alignment Systems — The Arcade Project, pp. 369–388. Kluwer Academic Publisher, Dordrecht (2000)
4. Simard, M.: Translation spotting for translation memories. In: HLT-NAACL 2003 Workshop on Building and using parallel texts: data driven machine translation and beyond, Edmonton, Canada, pp. 65–72 (2003)
5. Brown, P., Della Pietra, V., Della Pietra, S., Mercer, R.: The mathematics of statistical machine translation: parameter estimation. Computational Linguistics 19(2), 263–311 (1993)
6. Freund, Y., Schapire, R.: Large margin classification using the perceptron algorithm. Machine Learning 37(3), 277–296 (1999)
7. Collins, M.: Discriminative training methods for hidden markov models: theory and experiments with perceptron algorithms. In: EMNLP 2002, Philadelphia, PA, USA, pp. 1–8 (2002)
8. Liang, P., Bouchard-Côté, A., Klein, D., Taskar, B.: An end-to-end discriminative approach to machine translation. In: 21st COLING and 44th ACL, Sydney, Australia, pp. 761–768 (2006)
9. Chenna, R., Sugawara, H., Koike, T., Lopez, R., Gibson, T.J., Higgins, D.G., Thompson, J.D.: Multiple sequence alignment with the Clustal series of programs. Nucleic Acids Research 31(13), 3497–3500 (2003)
10. Saiou, N., Nei, M.: The neighbor-joining method: A new method for reconstructing phylogenetic trees. Molecular Biology and Evolution 4(4), 406–425 (1987)
11. Och, F.J., Ney, H.: A systematic comparison of various statistical alignment models. Computational Linguistics 29(1), 19–51 (2003)
12. Fleiss, J.L., Levin, B., Pai, M.C.: Statistical Methods for Rates and Proportions, 3rd edn. Wiley Interscience, Hoboken (2003)
13. Callisson-Burch, C., Bannard, C., Schroeder, J.: A compact data structure for searchable translation memories. In: 10th European Conference of the Association for Machine Translation (EAMT), Budapest, Hungary, pp. 59–65 (2005)
14. Vogel, S., Ney, H., Tillmann, C.: HMM-based word alignment in statistical translation. In: 16th conference on Computational linguistics, Copenhagen, Denmark, pp. 836–841 (1996)

Financial Forecasting Using Character N-Gram Analysis and Readability Scores of Annual Reports

Matthew Butler and Vlado Kešelj

Faculty of Computer Science, Dalhousie University
{mbutler,vlado}@cs.dal.ca

Abstract. Two novel Natural Language Processing (NLP) classification techniques are applied to the analysis of corporate annual reports in the task of financial forecasting. The hypothesis is that textual content of annual reports contain vital information for assessing the performance of the stock over the next year. The first method is based on character n-gram profiles, which are generated for each annual report, and then labeled based on the CNG classification. The second method draws on a more traditional approach, where readability scores are combined with performance inputs and then supplied to a support vector machine (SVM) for classification. Both methods consistently outperformed a benchmark portfolio, and their combination proved to be even more effective and efficient as the combined models yielded the highest returns with the fewest trades.

Keywords: automatic financial forecasting, n-grams, CNG, readability scores, support vector machines.

1 Introduction

The Securities and Exchange Commission (SEC) requires that each year all publicly-traded companies supply a third-party audited financial report, which states the company's financial position and performance over the previous year. Contained in these annual reports, inter alia, are financial statements, a letter to the share-holders, and management discussion and analysis. Over the years several research endeavours have been focused on the numbers contained in the financial statements, computing a variety of ratios and price projections without considering textual components of the reports. Peter Lynch, a famous investment "guru," once said that "charts are great for predicting the past," pointing out that there is more to making good investments than just processing the numbers. The textual components give insight into the opinions of the senior management team and provide a direction of where they feel the company is going. This information should not be trivialized or overlooked; it should be processed in a similar way to processing quantitative information, to extract meaningful information to aid in the forecasting process. Up until recently an analyst would have to read an annual

Y. Gao and N. Japkowicz (Eds.): Canadian AI 2009, LNAI 5549, pp. 39–51, 2009.

report and use their expertise to determine if the company is going to continue to do well or if there is trouble ahead. They would apply their skill and judgment to interpret what the Chief Executive Officer (CEO) is saying about the company and its direction for the future. This process can be very time consuming and it is a somewhat heuristic approach, considering that two experienced analysts could read the same report and have a different feeling about what it is saying. If an analyst has several companies to consider and even more annual reports to read it could be difficult to take in all the relevant information when it is most likely surrounded by noise and other erroneous information that has no effect on the stock price. Most numeric calculations can be automated to remove human error, and complex data mining and machine learning algorithms can be applied to extract meaningful relationships from them. It would be extremely valuable if the same could be done for the textual components, having a quick, efficient and accurate tool to analyze an annual report and make recommendations on its implications for the stock price over some given time period. This could erase some of the subjective judgments that arise from an individual's interpretation of the report, which could change from person to person. Also, given the sheer amount of annual reports that are produced each year, one would be able to analyze a larger number of companies and have a greater opportunity to find good investments.

In this paper an attempt is made at achieving this goal: two novel approaches to analyzing the text are put forward and then a combined model is also analyzed to see if a union of these approaches is more robust. The first novel technique is to convert the textual components to n-gram profiles and use the CNG distance measure [1] as proposed by Kešelj *et al.* to classify reports. The second is to generate three readability scores (Flesch, Flesch-Kincaid and Fog Index) for each report, and after combining with the previous year's performance, make class predictions using a support vector machine (SVM) method. The combined model will only make a recommendation on a particular annual report when the two models are in agreement; otherwise, the model outputs no decision. The models make predictions whether a company will over- or under-perform S&P 500 index over the coming year. This is an appropriate benchmark as all the companies being analyzed are components of this index. We believe that this is a very meaningful comparison. In some published results, performance of an algorithm was evaluated by measuring how accurately one can predict increase or decrease of a stock price. This evaluation approach may lead us to believe that an algorithm has a good performance, while it may be worse than the index performance. Hence it would be useless to an investor, who could simply invest in the index, achieve higher return, and be exposed to lower risk.

2   Related Work

As text processing techniques become more sophisticated its ability to work in the financial domain becomes more attractive. There has been a few publications in which textual information was analyzed in relation to financial performance. In comparison, the novelty of our approach is in applying character n-gram analysis and readability scores with the SVM method to the annual reports in making long-term predictions. Pushing the time-horizon for making predictions creates a more practical model, and thus it has a wider appeal in the investment industry. In [2], the effects of news articles on intra-day stock prices are analyzed. The analysis was conducted using vector space modeling and tfidf term weighting scheme, then the relationship between news stories and stock prices was defined with a support vector machine [2]. The experiments produced results with accuracy as high as 83% which translated to 1.6 times the prediction ability when compared to random sampling. Similarly, Chen and Schumaker (2006) [3] compared three text processing representations combined with support vector machines to test which was the most reliable in predicting stock prices. They analyzed the representations based on bag-of-words, noun phrases and named entities, and all of the models produced better results than linear regression; however named entities proved to be the most robust[3]. Other intra-day predictions facilitated through text mining were done by Mittermayer (2004) [4], where he created NewsCATS— an automated system that could day-trade the major American stock indexes. The model was created to automate the trading decisions based on news articles immediately after they are released. Kloptchenko *et al.* (2002) [5] focused on clustering quarterly financial reports in the telecom industry. They were not making predictions on future performance but attempting to use prototype-matching text clustering and collocational networks to visualize the reports. The collocational networks cut down the time required by an analyst to read the report and identify important developments [5]. This work was improved upon for making predictions and the new results (Kloptchenko *et al.* 2004) [6] were released new results, in which prototype-matching text clustering for textual information was combined with self-organizing maps for quantitative analysis. Their analysis was performed on quarterly and annual financial reports from three companies in the telecom industry. The results implied that some indication about the financial performance of the company can be gained from the textual component of the reports; however, it was also noted that the clusters from quantitative and qualitative analysis did not coincide. They explained this phenomenon by stating that the quantitative analysis reflects past performance and the text holds information about future performance and managerial expectations. Before complex text mining methods were developed, the work done by Subramanian, Insley, and Blackwell [7] in 1992 showed that there was a clear distinction between the readability scores of profitable and unprofitable companies. In more recent work by Li [8], he examined the relationship between annual report readability combined with current earnings and earnings persistence, with a firm's earnings. His conclusion was that firms with lower earnings had reports which were more difficult to read and longer.

3   Data Pre-processing

3.1   Data Collection

There are no known publicly available data sets that would contain a preprocessed sample of annual reports to analyze, so the data set was created from scratch. To facilitate this, the website of each company considered was visited and the relevant annual reports were downloaded from the investor relations section. Prior to downloading, every report's security features were checked to ensure the PDF was not protected; if it was, then it was discarded as the file could not be converted to text (text format is required to apply n-gram and readability programs). Once a sufficient sample size of annual reports was collected, they were converted to text using a Perl script with program `pdftotext`.

3.2   Data Labeling

The most sensitive and time consuming process of the experiment was class labeling of the training and testing data. It is not mandated by the SEC that companies file their annual reports at the same time, so as a result, each performance measure has to be individually calculated for each company, based on different months. To expedite this process, a matrix of relative returns was created based on monthly closing prices for each stock from data obtained from Yahoo! Finance [9]. The returns for each month were calculated as a numeric figure, and introduced as a class attribute as either over or under performing the S&P 500 over the trailing 12 month period. Next, the filling date for the reports was captured from the SEC website and the appropriate text file is labeled. This was done manually for each report.

3.3   Generating N-Gram Profiles

The n-gram profiles were created as defined by the CNG method [1] using the Perl n-gram module Text::Ngrams developed by Keselj [10]. The character six-grams and word tri-grams were used, and various profile lengths up to 5000 unique, normalized, most-frequent n-grams from an annual report were used.

3.4   Generating Readability Scores

A Perl script was created that generated the three readability scores from source code developed by Kim Ryan [11] and made publicly at CPAN [12]. The scores for each annual report are combined with the underlying securities' 1-year past performance to form the input attribute set for the SVM. The previous year's performance was represented in two ways: first by its relative performance to the S&P 500, and by an indicator whether or not it decreased or increased in value over the last year. To make the data appropriate for the SVM it was scaled between 0 and 1 to cut down on computation size and transformed into the required format. The three readability scores considered where the Gunning Fog Index, Flesch Reading Ease, and Flesch-Kincaid Grade Level. The Gunning Fog Index developed by Robert Gunning in 1952 is a measure of readability of an English sample of writing, the output is a reading level that indicates the number of years of formal education required to understand the text, and the equation is as follows:

$$Gunning \ Fog \ Index = 0.4 \cdot \left(\frac{\#words}{\#sentences} + 100 \cdot \frac{\#complex \ words}{\#words} \right)$$

where *#words* is the number of words in text, *#sentences* number of sentences, and *#complex words* number of words that are not proper nouns and have three or more syllables. The Flesch Reading Ease (FRE) and Flesch-Kincaid Grade Level (FKL) were both created by Rudolph Flesch. The higher the FRE score the simpler the text and the output for the FKL is similar to the Gunning Fog Index, where it generates a Grade Level that reflects the number of years of formal education required to understand it. The two scores are imperfectly correlated and therefore it is meaningful to consider them both. Their respective equations are given below:

$$Flesch\ Reading\ Ease = 206.835 - 1.015 \cdot \frac{\#words}{\#sentences} - 84.6 \cdot \frac{\#syllables}{\#words}$$

$$Flesch\text{-}Kinkaid\ Grade\ Level = 0.39 \cdot \frac{\#words}{\#sentences} + 11.8 \cdot \frac{\#syllables}{\#words} - 15.59$$

The algorithm for syllable count was implemented as the Perl module Lingua::EN::Syllable [13], with estimated accuracy of 85–90%.

4   CNG Classification of N-Gram Profiles

The n-gram classification technique was inspired by work done by Kešelj, Peng, Cercone and Thomas [1], where n-gram profiles were used, with a high degree of accuracy, to predict author attribution for a given unlabeled sample of writing. A generalized profile for a given author was generated and then used to gauge a distance calculation from new testing documents. For financial forecasting a general n-gram profile was created from all of the company annual reports for a given class. The classifier would concatenate all the files from one class or another and then generate one overall n-gram profile with the same settings as discussed in the data pre-processing subsection. For each testing year x the training profiles would be generated from years $x - 1$ and $x - 2$. Once the two generalized profiles are created, one for over-preforming and one for under-performing stocks, the profiles of documents from the testing year are compared with the training profiles using the CNG distance measure:

$$\Sigma_{s \in profiles} \left(\frac{f_1(s) - f_2(s)}{\frac{f_1(s) + f_2(s)}{2}} \right)^2$$

where s is any n-gram from one of the two profiles, $f_1(s)$ is the frequency of the n-gram in one profile, or 0 if the n-gram does not exist in the profile, and $f_2(s)$ is the frequency of the n-gram in the other profile.

5   SVM Classification with Readability Scores

The input attributes to the SVM method where vector representations of the annual reports that contained the three readability scores and the stock's performance over the previous year. An SVM is a very robust classifier that has proven effective when dealing with highly complex and non-linear data, which is indicative of data found in the financial domain. SVM's had been widely experimented with financial forecasting in both classification [14,15,16] and level estimation or regression [17] domains. Because the scores are not time sensitive and the SVM does not take into account any time dependencies when evaluating the data, all of the vector representations were used to train the system, except for the particular year it was tested on at any given time. The Support Vector Machine environment utilized was LIBSVM [18]—a very powerful integrated software for support vector classification and regression. It utilizes an SMO-type algorithm[14] for solving the Karush-Kuhn-Tucker (KKT) conditions. A polynomial kernel of degree 3 was used, with the c-SVM approach; i.e., the use of slack variables to allow for "soft" margin optimization.

Five input attributes are used in SVM classification: three readability scores from annual reports, and two performance measures in the previous year: one whether the stock over or under performed, and the second whether the stock price increased or decreased in the previous year.

6   Experimental Results

In general all three individual models and the two combinations preformed well and overall, they each outperformed the benchmark return in the testing period. To display the results, a special attention is given to the three criteria: overall accuracy, overperform precision, rate and investment return. Over-performing precision is a point of interest on its own as positive predictions classify a stock as a future over-performer, and therefore would initiate an investment in the market. This opens the portfolio up to potential losses since an actual position has been taken. However, when the model predicts an under-performing stock, it passes it over for investing and when the prediction is wrong it is only penalized by missing out on a return—an opportunity cost and not an actual dollar loss. Next, we look at each model's performance individually, and then on some comparisons between them and the benchmark. The benchmark portfolio consists of an equal investment in all available stocks in each of the testing periods. The S&P 500 was not used as the experiment sample did not include all underlying assets in the S&P 500 index.

Table 1 displays comparative models' performance year over year for percentage return, cumulative dollar returns and accuracy, and over- and under-performance precision of the model.

Character N-grams with CNG (C-grams) method outperformed the benchmark portfolio return overall and in five of the six years.

Word N-grams with CNG Classification (W-grams) model had superior accuracy and over-performance precision to that of the character n-gram model, and it also outperformed the benchmark return.

Readability Scores with SVM (Read) performed well, and in all but one year outperformed the benchmark and the n-gram model.

Combined Readability-scores with Character N-grams (Combo-char) makes a recommendation only when there is an agreement between the

Table 1. Detailed Experimental Results

Character N-gram Model

Year	Return (% and $)		Accuracy	Over-perf.	Under-perf.	No Decision
2003	-6.59%	$9341.18	61.91%	70.59%	25.00%	
2004	47.80%	$13806.26	60.87%	65.00%	33.33%	
2005	20.32%	$16611.11	53.12%	52.63%	53.85%	
2006	31.48%	$21839.65	51.28%	52.38%	50.00%	
2007	34.67%	$29410.73	63.41%	75.00%	58.62%	
2008	-10.33%	$26371.62	41.02%	26.67%	50.00%	
Overall	163.72%	$26371.62	55.27%	57.04%	45.13%	

Word N-gram Model

Year	Return (% and $)		Accuracy	Over-perf.	Under-perf.	No Decision
2003	-3.00%	$9700.00	71.43%	80.00%	50.00%	
2004	50.53%	$14601.35	56.52%	64.71%	33.33%	
2005	15.82%	$16911.02	50.00%	50.00%	50.00%	
2006	27.94%	$21636.71	53.85%	55.56%	47.62%	
2007	36.60%	$29555.75	70.73%	80.00%	65.38%	
2008	-9.29%	$26808.80	51.28%	41.18%	59.09%	
Overall	168.09%	$26808.80	58.97%	61.91%	50.90%	

Readability Model with SVM

Year	Return (% and $)		Accuracy	Over-perf.	Under-perf.	No Decision
2003	-2.42%	$9758.33	66.67%	81.82%	44.44%	
2004	30.07%	$12692.34	56.52%	66.67%	37.50%	
2005	25.23%	$15894.71	59.38%	61.54%	57.89%	
2006	48.06%	$23534.11	69.23%	75.00%	65.22%	
2007	19.33%	$28084.04	60.98%	59.26%	64.29%	
2008	-3.13%	$27206.41	64.10%	62.50%	64.52%	
Overall	172.06%	$27206.41	62.81%	67.80%	55.64%	

Combined Readability and Character N-grams

Year	Return (% and $)		Accuracy	Over-perf.	Under-perf.	No Decision
2003	-2.42%	$9,758.33	68.75%	83.33%	5.88%	5.60%
2004	27.69%	$12,460.64	64.29%	61.54%	25.49%	9.97%
2005	35.22%	$16,849.56	61.11%	66.67%	9.80%	7.72%
2006	73.50%	$29,233.98	78.57%	83.33%	8.82%	7.54%
2007	41.50%	$41,366.08	72.73%	90.00%	11.44%	9.09%
2008	39.00%	$57,498.85	55.56%	100.00%	1.04%	1.06%
Overall	474.99%	$57,498.85	66.83%	76.47%	62.48%	6.83%

Combined Readability and Word N-grams

Year	Return (% and $)		Accuracy	Over-perf.	Under-perf.	No Decision
2003	-3.55%	9,645.4545	72.22%	83.33%	50.00%	14.29%
2004	26.30%	12,182.2091	63.64%	60.00%	100.00%	52.17%
2005	32.50%	16,141.4270	58.82%	70.00%	42.86%	46.88%
2006	40.50%	22,678.7050	76.47%	66.67%	81.82%	75.86%
2007	43.08%	32,449.4471	78.26%	91.67%	63.64%	43.90%
2008	4.00%	33,747.4250	68.75%	100.00%	66.67%	58.97%
Overall	237.47%	33,747.4250	69.69%	76.47%	65.68%	48.68%

two combined methods. In addition to previously mentioned measures, for the combined models we also consider the percentage of cases with no decision due to the disagreement of the models.

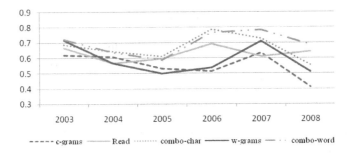

Fig. 1. Year over year accuracy

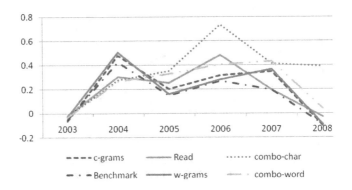

Fig. 2. Year over year % returns

Combined Readability-scores with Word N-grams (Combo-word) performed better than the benchmark, but significantly worse than the Combo-char model.

6.1 Model Results Comparison

To adequately compare the models we present in this subsection performances graphically on a combined plot. Figure 1 plots the year over year percentage accuracy of the five models. We can see that the word-combo model had better accuracy in all six years including 2008 when the market experienced a major trend shift. It is worth noting that the character-gram model slipped below the 50% margin in the last year during the trend change in 2007–2008. This was the only occurrence of any of the models performing below 50% accuracy.

Figures 2 and 3 chart the percentage return and overall dollar return respectively for the five models and the benchmark portfolio.

Comparing the plots between the models and the benchmark portfolio it appears that their trends all match a general shape, only that in the majority of the years the benchmark is the poorest performer. In 2008 the only models to

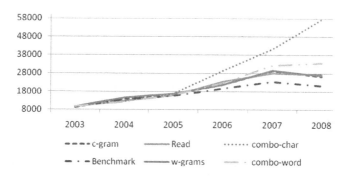

Fig. 3. Cumulative investment returns (in dollars, with initial investment $10,000)

produce a positive return were the combined models and this was achieved when the benchmark lost nearly 10%.

By a large margin the character n-gram combination model had the superior investment strategy. For the first three years all 4 portfolios were quite close but in 2006 the character n-gram combination model pulled away and in 2008 picked up its most significant relative gain. This 2008 return is a direct result from the benefit of having a perfect overperformance precision rate.

7   Discussion and Conclusions

In general, the endeavour put forth here is an attempt to automate the analysis of annual reports. The expected benefit is that one could quickly evaluate the textual component and remove some of the uncertainty that arises from analysts having different opinions. More specifically, two novel NLP techniques are applied to solving the aforementioned problem. This section details the results, and gives some explanations as to what worked and what did not.

7.1   N-Grams with CNG Classification

It has been shown that this methodology can be effective the problem of authorship attribution. In changing from the authorship attribution task to recognizing language indicative to one type of behaviour to another is a bit of a stretch. The belief is that certain language and phrases are used when the outlook is bleak and is measurably different than that when the outlook is positive. Overall, both the n-gram models were the weakest of the five models constructed, however they were still superior to the benchmark portfolio and that fact alone makes the experiment a success. The two n-gram based models had similar results, with the word-grams performing slightly better in overall accuracy and investment return. Although neither n-gram approach could capture all the information in the report, it was able to model a portion of it, such that, sufficient enough to give above average returns. The n-grams proved to be least effective when the market trend drastically shifted in 2007–2008. This may not necessarily be a short-coming of the n-grams themselves but the classification approach applied to them. It would be interesting to use a SVM for the n-gram profiles as a comparison to the CNG method. The overall accuracy of the models were about 55% and 59% for character-grams and word-grams respectively which is quite typical of investment models and is good evidence that it is better that random guessing.

7.2 Readability-Scores with SVM

As noted earlier, SVM's have proven very effective at producing robust investment models and dealing with the highly complex and non-linear data that is inherent in financial forecasting. Part of the success of this model could be attributed to the SVM choice of the classifier. Based on our preliminary tests, some other algorithms such as Artificial Neural Networks or Naïve Bayes could not achieve the same accuracy. Readability scores and their relation to stock performance have been well documented and the favourable results of this method are not unexpected as this model combined a proven linguistic analysis technique with a powerful classification algorithm. This model outperformed the n-grams technique and the benchmark portfolio on investment return (percentage and dollars) and in overperform precision, which made for more efficient trades. The overall accuracy and over-performance precision was 62.81% and 67.80% respectively, giving evidence that the model was more than just random guessing. This technique also demonstrated an ability to partly understand the text in the annual reports and learn what it indicated for future performance.

7.3 Combined Models

Choosing to only make decisions when the models agreed proved to be a valuable approach. This approach could be characterized as an ad hoc ensemble approach. It is evident that the three individual models were each able to explain part of the relationship between performance and the textual components of the annual reports and that what they learned was not completely overlapping. The combined models consistently outperformed the individual models and the benchmark portfolio. The combined models were also the most efficient as they made only about half the number of trades as the other three. This fact is evident from the "no decision" figures in table 1, where on average 40% (character n-grams combo) and 48% (word n-gram combo) of the time the two models did not agree and therefore no position was taken. Having the two models agree introduced a further confidence factor into the combined model which makes it more robust to noise in the market. In the majority of the years and overall the combined models proved superior in terms of investment return (dollar and percentage), overperformance precision, accuracy, and efficiency of investments. The most significant difference came in 2008 when the other three portfolios posted negative returns and the combined models made a positive gain of 39% (character n-gram combo) and 4% (word n-gram combo). It is also interesting that in this year the character combined model was not as accurate as the Readability model but it did, like the word n-gram combo, have a perfect 100% for overperform precision and therefore made no poor

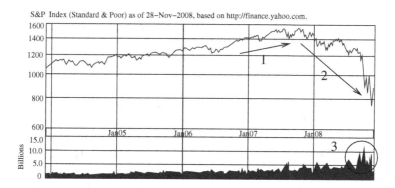

Fig. 4. S&P 500 Index Returns

choices when an actual position in the market was taken. This abnormal investment return in 2008 is a bit of an anomaly and is not entirely realistic and will be discussed in the next section.

7.4   The 2008 Investment Anomaly

An overperformance precision of 1 and an investment return of 39% or 4% when the market losses almost 10% seems very good, however the problem is the models are suppose to build a portfolio of investments to spread the risk. Due to the volatile nature of the markets in 2007-2008 the two models were only able to agree once on an over-performer and therefore only made one investment each in the market. In reality an investment manager would most likely not have accepted this response and either moved some of the assets to the money market or conducted further analysis on the companies to find other suitable investments. The annual reports that the 2008 returns are calculated from are the 2006 annual reports produced sometime in 2007. Figure 4 illustrates the massive shift of market momentum in 2007.

The arrow labeled '1' represents the time period when the 2006 annual reports were being published and arrow '2' represents the time period when the actual performance was being evaluated. It is quite clear that the market environment drastically changed between those two time periods and the increase volatility is supported by the large increase in market volume highlighted by the circle labeled '3'.

8   Drawbacks, Limitations, and Future Work

Although the results are persuasive that the techniques presented are effective at analyzing annual reports, there still is a need for more thorough testing with an expanded data set that contains more of the companies in the S&P 500 index. The n-gram profiles were set size 6 and 3 for the character grams and word grams respectively taking up to the top 5000, these settings are most likely a local optimum and require fine tuning to optimize the model. With all the recent turmoil and volatility in the financial markets it will be worth applying the models to the newly released annual reports over the coming year to see how the models hold up under such extreme conditions. There is also a lot of information that is generated and can be learned from the experiment and deeper drilling down through the data could reveal more interesting information. For example, it would be interesting to know if there are some companies that produce more easily read annual reports making them more transparent, and therefore a safer investment, or if the distance score that the CNG classifier reports is an indication of how sure the model is and could a threshold be introduced to improve overall accuracy and overperform precision. Finally the labeling process should be automated to cut down on pre-processing time and human error.

References

1. Keselj, V., Peng, F., Cercone, N., Thomas, C.: N-gram-based author profiles for authorship attribution. In: Proceedings of the Conference Pacific Association for Computational Linguistics, PACLING 2003, Dalhousie University, Halifax, Nova Scotia, Canada, pp. 255–264 (August 2003)
2. Falinouss, P.: Stock trend prediction using news articles. Master's thesis, Lulea University of Technology (2007) ISSN 1653-0187
3. Schumaker, R., Chen, H.: Textual analysis of stock market prediction using financial news articles. In: Proc. from the America's Conf. on Inform. Systems (2006)
4. Mittermayer, M.: Forecasting intraday stock price trends with text mining techniques. In: Proc. of the 37th Hawaii Int'nal Conf. on System Sciences (2004)
5. Kloptchenko, A., Magnusson, C., Back, B., Vanharanta, H., Visa, A.: Mining textual contents of quarterly reports. Technical Report No. 515, TUCS (May 2002) ISBN 952-12-1138-5
6. Kloptchenko, A., Eklund, T., Karlsson, J., Back, B., Vanharanta, H., Visa, A.: Combined data and text mining techniques for analysing financial reports. Intelligent Systems in Accounting, Finance and Management 12, 29–41 (2004)
7. Subramanian, R., Insley, R., Blackwell, R.: Performance and readability: A comparison of annual reports of profitable and unprofitable corporations. The Journal of Business Communication (1993)
8. Li, F.: Annual report readability, current earnings, and earnings persistence. Journal of Accounting and Economics (2008)
9. Yahoo! Inc.: Yahoo! finance, http://ca.finance.yahoo.com/ (last access 2008)
10. Kešelj, V.: Text:Ngrams Perl module for flexible ngram analysis (2003–2009), http://www.cs.dal.ca/~Evlado/srcperl/Ngrams/Ngrams.html Ver, 2.002. Avail. at CPAN
11. Ryan, K.: Lingua:EN: Fathom Perl module for measuring readability of english text. Available at CPAN (2007)
12. CPAN Community: CPAN—Comprehensive Perl Archive Network (1995–2009), http://cpan.org
13. Fast, G.: Lingua:EN: Syllable Perl module for estimating syllable count in words. Available at CPAN (1999), http://search.cpan.org/perldoc?Lingua:EN:Syllable

14. Fan, R., Chen, P., Lin, C.: Working set selection using second order information for training SVM. Journal of Machine Learning Research 6, 1889–1918 (2005)
15. Fan, A., Palaniswami, M.: Stock selection using support vector machines. In: Proceedings of IJCNN 2001, vol. 3, pp. 1793–1798 (2001)
16. Huang, W., Nakamori, Y., Want, S.-Y.: Forecasting stock market movement direction with support vector machine. Computers and Operations Research 32, 2513–2522 (2005)
17. Kim, K.: Financial time series forecasting using support vector machines. Neurocomputing 55, 307–319 (2003)
18. Chang, C.C., Lin, C.J.: LIBSVM: a library for support vector machines (2001), http://www.csie.ntu.edu.tw/~7Ecjlin/libsvm

Statistical Parsing with Context-Free Filtering Grammar

Michael Demko and Gerald Penn

Department of Computer Science
University of Toronto, Ontario, M5S 3G4
{mpademko,gpenn}@cs.toronto.edu

Abstract. Statistical parsers that simultaneously generate both phrase-structure and lexical dependency trees have been limited to date in two important ways: detecting non-projective dependencies has not been integrated with other parsing decisions, and/or the constraints between phrase-structure and dependency structure have been overly strict. We introduce context-free filtering grammar as a generalization of a lexicalized factored parsing model, and develop a scoring model to resolve parsing ambiguities for this new grammar formalism. We demonstrate the new model's flexibility by implementing a statistical parser for German, a freer-word-order language exhibiting a mixture of projective and non-projective syntax, using the TüBa-D/Z treebank [1].

1 Context-Free Filtering Grammar

The factored parsing model of Klein & Manning [2] is an interesting framework in which phrase structure and dependency parsing are interleaved. Unfortunately, the framework as presented therein does not permit the use of linguistically plausible dependency trees for free word-order languages: the dependency trees are constrained to match the phrase structure very closely, one consequence of which is that the dependency trees must be projective.

Context-free filtering grammar (CFiG) provides more flexibility by permitting more than one token of a constituent to behave like a head in the sense that it may govern any token within the constituent and depend on tokens outside the constituent. The parsing model of [2] is a special case of CFiG.

Our model uses a PCFG to score phrase structure trees and McDonald's [3] linear model to score dependency trees. Combined trees are scored by summing the log PCFG score and the dependency score. Like [2], we use an A* search to find the model-optimal combined parse. A* requires an admissible and monotonic scoring heuristic to predict the optimal score of any full solution incorporating some partial result. We use outside scores to build a heuristic for the PCFG model and a maximum-weight incident edge heuristic for the dependency model.

There are, of course, a multitude of linguistic formalisms that effectively generate dependency representations while accommodating unbounded dependencies and non-projective structures, e.g., CCG, HPSG and LFG. When the mapping to dependencies is left suitably non-deterministic, these can serve as the basis

Y. Gao and N. Japkowicz (Eds.): Canadian AI 2009, LNAI 5549, pp. 52–63, 2009.

for statistical parsers, e.g., as by [4]. These formalisms are much more ambitious, however, in their rejection of classical context-free phrase structure as the goal representation of parsing. While this rejection is not without merit, the question nevertheless remains whether there might be a more conservative extension of CFGs that can cope with freer word order. The search for and experimentation with such extensions, including CFiG, allows us to investigate the shortcomings of CFGs in a more controlled fashion, which in turn may inform these more ambitious attempts.

1.1   Definitions

Formally, a CFiG G is a 4-tuple $\langle \mathcal{N}, \mathcal{T}, \mathcal{P}, S \rangle$, where $\mathcal{N}$ is a set of non-terminals, $\mathcal{T}$ is a set of pre-terminals (parts of speech), $\mathcal{P}$ is a set of production rules, $S \in \mathcal{N}$ is the start symbol. Each production rule in $\mathcal{P}$ has the form: $L \leftarrow R_0^{e_0} R_1^{e_1} ... R_n^{e_n}$, where $L \in \mathcal{N}$, $R_i \in \mathcal{N} \cup \mathcal{T}$ and $e_i \in \{Hidden, Exposed, Unlinked\}, \forall i$ from $0..n$. At least one of $e_0, e_1, ..., e_n$ must have the value $Exposed$ or $Unlinked$.

Informally, $(L \leftarrow R_0^{e_0} R_1^{e_1} ... R_n^{e_n})$ is the same as a context-free production, except that each right-hand-side symbol is marked as either exposing or hiding its tokens. In a corresponding dependency tree, tokens inside $Exposed$ children may govern tokens of their siblings and may depend on tokens outside the parent constituent. Tokens inside $Hidden$ children may govern tokens only within the same child and may depend only on tokens within the parent constituent. As we will see in Section 1.4, the label $Exposed$ can be thought of as a generalization of the label $Head$ from more traditional lexicalized phrase structure formalisms. Dependency trees can be thought of here as 'skeletons' and phrase structure trees as 'shells' containing these skeletons, restricting their possible attachments. $Hidden$ is discussed in Section 1.2, and $Unlinked$, in Section 1.3.

Figure 1 shows an example of the TüBa-D/Z annotation. A small fragment of the CFiG stripped from TüBa-D/Z is shown in Figure 2, and will be used for some motivational examples. We will adopt the familiar notation for phrase structure derivations: $(D : S \rightarrow_G^* \sigma)$ of the string σ from S according to G. Restricting ourselves to left-most depth-first derivations, we can also speak of (labeled) phrase structure trees produced by CFiGs.

Constituent C_2 *strictly dominates* C_1 $(C_2 \triangleright C_1)$ iff C_1 is contained within C_2 (and $C_1 \neq C_2$). Abusing notation, we say that a pre-terminal constituent

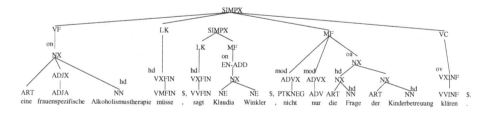

Fig. 1. A disconnected phrase structure tree from TüBa-D/Z

$$\mathcal{T} = \{NN, PRELS, PIDAT, VMFIN, VAFIN, ADV, PIS, VVINF, PPER, VVFIN\}$$
$$\mathcal{N} = \{NX, VF, SIMPX, MF, C, VXFIN, LK, ADVX, VXINF, R\ SIMPX, VC, NF\}$$
$$S = SIMPX$$
$$\mathcal{P} = \{SIMPX \leftarrow VF^H LK^E MF^E VC^E NF^H, R\ SIMPX \leftarrow C^H MF^H VC^E,$$
$$C \leftarrow NX^E, VF \leftarrow NX^E, LK \leftarrow VXFIN^E, MF \leftarrow NX^E, MF \leftarrow ADVX^E NX^E,$$
$$MF \leftarrow NX^E NX^E, VC \leftarrow VXFIN^E, VC \leftarrow VXINF^E, NF \leftarrow R\ SIMPX^E,$$
$$NX \leftarrow NX^E R\ SIMPX^H, NX \leftarrow NN^E, NX \leftarrow PRELS^E, NX \leftarrow PPER^E,$$
$$NX \leftarrow PIDAT^H NN^E, VXFIN \leftarrow VMFIN^E, VXFIN \leftarrow VAFIN^E,$$
$$VXFIN \leftarrow VVFIN^E, VXINF \leftarrow VVINF^E, ADVX \leftarrow ADV^E\}$$

Fig. 2. German CFiG fragment (TüBa-D/Z)

dominates its matching terminal, although G (as a CFG) only generates pre-terminals. If two constituents C_1 and C_2 are in P, then we refer to their lowest common ancestor C^* as their *join* ($C^* = C_1 \sqcup_P C_2$). If p is a phrase structure tree with terminal x in its yield, then $envelope_p(x)$ refers to the smallest *Hidden* constituent in p that dominates x, or the root constituent if x is not contained in any *Hidden* constituent.

1.2 Constraining Dependency Structure with Constituent Hiding

We will represent dependency trees as connected DAGs. Each vertex is labeled with a terminal (word token), pre-terminal and linear position (from string σ). Each edge represents a dependency, pointing from the governor to the dependent. One vertex, the root, has no incoming edges; all others have exactly one incoming edge. In figures, we denote the root by an incoming edge from outside the graph.

Let σ be a string, G be a CFiG, and p be the parse tree induced by $D : S \rightarrow^*_G \sigma$, and $g = \{\mathcal{V}, \mathcal{E}\}$ be a dependency tree of σ. Then, together, p and g are a *parse of* σ iff: **(1)** $(u, v) \in \mathcal{E} \Rightarrow envelope_p(u) \rhd v$ (i.e., tokens in *Hidden* constituents may only govern tokens within the same constituent, since dependency edges may not leave a constituent marked *Hidden*); and **(2)** $(u, v) \in \mathcal{E} \Rightarrow \not\exists i.u \sqcup_p v \rhd i \rhd envelope_p(v)$ (i.e., a token in a *Hidden* constituent may only depend on a token within its immediate parent, since dependency edges may not enter a *Hidden* constituent from anywhere other than within the parent).

1.3 Unlinking

The experiment described in Section 4 is conducted on a modified version of the TüBa-D/Z treebank. TüBa-D/Z is a phrase structure treebank, in which some trees contain disconnected constituents (Figure 1). We factor these into pairs of phrase structure and dependency trees, and re-connect the phrase structure with some simple heuristics. Connecting the dependency trees is considerably more difficult, in general, so we left these disconnected. In other words, there may be multiple root governors in the dependency structure. To allow for these disconnected dependency structures, we need the third exposure value, *Unlinked*. Each token exposed within an unlinked child must either be governed by another token exposed within the child, or must be ungoverned (a root). Tokens from an *Unlinked* constituent may not take part in dependency relations with tokens from sibling constituents.

To simplify parsing, we will want that *only* tokens of *Unlinked* constituents be allowed as roots. The easiest way to achieve this is to add a new start symbol S', and the rule $S' \leftarrow S^{Unlinked}$.

1.4 Factored Lexicalized PCFG as a Special Case

The CFiG approach is best seen as a generalization of the factored parsing model of Klein & Manning [2]. They propose this model in order to disentangle statistical models of phrase structure and lexical headedness within a lexicalized phrase structure parser. They recommend simultaneously searching for a phrase structure tree and a compatible dependency tree using A* search. Because their starting point is lexicalized PCFG parsing, their notions of 'dependency tree' and 'compatibility' are fairly limited. In particular, each phrase structure constituent must have exactly one head child, the lexical head of the constituent is the lexical head of that child, and the lexical heads of all other children of a constituent must be immediately governed by the head child's lexical head.

Consider the grammar in Figure 3, in which head symbols are marked with an asterisk. With this grammar, we can find a lexicalized parse tree for "John/NP drove/V Alana/NP To/TO School/NP" as shown in Figure 4(a). We can trivially express this grammar, and any context-free grammar with one marked head child per rule, as a CFiG, as shown in Figure 3, for which the corresponding parse tree for "John/NP drove/V Alana/NP To/TO School/NP" is shown in Figure 4(b). Thin arrows indicate dependency edges. Attached subconstituents are enclosed either with a solid line to indicate that their contents have been hidden by their parent, or a dashed line to indicate that their contents have been exposed by their parent. Unattached subconstituents are drawn with a dotted line: a constituent's exposure is only defined in relation to its parent constituent.

In fact, the factored grammars of [2] are precisely the subset of CFiGs in which only one child is exposed in each constituent. In particular, the one *exposed* child in each constituent is the *head* child of the factored model, only one token is exposed within each constituent (provable by induction on the height of the phrase structure) — this token is the *lexical head* of the constituent under the factored model — and since only one token is exposed within a constituent, that one token is the only candidate governor for the lexical heads of the other children. It follows that the lexical heads of the other children will all be immediate dependents of that token (the head child's lexical head).

$$\mathcal{T} = \{V, NP, TO\} \quad \mathcal{N} = \{S, VP, PP\}$$
$$\mathcal{P} = \{S \leftarrow NP\ VP^*, PP \leftarrow TO^*\ NP,$$
$$VP \leftarrow V^*\ NP\ PP\}$$
$$(a)$$

$$\mathcal{T} = \{V, NP, TO\} \quad \mathcal{N} = \{S, VP, PP\}$$
$$\mathcal{P} = \{S \leftarrow NP^H VP^E, PP \leftarrow TO^E NP^H,$$
$$VP \leftarrow V^E NP^H PP^H\}$$
$$(b)$$

Fig. 3. (a) A lexicalized grammar and (b) its equivalent CFiG

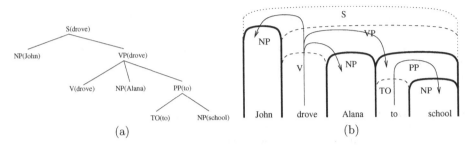

Fig. 4. (a) A simple lexicalized parse tree and (b) its factored representation

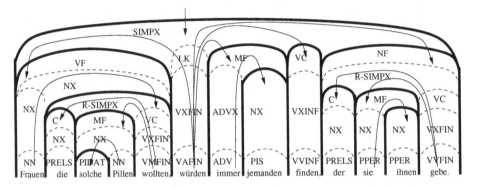

Fig. 5. A phrase structure / dependency structure pair à la Klein & Manning [2]

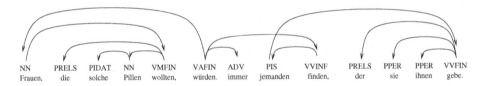

Fig. 6. A better dependency tree

As to how this difference bears on German, Figure 5 shows an example phrase structure annotation under the Klein & Manning factored model, which does not admit a linguistically plausible dependency structure (such as Figure 6). Three problems are evident. First, German topological fields do not always have lexical heads. The Mittelfeld (`MF`) constituent in particular contains two tokens properly in dependency relations with tokens from other fields. Second, although the main clause can be said to have a head ('würden'), not all tokens exposed within the clause are proper dependents of it, e.g. 'jemanden', which should be governed by 'finden'. Third, a good dependency tree for this sentence is not projective: the relative clause is governed by the token 'jemanden,' and this dependency must cross the relation between 'würden' and 'finden.'

All of these problems can be overcome with a CFiG. The parse tree shown in Figure 7 is derived from the grammar of Figure 2.

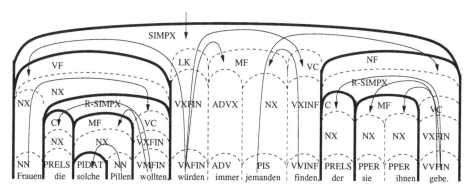

Fig. 7. A better phrase-filtered dependency tree licensed by a CFiG

2   Parsing

As with traditional lexicalized grammars we can use a bottom-up chart parser
for CFiGs. The additional flexibility of the new formalism comes at the cost
of greater interconnectedness within its parse trees, however. In a traditional
lexicalized phrase structure tree, each constituent is characterized by a *phrase
category* label and a *single exposed token*. Constituents with the same category
and head are interchangeable syntactically. This interchangeability permits the
development of an efficient chart parser.

In CFiGs,[1] passive arc signatures consist of a phrase-label and a *set* of tokens
exposed within the constituent spanned by the arc. Passive arcs store both the
internal phrase structure of the constituent and all of the dependency relations
involving hidden tokens spanned by the arc. Constituents are interchangeable
syntactically only if their category and entire set of exposed tokens match. If a
grammar permits the creation of constituents with arbitrarily large exposed sets,
then dynamic programming alone may not guarantee efficient parsing. Active
arcs match lists of passive arcs to prefixes of production rule right-hand-sides.
This matching includes an assignment of exposure to each passive arc.

Fortunately it is possible, within the framework, to limit this production of
arbitrarily large exposed sets to a small group of phrase categories. A phrase of
category L can expose more than one token only if either a rule with L in its
left-hand-side marks more than one right-hand-side symbol as exposed, or a rule
with L in its left-hand-side marks as exposed a symbol of another category that
can expose more than one token.

We can therefore construct a grammar for which many phrase categories are
as well-behaved in complexity as they would be in a stricter lexicalized grammar
— only a small number of categories have the greater freedom required to license
plausible linguistic structures. In the case of TüBa-D/Z , German treebank used
in our experiment, phrase constituents like noun-phrases will only ever expose a
single head-token, whereas some topological constituents, notably the Mittelfeld,

---

[1] The complete details of the parser and its chart actions can be found in XXX (2007).

Table 1. Dependency edge scoring features — c-w: child (dependent) word type, c-p: child POS, p-w, p-p: parent (governor) word and POS, dist: position of child minus position of parent, dir: sign of dist, b-p: POS of a token between parent and child, p-p+1, p-p-1, c-p+1, c-p-1: POSs of tokens (immediately) after parent, before parent, after child and before child

base features

dist, p-w, p-p, c-w, c-p	dist, p-w, c-w
dist, p-w, p-p, c-w	dist, p-w, c-p
dist, p-w, p-p, c-p	dist, p-p, c-w
dist, p-w, p-p,	dist, p-p, c-p
dist, p-w	**between features**
dist, p-p	dir, p-p, b-p, c-p
dist, p-p, c-w, c-p	**context features**
dist, p-w, c-w, c-p	dir, p-p, p-p+1, c-p, c-p+1
dist, c-w, c-p	dir, p-p-1, p-p, c-p, c-p+1
dist, c-w	dir, p-p, p-p+1, c-p-1, c-p
dist, c-p	dir, p-p-1, p-p, c-p-1, c-p

will expose multiple tokens. Clauses also need to expose multiple tokens, so that tokens from more than one field can act as governors within the clause, but the clauses themselves will normally be *Hidden* when embedded in other constituents. This limits the proliferation of exposed tokens.

3   Statistical Parametrization

Following [2], we use a factored scoring model for CFiG parses: a function g_p numerically scores each possible phrase structure tree, another function g_d scores each possible dependency tree, and the score of a pair of trees is defined as $g_p + g_d$.

For this factored model to be of any use, we need a search method that will find the best combined score over all allowed pairs of phrase structure and dependency trees, but as mentioned in Section 2, the search space of our chart parser is potentially exponential in size, so an exhaustive search is not feasible. Fortunately, factored parsing models can be amenable to A* search.

Following McDonald et al. [3,5], we take g_d to be the sum of the independent scores of the individual dependency edges in the tree. For the sake of uniformity, root tokens will be scored as if they were dependent on a special hidden 'ROOT' node. The score of an individual edge e is given by: $score_e = \mathbf{f_e} \cdot \mathbf{w}$, where $\mathbf{f_e}$ is a vector of binary *feature values* (each value is 1 if the dependency has the feature, 0 if not) and $\mathbf{w}$ is a vector of *feature weights*. Every combination of values seen in the training data that matches the fields of one of the feature patterns in Table 1 is considered one feature. ($p\ w = haben, c\ p = PPER, dist = -1$), for example, could be a feature matching the pattern ($p\ w, c\ p, dist$), as could be ($p\ w = Mai, c\ p = APPRAT, dist = -2$). These are only a subset of McDonald's [3,5] features, however; they use not only word-based fields (p_w, c_w) in pattern construction, but also 5-character-prefix-based fields.

Observe that every input token is the destination of one edge. So, an admissible heuristic for g_d on some chart arc is the sum the scores of dependency edges over all tokens for which an edge has been chosen, plus the maximum score over all possible edges that could be chosen for the remaining tokens.

For g_p, like [2], we use the log joint probability of the production rules from the chart arc's derivation, plus the outside score of the arc's constituent. To improve the function's granularity, we use parent encoding and Markovize the resulting PCFG so that children are generated individually, conditioned on their head and (if distinct) inwardly adjacent sibling. Unlike [2], we also condition on the inwardly adjacent sibling's exposure value. In any rule with exactly one exposed child, that child is chosen as the head for scoring; otherwise the leftmost exposed or (for top-level rules) unlinked child is selected.

3.1   Factored Training

As with regular PCFGs, we count the number of occurrences of each rule (keeping in mind that rules may be distinguished by differing exposures, even if the right-hand-side symbols are otherwise identical), and use the counts to compute estimates for all Markovized rule generation probabilities.

TüBa-D/Z is not annotated with constituent exposure settings, however, so these must be inferred during grammar extraction. TüBa-D/Z also includes topological field constituents; all children of each topological field were marked as *Exposed*. Considering all tokens in the yield of any other constituent c, if no token is dependent on a token outside c, then c must be *Unlinked*. If any token governs a dependent outside c or any token is the dependent of another token not in the immediate parent of c, then c must be *Exposed*. Otherwise c is *Hidden*. In some cases, it may be preferable to assign exposures based on linguistic knowledge about the category of c instead.

To train the dependency factor, we must keep in mind that factored parsing must train a model to select among dependency trees licensed by high-scoring phrase structure trees, not among all dependency trees. This is reflected in both the selection of the training tree T and the update procedure used in training to achieve this. Rather use an MST dependency parser, we must choose T using A* search over the full factored scoring model.

In the case where the phrase structure (call it P) returned with T licenses the correct dependency tree T^*, or is outscored by a phrase structure P^* that does, then we follow McDonald et al. [3,5] by using MIRA [6] to estimate the vector of weights, modified to average the weight vector's values over a full course of training [7]. Otherwise, we find the highest scoring phrase structure (call it P_h) that does license the tree T^* and we update by minimizing $| \mathbf{w}' - \mathbf{w} |$ subject to the constraint: $\mathbf{w}' \cdot \mathbf{f}_{T^*} - \mathbf{w}' \cdot \mathbf{f}_T \geq 1.0 + g_p(P) - g_p(P_h)$. In the first case, the phrase structure found by the factored parse is correct — g_p has contributed as much as it can. In the second case, g_p predicts the wrong phrase structure, and g_d must be modified to compensate for the error. The constant factor of 1.0 ensures that the new classifier will give a better score to the correct tree than the found tree, rather than simply an equal score.

4   Experiment

CFiG is of interest for future research because: (1) it permits a parser to reliably generate more useful syntactic analyses than was previously possible; and (2) it improves parsing reliability on tasks of which already-existing systems are capable. So far, we have dwelt on (1) — context-free filtering grammar does allow a simultaneous and complementary phrase structure and dependency analysis that is not possible in previous statistical parsing systems. The experiment described here aims to establish (2). As this section demonstrates, the new system provides this enhanced output without compromising accuracy compared to either pure phrase structure or pure dependency parsers.

Before development of the parsers began, TüBa-D/Z was split into three subsets: a training set comprising roughly 90% of sentences (19,856 of them), a development set comprising roughly 5% of sentences (1,109 of them) and an test set comprising roughly 5% of sentences (1,126 of them). Because of computing resource (memory) limitations, only sentences with 40 or fewer tokens, including punctuation, were used (1,070 in the development set and 1,086 in the test set). The split was done randomly, each sentence being assigned to one of the three subsets independently of all other sentences.

The dependency treebank used in this experiment was extracted from TüBa-D/Z by a rule-based script. In designing this script, we tried as much as possible to rely directly on phrase structure and the edge-labels from the original data.

In addition to the proper dependency trees used in the experiment, we also need trees of linked phrase heads. These phrase-head trees are required to train the Markovized PCFG, against which the experimental model is compared.

The parsing system used was written specifically for this experiment. In total, three parsing modes are supported: a pure phrase structure parser, a pure dependency parser, and a factored parser. In order to limit as much as possible the differences between the algorithms, the factored mode re-uses as many of the components of the two pure modes as possible.

The factored system does not use a pure A* algorithm: the system prunes arcs to avoid running out of memory on difficult sentences. In training, we limit to 500 the number of arcs considered for each possible span and for each left-hand-side category; in evaluation we relax the limit to 1600.[2]

All of the parsers require POS-annotated input; in this experiment we provided gold-standard parts of speech in the input. None of the parsing systems employ parameter smoothing.

4.1   Experimental Hypothesis

We would expect the experimental model to outperform a pure phrase based parser or a pure dependency parser, based upon the original finding of [2] that

---

[2] Increasing the pruning limit beyond 1600 had no impact on the parser output against development data. It would appear that at this level only arcs that will never be part of an optimal solution are pruned.

Table 2. Final results: perfect parse scores, non-root governor precision and recall, and labeled constituent precision/recall with mean crossing brackets per sentence

parser	dependency perfect	phrase perfect
PCFG parser	–	0.361
dependency parser	0.355	–
KM'02 parser	–	0.371
experimental parser	0.368	0.381

parser	NRG precision	NRG recall
dependency parser	0.885	0.884
experimental parser	0.884	0.885

parser	LC precision / recall	x-brackets
PCFG parser	0.909 / 0.900	0.963
KM'02 parser	0.914 / 0.906	0.861
experimental parser	0.917 / 0.909	0.804

their factored model outperforms its component phrase and dependency models on the Penn treebank, and upon the finding by Kuebler et al. [8] that the same factored model outperforms its phrase structure component on TüBa-D/Z .

Having conducted some preliminary tests to tune the parameters of the model on the development set, we ran tuned versions of (1) a pure phrase structure parser, (2) a pure dependency parser, (3) a factored parser à la [2], and a factored parser with our experimental model, on the test set.

4.2   Experimental Results

Tuning determined that the best pure phrase structure parser uses a grammar including grandparent encoding and two siblings worth of horizontal context. This result is consistent with the findings of Klein & Manning [9] on the English Penn treebank. Also, the best dependency parser makes 5 training passes using all features seen at least once in the corpus. It is a bit surprising that restricting features to those seen a minimum number of times does not produce better results. With many learning methods, a glut of rare features could be expected to lead to overfitting; but with the averaged-perceptron-style algorithm we use, the extra features may be reducing overfitting by introducing slack into the model. On the other hand, the best parser à la [2] simply combines the best phrase-structure model with a lexical heads dependency model trained over 8 passes with all available features. No additional training needs to be done after the two models are combined. The best parser using our experimental model is achieved by combining the best phrase structure model with the best dependency model, and then performing an additional 3 training passes.

The final results on evaluation data for these four systems are shown in Table 2. The phrase structure parsing metrics are from PARSEVAL [10]. The experimental model's accuracy does not differ significantly from that of the

Table 3. Comparison with Kuebler et al. [8]

	Kübler et al.	**PCFG**	**experimental**
unlabeled precision	0.923	0.931	0.936
unlabeled recall	0.909	0.923	0.926
labeled precision	0.899	0.914	0.917
labeled recall	0.885	0.906	0.909

pure dependency parser. Both the Klein & Manning model and the experimental model produce significantly ($p > 0.01$) better results than the unlexicalized PCFG model with respect to the PARSEVAL metrics.[3] There is no significant difference between the experimental model and the Klein & Manning model.

How do the parsers of this experiment compare with similar studies on this corpus? Unfortunately, there have been almost no publications on full statistical parsing with TüBa-D/Z . The most comparable we know of is Kuebler et al. [8]. Table 3 compares their best results with ours, according to precision and recall on labeled and unlabeled constituents. Naturally, no definite conclusions can be drawn because of the large differences in experimental setups. The dependency results cannot be compared with other studies, because the heuristically extracted trees have only ever been used for this experiment.

5   Future Work

This first serious attempt to put context-free filtering grammar into large-scale practice has shed light on several very interesting directions of future research. The first would be to apply online large-margin learning to a phrase structure parser. Taskar et al. [11] have already shown that features can be chosen for such a model, but they did not use a computationally efficient learning approach to learn the model weights. McDonald et al. [3] did. A homogeneous linear large-margin factored model for dependency and phrase structure would then be immediately feasible. The training of such a model would be far less awkward than that of the model used in this paper.

Investigation subsequent to the experiment presented here suggests that pruning arcs during A* search has a significant impact on parsing accuracy. It would be beneficial to reduce the memory requirements of the parser, both by optimizing the CKY outside-score precomputation and by using a tighter dependency scoring heuristic.

Finally, it would be valuable to extend the error analysis made here by differentiating among errors made because of flaws in the models, because of pruning and because of inconsistencies in the corpus annotation.

---

[3] As determined by a Monte-Carlo estimate of an exact paired permutation test.

References

1. Telljohann, H., Hinrichs, E.W., Kübler, S., Zinsmeister, H.: Stylebook for the Tübingen treebank of written German (TüBa-D/Z). Technical report, Seminar für Sprachwissenschaft, Universität Tübingen, Germany (2005)
2. Klein, D., Manning, C.D.: Fast exact inference with a factored model for natural language parsing. In: 16th NIPS, pp. 3–10 (2002)
3. McDonald, R., Crammer, K., Pereira, F.: Online large-margin training of dependency parsers. In: 43rd ACL, pp. 91–98 (2005)
4. Briscoe, T., Carroll, J.: Evaluating the accuracy of an unlexicalized statistical parser on the parc depbank. In: 44th ACL, pp. 41–48 (2006)
5. McDonald, R., Pereira, F., Ribarov, K., Hajič, J.: Non-projective dependency parsing using spanning tree algorithms. In: HLT/EMNLP, pp. 523–530 (2005)
6. Crammer, K., Dekel, O., Singer, Y., Shalev-Shwartz, S.: Online passive-aggressive algorithms. In: Thrun, S., Saul, L., Schölkopf, B. (eds.) Advances in Neural Information Processing Systems, vol. 16, pp. 1229–1236. MIT Press, Cambridge (2004)
7. Collins, M.: Discriminative training methods for hidden markov models: theory and experiments with perceptron algorithms. In: EMNLP, pp. 1–8 (2002)
8. Kübler, S., Hinrichs, E.W., Maier, W.: Is it really that difficult to parse German? In: EMNLP, pp. 111–119 (2006)
9. Klein, D., Manning, C.D.: Accurate unlexicalized parsing. In: 41st ACL, pp. 423–430 (2003)
10. Black, E., Abney, S., Flickenger, S., Gdaniec, C., Grishman, C., Harrison, P., Hindle, D., Ingria, R., Jelinek, F., Klavans, J., Liberman, M., Marcus, M., Roukos, S., Santorini, B., Strzalkowski, T.: Procedure for quantitatively comparing the syntactic coverage of English grammars. In: HLT, pp. 306–311 (1991)
11. Taskar, B., Klein, D., Collins, M., Koller, D., Mannning, C.: Max-margin parsing. In: EMNLP, pp. 1–8 (2004)

Machine Translation of Legal Information and Its Evaluation

Atefeh Farzindar[1] and Guy Lapalme[2]

[1] NLP Technologies Inc.
3333 Queen Mary Road, suite 543
Montréal, Québec, Canada, H3V 1A2
farzindar@nlptechnologies.ca
[2] RALI-DIRO
Université de Montréal
Montréal, Québec, Canada, H3C 3J7
lapalme@iro.umontreal.ca

Abstract. This paper presents the machine translation system known as TransLI (**Trans**lation of **L**egal **I**nformation) developed by the authors for automatic translation of Canadian Court judgments from English to French and from French to English. Normally, a certified translation of a legal judgment takes several months to complete. The authors attempted to shorten this time significantly using a unique statistical machine translation system which has attracted the attention of the federal courts in Canada for its accuracy and speed. This paper also describes the results of a human evaluation of the output of the system in the context of a pilot project in collaboration with the federal courts of Canada.

1 Context of the Work

NLP Technologies[1] is an enterprise devoted to the use of advanced information technologies in the judicial domain. Its main focus is DecisionExpress™ a service utilizing automatic summarization technology with respect to legal information. DecisionExpress is a weekly bulletin of recent decisions of Canadian federal courts and tribunals. It is an tool that processes judicial decisions automatically and makes the daily information used by jurists more accessible by presenting the legal record of the proceedings of federal courts in Canada as a table-style summary (Farzindar et al., 2004, Chieze et al. 2008). NLP Technologies in collaboration with researchers from the RALI[2] at Université de Montréal have developed TransLI to translate automatically the judgments from the Canadian Federal Courts. As it happens, for the new weekly published judgments, 75% of decisions are originally written in English and 25% in French. By law, the Federal Courts have to provide a translation in the other official language of Canada.

The legal domain has continuous publishing and translation cycles, large volumes of digital content and growing demand to distribute more multilingual information. It is necessary to handle a high volume of translations quickly.

---

[1] http://www.nlptechnologies.ca
[2] http://rali.iro.umontreal.ca

Y. Gao and N. Japkowicz (Eds.): Canadian AI 2009, LNAI 5549, pp. 64–73, 2009.
© Springer-Verlag Berlin Heidelberg 2009

Currently, a certified translation of a legal judgment takes several months to complete. Afterwards, there is a significant delay between the publication of a judgment in the original language and the availability of its human translation into the other official language.

Initially, the goal of this work was to allow the court, during the few months when the official translation is pending, to publish automatically translated judgments and summaries with the appropriate caveat. Once the official translation would become available, the Court would replace the machine translations by the official ones. However, the high quality of the machine translation system obtained, developed and trained specifically on the Federal Courts corpora, opens further opportunities which are currently being investigated: machine translations could be considered as first drafts for official translations that would only need to be revised before their publication. This procedure would thus reduce the delay between the publication of the decision in the original language and its official translation. It would also provide opportunities for saving on the cost of translation.

We evaluated the French and English output and performed a more detailed analysis of the modifications made to the translations by the evaluators in the context of a pilot study to be conducted in cooperation with the Federal Courts.

This paper describes our statistical machine translation system, whose performance has been assessed with the usual automatic evaluation metrics. We also present the results of a manual evaluation of the translations and the result of a completed translation pilot project in a real context of publication of the federal courts of Canada. To our knowledge, this is the first attempt to build a large-scale translation system of complete judgments for eventual publication.

2 Methodology

NLP Technologies' methodology for machine translation of legal content consists of the following steps:

- Translated judgments are gathered;
- The HTML markup is removed from the judgments, which are then aligned at the level of the sentence;
- a translation model is created using the pairs of translated sentences;
- The court tests the usability of the Statistical Machine Translation (SMT) in the context of a pilot project;
- The SMT is then deployed.

In the context of our project, NLP Technologies in collaboration with RALI used the existing translated judgments from the Federal Court of Canada as a training corpus for our SMT system. The next section provides more details on the translation system:

3 Overview of the System

We have built a phrase-based statistical translation system, called TransLI (Translation of Legal Information), that takes as input judgments published (in HTML) on the

Federal Courts web site and produces an HTML file of the same judgment in the other official language of Canada. The architecture of the system is shown in Figure 1.

The first phase (semantic analysis) consists in identifying various key elements pertaining to a decision, for instance the parties involved, the topics covered, the legislation referenced, whether the decision was in favor of the applicant, etc. This step also attempts to identify the thematic segments of a decision: **Introduction**, **Context**, **Reasoning** and **Conclusion** (see section Evaluation in a pilot project). During this phase, the original HTML file is transformed into XML for internal use within NLP Technologies in order to produce DecisionExpress™ fact sheets and summaries. We extract the source text from these structured XML files in which sentence boundaries have already been identified. This is essential, since the translation engine works sentence by sentence.

The second phase translates the source sentences into the target language using SMT. The SMT module makes use of open source modules GIZA++ (Och and Ney, 2003) for creating the translation models and SRILM for the language models. We considered a few phrase-based translation engines such as Phramer (Olteanu et al, 2006), Moses (Koehn et al., 2007), Pharaoh (Koehn, 2004), Ramses (Patry et al., 2006) and Portage (Sadat et al., 2005). Moses was selected because we found it to be a state-of-the-art package with a convenient open source license for our testing purposes.

The last phase is devoted to the rendering of the translated decisions in HTML. Since the appropriate bookkeeping of information has been maintained, it is possible to merge the translation with the original XML file in order to yield a second XML file containing a bilingual version of each segment of text. This bilingual file can then be used to produce an HTML version of the translation, or for other types of processing, like summarization.

Indeed, since summaries of judgments produced by NLP Technologies are built by extracting the most salient sentences from the original text, producing summaries in both languages should be as simple as selecting the translation of every sentence retained in the source-language summary.

Gotti et al. (2008) describe the development and the testing of the TransLI statistical machine translation system. The final configuration is a compromise between quality, ease of deployment and maintenance and speed of translation with the following features: a distance based reordering strategy, a tuning corpus based on recent decisions; a large training corpus and the integration of specialized lexicons.

Although these types of texts employ a specialized terminology and a specific cast of sentences, the availability of large amounts of high quality bilingual texts made it possible to develop a state-of-the-art SMT engine. These excellent results prompted us to perform a human evaluation also described in (Gotti et al. 2008) on 24 randomly selected sentences from our test set. This evaluation centered on the quality of the produced translation and on its fidelity, i.e. to what extent the SMT conveys all the semantic content of the original.

A key element in the success of an SMT system lies in the availability of large corpora of good quality. In the Canadian judicial domain, we are fortunate enough to have access to public web sites providing translations of excellent quality for almost all judgments of the most important Canadian courts. For our work, we built a set of corpora, the characteristics of which are shown in Table 1.

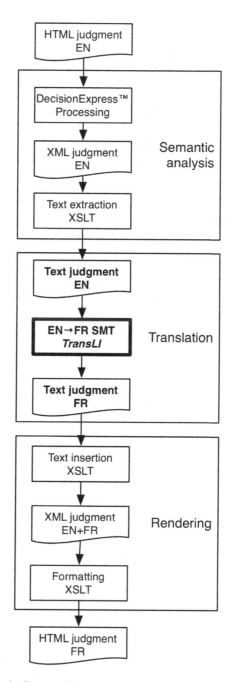

Fig. 1. The translation pipeline translates an HTML court decision written in English into a French decision (also in HTML). A similar pipeline performs translations from French to English.

Table 1. Corpora used for developing TransLI

Corpus name	# sent pairs	# en words	# fr words
principal	245,000	6,510,000	7,510,000
train	244,000	6,500,000	7,500,000
tune-1	300	8,000	9,000
test	1300	28,000	33,000
tune-recent	400	8,000	10,000
train-lexum	1,000,000	22,340,000	25,720,000

principal: we downloaded 14,400 decisions in HTML from the Federal Court of Canada web site[3] from which we extracted the text. Because many judgments did not have a translation or could not be parsed automatically with our tools because of inconsistent original formatting, we ignored them and we were left with 4500 valid judgment pairs. From these pairs, we extracted the sentences and aligned them to produce a bi-text of around 260,000 sentence pairs. A number of them had English citations in the French text and vice-versa. Once these cases were filtered out, we were left with 245,000 sentence pairs.

train: 99% of the sentences from principal, used to train the SMT system.

tune-1: 1% of principal used to adjust the parameters of the system. There is no overlap with train.

test: 13 recent decisions that were published after the decisions occurring in principal. This better simulates the application context for our system, which will be used for translating recent decisions.

tune-recent: 6 recent decisions that were published after the decisions in principal.

train-lexum: Since the RALI has a long experience in dealing with judicial texts in collaboration with the Lexum[4] at the Université de Montréal in the context of the TransSearch[5] system, we decided to add 750,000 bilingual sentence pairs from our existing bilingual text database. These sentences are taken from decisions by the Supreme Court, the Federal Court, the Tax Court and the Federal Court of Appeal of Canada.

For the quality of language, we asked three evaluators to assign each of the 24 passages a score: 1 (unacceptable), 2 (bad), 3 (fair), and 4 (perfect), according to whether they found it to be in a correct and readable target language, independently of the source language. This would correspond to the case where a non-French speaking person wanted to consult an English translation of a French text. Our evaluators

---

[3] decisions.fct-cf.gc.ca/en/index.html
[4] www.lexum.ca
[5] www.tsrali.com

did not know which translations had been produced by a human or which were produced by a machine.

The same three evaluators were given groups of two or three sentences containing the source French text and the English translation produced either by TransLI or by a human translator (the reference text). The evaluators were asked to modify them in order to make them good enough for publication. Overall they took an average of 27 minutes to revise 8 TransLI texts (475 words), which corresponds to 1070 words/hour. That would amount to 8000 words per day compared to the average of about 6000 often used in the industry for revision (4 times the productivity of 1500 words translated per day per translator).

4   Evaluation in a Pilot Project

Although still not of publishable quality, the translations of the TransLI system that we developed in this project can be readily used for human revision, with promising productivity gains. Following those encouraging results on a small sample of a few sentences, we conducted a pilot study with the Federal Courts of Canada in which we translated a certain number of complete judgments from French to English and from English to French. We herein set out the more detailed evaluation of the revision process that we performed on a randomly selected set of 10 decisions (6 from French to English and 4 from English to French).

We also describe how we evaluate the quality of our current automatic judgment translations and the effort needed to revise them so that they can be published. As the summarization system of *NLP Technologies* already divides a judgment into four main thematic segments: **Introduction**, **Context**, **Reasoning** and **Conclusion**, we describe the evaluation using those divisions. In order to give an idea of the source text, of the raw SMT translation produced and of the revised output judged acceptable for publication, Table 2 shows a few sentences from each division.

The thematic segmentation is based on specific knowledge of the legal field. According to our analysis, legal texts have a thematic structure independent of the category of the judgment (Farzindar and Lapalme, 2004) Textual units dealing with the same subject form a thematic segment set. In this context, we distinguish four themes, which divide the legal decisions into thematic segments, based on the work of judge Mailhot (1998):

- **Introduction** describes the situation before the court and answers these questions: who did what to whom?
- **Context** explains the facts in chronological order: it describes the story including the facts and events related to the parties and it presents findings of credibility related to the disputed facts.
- **Reasoning** describes the comments of the judge and the finding of facts, and the application of the law to the found facts. This section of the judgment is the most important part for legal experts because it presents the solution to the problem between the parties and leads the judgment to a conclusion.
- **Conclusion** expresses the disposition, which is the final part of a decision containing the information about what is decided by the court.

Table 2. Sentences from a decision (2008fc1224 from http://decisions.fct-cf.gc.ca/fr/2008/ 2008cf1224/2008cf1224.html). The first column indicates the theme in which the summarization system has classified the source sentence of the fourth column; the second column indicates the Levenshtein edit distance in terms of space separated tokens between the original SMT output (fifth column) and the revised output (sixth column). Replacement of tokens is shown in bold in the original and revised. Insertion in the revision is underlined and the insertion point is shown with a bullet in the original. Deletion of original is indicated by a strike-through the deleted text in the original. Because the sentences are tokenized at blank spaces, some indications may not reflect accurately the minimum distance or the sequence of editing operations performed by the revisor: for example, in paragraph [8] «officer» and «officer,» were considered as distinct tokens so the editing sequence is reported as an insertion of «officer,» and a replacement of «officer» by «i.e.».

	dist	ODS	French original	SMT Fr->En output	Post edited version
Introduction	1	1	[1] Il s'agit d'une requête visant à obtenir un sursis d'exécution de l'ordonnance de déportation émise contre le demandeur prévue pour le 3 novembre 2008 à 18 h 30.	[1] This is a motion for a stay of execution of the deportation order issued against the applicant scheduled for November 3, 2008 at 6:**30**.	[1] This is a motion for a stay of execution of the deportation order issued against the applicant scheduled for November 3, 2008 at 6:**30p.m.**
Context	5	5	[8] Le 13 avril 2007, le demandeur s'est prévalu d'un Examen des risques avant renvoi (« ERAR ») et, le 16 mai 2007, il présentait une deuxième demande de résidence permanente pour raisons humanitaires. Ces deux dernières demandes furent entendues par le même agent i.e. Patricia Rousselle, laquelle, par décision du 31 juillet 2008, rejetait les deux demandes.	[8] On April 13, 2007, the Applicant availed • of a pre-removal risk assessment ("PRRA") and, on May 16, 2007, he submitted a second application for permanent residence on humanitarian and compassionate grounds. These last two applications were heard by the same • **officer** Patricia Rousseau, ~~i.e.~~ **that,** by decision dated July 31, 2008, dismissed both applications.	[8] On April 13, 2007, the Applicant availed *himself* of a pre-removal risk assessment ("PRRA") and, on May 16, 2007, he submitted a second application for permanent residence on humanitarian and compassionate grounds. These last two applications were heard by the same officer, **i.e.** Patricia Rousseau, **who,** by decision dated July 31, 2008, dismissed both applications.
Reasoning	5	3	[16] Quant au préjudice allégué, il s'agit de celui subi par tous ceux qui sont l'objet d'une ordonnance de renvoi.	[16] As to the harm • **alleged** is that • by all those who are subject to a removal order.	[16] As to the harm alleged, **it** is that which is experienced by all those who are subject to a removal order.
Reasoning	4	3	[17] En conséquence, les conditions exigées par l'arrêt Toth n'étant pas respectées, la demande de sursis ne peut être accueillie.	[17] Accordingly, the conditions required by • Toth **is** not • met, the application for a stay cannot be allowed	[17] Accordingly, the conditions required by the Toth **case** not having been met, the application for a stay cannot be allowed.
Conclusion			[18] Pour ces motifs la Cour ordonne que la demande de sursis de la mesure de renvoi soit rejetée.	[18] For these reasons, the Court orders that the application for • stay of removal is dismissed.	[18] For these reasons, the Court orders that the application for a stay of removal is dismissed.

In order to evaluate the results of the automatic translation, we computed two automatic measures over the space-separated tokens of a sentence. A token is thus a word plus any accompanying punctuation or symbols. A token can also be any sequence of contiguous non-space characters:

- **Edit distance:** the number of tokens that differ in the source and revised text as computed by the classical Levenshtein distance algorithm (Levenshtein, 1966).
- **Number of operations:** the number of consecutive insertion, deletion and replacement operations to transform the source into the revised text. For example, replacing 5 consecutive words would count as 5 in the edit distance but for only one operation. This measure approximates the number of cut and paste operations needed to revise an SMT translation.

Table 3. Number and percentage of editing operations and tokens in each division over ten judgments

Theme	English to French (4 texts)				French to English (6 texts)			
	Nb ops		Nb tokens		Nb ops		Nb tokens	
Introduction	31	5%	397	8%	13	2%	350	5%
Context	297	47%	2046	41%	154	18%	1246	19%
Reasoning	281	44%	2243	45%	646	76%	4642	69%
Conclusion	28	4%	298	6%	38	4%	457	7%
Total	637	100%	4984	100%	851	100%	6695	100%

Table 4. Edit distance in tokens for each division, the percentages are taken over the number of tokens given in the fourth and eighth column of Table 3

Edit distance	English to French		French to English	
Introduction	51	13%	19	5%
Context	626	31%	263	21%
Reasoning	518	23%	1213	26%
Conclusion	43	14%	68	15%
Overall	1238	25%	1563	23%

Table 2 shows examples of values of these measures on a few sentences. Even though, the exact values of the number of operations might differ from what a careful reviewer might do, we think this value is a good approximation of the work needed for revision.

Table 3 shows that for both translation directions, the number of editing operations is roughly equivalent to the number of tokens in each division. Table 4 shows that the global proportion of differences is similar for both directions of translation. The results are slightly better on the French to English direction, which is expected due to the complexity of the French language (with the accents and exceptions) bringing more complications to the machine translations. When we compare the different themes, we see that the Introduction and Conclusion themes require significantly less editing than the Context or the Reasoning themes. The type of text used in these themes in part explains these differences. In the legal field, the sentences used for the Introduction and Conclusion of the judgments often use the same expressions while the Context and Reasoning contain more sentences which are seldom seen in multiple judgment. Sentences from the Context that explain the litigation events are more variable.

5   Conclusion

The volume of legal content is growing rapidly. In Canada it is even more problematic because it is created in two languages and different formats. As a result, the amount of data that must be translated in short time has grown tremendously, making it difficult to translate and manage.

Legal organizations need solutions that enable them to handle quickly a high volume of translations. Our goal was to study the ability to train translation systems on a specific domain or subject area like the legal field so as to radically increase translation accuracy. This process recycles existing translated content to train the machine on the terminology and style of the requested domain.

To our knowledge this is one of the first times that an SMT engine has been developed specifically for judicial texts and evaluated in a pilot study. We managed to establish that an SMT engine trained on an appropriate corpus can produce a cost-effective revisable text.

6   Future Work

An interesting aspect of our findings is that review and post-editing of judicial translations are an important part of an SMT-integrated work flow. Reviewers with subject knowledge need to have direct access to the translation process in order to provide a feedback loop to the SMT training process.

We will therefore continue further investigation into an optimization of the post-editing and reviewing process, specifically with a focus on quantifying the distance, measured in number of operations and edits, to arrive at a fully acceptable translation.

As part of an ongoing collaboration with Palomino System Innovations Inc., a Canadian web content management system provider, we will evaluate integration of TransLI SMT into a translation work flow system – with a view to apply SMT to generic web content in the future.

Acknowledgments

We thank the Precarn Program for partially funding this work and the Federal Courts for their collaboration and feedback. We sincerely thank our lawyers for leading the human evaluation: Pia Zambelli and Diane Doray. The authors thank also Fabrizio Gotti and Jimmy Collin for technical support of experiments. We also thank Elliott Macklovitch, the Coordinator of the RALI and Markus Latzel, CEO of Palomino System Innovations Inc. for the support and many fruitful discussions.

References

Chieze, E., Farzindar, A., Lapalme, G.: Automatic summarization and information extraction from Canadian immigration decisions. In: Proceedings of the Semantic Processing of Legal Texts Workshop, LREC 2008 (2008)

Farzindar, A., Lapalme, G.: LetSum, an automatic Legal Text Summarizing system. In: Gordon, T.F. (ed.) Legal Knowledge and Information Systems, Jurix 2004: the Sevententh Annual Conference, pp. 11–18. IOS Press, Berlin (2004)

Farzindar, A., Lapalme, G., Desclés, J.-P.: Résumé de textes juridiques par identification de leur structure thématique. Traitement automatique de la langue (TAL) 45(1), 39–64 (2004)

Gotti, F., Farzindar, A., Lapalme, G., Macklovitch, E.: Automatic Translation of Court Judgments. In: AMTA 2008 The Eighth Conference of the Association for Machine Translation in the Americas, Waikiki, Hawai'i, pp. 1–10 (October 2008)

Koehn, P., Hoang, H., Birch, A., Callison-Burch, C., Federico, M., Bertoldi, N., Cowan, B., Shen, W., Moran, C., Zens, R., Dyer, C., Bojar, O., Constantin, A., Herbst, E.: Moses: Open Source Toolkit for Statistical Machine Translation. In: Annual Meeting of the Association for Computational Linguistics (ACL), demonstration session, Prague, Czech Republic (2007)

Koehn, P., Shroeder, J.: Experiments in Domain Adaptation for Statistical Machine Translation. In: Proceedings of the 2nd Workshop on SMT, Prague, Czech Republic (2007)

Koehn, P.: Pharaoh: A beam search decoder for phrase-based statistical machine translation models. In: Frederking, R.E., Taylor, K.B. (eds.) AMTA 2004. LNCS, vol. 3265, pp. 115–124. Springer, Heidelberg (2004)

Levenshtein, V.I.: Binary codes capable of correcting deletions, insertions, and reversals. Soviet Physics Doklady 10, 707–710 (1966)

Olteanu, M., Davis, C., Volosen, I., Moldovan, D.: Phramer, An Open Source Statistical Phrase-Based Translator. In: Proceedings of the Workshop on Statistical Machine Translation, pp. 146–149 (2006)

Mailhot, L.: Decisions, Decisions: a handbook for judicial writing. Editions Yvon Blais, Québec, Canada (1998)

Och, F.J., Ney, H.: A Systematic Comparison of Various Statistical Alignment Models. Computational Linguistics 29(1), 19–51 (2003)

Patry, A., Gotti, F., Langlais, P.: Mood at work: Ramses versus Pharaoh. In: Workshop on Statistical Machine Translation, HLT-NAACL, New-York, USA (2006)

Sadat, F., Johnson, H., Agbago, A., Foster, G., Kuhn, R., Martin, J., Tikuisis, A.: Portage: A Phrase-based Machine Translation System. In: ACL 2005 Workshop on Building and Using Parallel Texts: Data-Driven Machine Translation and Beyond, Ann Arbor, Michigan, USA, June 29-30, 2005, pp. 133–136 (2005)

An Iterative Hybrid Filter-Wrapper Approach to Feature Selection for Document Clustering

Mohammad-Amin Jashki, Majid Makki, Ebrahim Bagheri,
and Ali A. Ghorbani

Faculty of Computer Science
University of New Brunswick
{a.jashki,majid.makki,ebagheri,ghorbani}@unb.ca

Abstract. The manipulation of large-scale document data sets often involves the processing of a wealth of features that correspond with the available terms in the document space. The employment of all these features in the learning machine of interest is time consuming and at times reduces the performance of the learning machine. The feature space may consist of many redundant or non-discriminant features; therefore, feature selection techniques have been widely used. In this paper, we introduce a hybrid feature selection algorithm that selects features by applying both filter and wrapper methods in a hybrid manner, and iteratively selects the most competent set of features with an expectation maximization based algorithm. The proposed method employs a greedy algorithm for feature selection in each step. The method has been tested on various data sets whose results have been reported in this paper. The performance of the method both in terms of accuracy and Normalized Mutual Information is promising.

1 Introduction

Research on feature selection has recently gained considerable amount of attention due to the increasing complexity of objects within the target domains of interest whose data sets consist of hundreds of thousands of features. These target domains cover areas such as the analysis and understanding of corporate and publicly available documents, the detection of the most influential genes based on DNA micro-array experiments, and experiments in combinatorial chemistry, among others [8]. Document clustering has specially been a fertile area for the employment of feature selection techniques due to its wide variety of emerging applications including automatic email spam detection, News article categorization, document summarization, and others. In these domains, documents are usually represented by 'bag-of-words', which is a vector equal in dimension to the number of existing vocabulary in the domain of discourse [4]. Studies have revealed that document collections whose vocabulary domain size are between 5,000 and 800,000 are common, whose categorization would hence require scalable techniques that are able to operate over a feature size of this magnitude [13]. It is clear that this representation scheme possesses two main characteristics: 1) high dimensionality of the feature space; and 2) inherent sparsity of each

Y. Gao and N. Japkowicz (Eds.): Canadian AI 2009, LNAI 5549, pp. 74–85, 2009.

vector over all terms, which are detrimental to the achievement of the optimal performance for many learning techniques.

Despite these problems, the bag-of-words representation usually performs well with small enhancements on textual data sets due to some of their inherent traits: First, since the bag-of-words representation is oblivious to the word appearance sequence in a document, neighboring terms do not have to necessarily co-exist in the bag-of-words vector. Second, investigations have shown that the large feature space developed by the bag-of-words representation typically follows the Zipf-like distribution, i.e., there are a few very common frequently seen terms in the document set along with many very unfrequent terms [5]. Third, even among the very common frequent terms, most of them do not possess a high discrimination power (frequent-but-non-discriminant terms). This may be due to the fact that such terms have a similar occurrence pattern in all existing document classes. Stop words and auxiliary verbs are examples of such terms. This can also be analyzed within the context of relevant but redundant terms. Therefore, feature selection techniques can be applied on features of the bag-of-words representation in document categorization in order to select a highly discriminant subset such that 1) the prediction performance of the classifiers/clusterers are enhanced; 2) classifiers/clusterers are learnt faster and more cost effectively; and 3) the underlying concepts behind the available corpus of data are revealed and understood.

In this paper, we propose an iterative feature selection scheme, which greedily selects the best feature subset from the bag-of-words that best classify the document set in each step. The method employs an Expectation Maximization (EM) approach to feature selection and document clustering due to the restriction that supervised feature selection techniques cannot be directly applied to textual data because of the the unavailability of the required class labels. Briefly explained, our method initially labels the documents with random labels. Based on these labels, it then greedily chooses the best representative subset from the feature space. The selected features are then employed for clustering the documents using the k-Means algorithm. This iterative process of feature selection and document clustering is repeated until the satisfaction of a certain stopping criterion. It is important to mention that the proposed greedy algorithm evaluates the suitability of the features locally within the context of each individual cluster. This local computation allows the greedy algorithm to find the most discriminative features. Our approach in combining the EM algorithm with a greedy feature selection method shows improved performance with regards to *accuracy* and *Normalized Mutual Information (NMI)* compared with some existing techniques for document clustering.

The paper is organized as follows: the next section, reviews some of the existing techniques for feature selection and document clustering. Section 3 introduces our proposed iterative feature selection and document clustering technique. The results of the evaluation of the performance of the proposed technique is given in Section 4. The paper is then concluded in Section 5.

2   Related Work

The work on the manipulation of the feature space for text categorization has been three-fold: 1) feature generation; 2) feature selection; and 3) feature extraction [13]. In feature generation, researchers have focused on employing the base features available in the initial feature space to create suitable discriminant features that best reflect the nature of the document classes. For instance, some approaches consider different word forms originating from the same root as one term in the feature space and employ stemming techniques to extract such features. Here, all of the terms such as *fishing, fished, fishy,* and *fisher* would be represented by their root word, *fish,* in the feature space. Along the same lines, some other techniques have gone further in exploring similarity between words by using thesaurus and ontologies to create groups of features, which would be considered as a single feature in the feature space. In this approach, terms such as *office, bureau,* and *workplace* would be all grouped into one feature in the feature space [2]. Furthermore, as was mentioned earlier, the bag-of-words approach fundamentally neglects the sequence of word occurrences in the document. To alleviate this, some researchers have made the observation that the use of word *n-grams*[1] for creating word phrase features can provide more specificity. In this approach, n-grams are mined from the documents (2-word phrases are most popular) and are used as representative features in the feature space. It is important to notice that although the number of potential n-grams increases exponentially with the value of n, there are only a small fraction of phrases that receive considerable probability mass and are found to be predictive; therefore, the size of the feature space would not grow very large. Other methods such as using unsupervised techniques for clustering the terms in the feature space can be used to create classes of similar features. These classes can themselves be used as a complex features in the new feature space.

There are three main approaches within the realm of feature selection [10] that have been introduced in the following:

1. *Filter methods* assess each feature independently, and develop a ranking between the members of the feature space, from which a set of top-ranked features are selected. The evaluation of each feature is performed on their individual predictive power. This can be done for instance using a classifier built using that single feature, where the accuracy of the classifier can be considered as the fitness of the feature. Here, a limitation for such an evaluation is the absence of the required class labels in document categorization data sets; therefore, unsupervised methods need to be employed to evaluate the features. Some filter methods assess each feature according to a function of its *Document Frequency (DF)* [11], which is the number of documents in which a term occurs in the data set. Other unsupervised measures have been defined on this basis such as *Term Strength (TS)* [11], which is based on the conditional probability that a feature occurs in the second half of a pair of

---

[1] Or character n-grams for languages such as Chinese and Japanese that do not have a space character.

related documents given that it appeared earlier, *Entropy-based Ranking* [3] that is measured by the amount of entropy reduction as a result of feature elimination, and *Term Contribution (TC)*, which is a direct extension of DF that takes term weights in to account. In addition, Liu et al [11]. have shown how several supervised measures like *Information Gain* and χ^2 *statistic* can be used for the purpose of feature ranking in document clustering. However, the computation time of this method is still under question. Filtering methods are computationally and statistically scalable since they only require the computation of n scores for n features, and also they are robust to overfitting, since although they increase bias, they may have less variance; however at the same time, they may be exposed to the selection of redundant features.

2. *Wrapper methods* employ AI search techniques such as greedy hill climbing or simulated annealing in order to find the best subset of features from the feature space. Different feature subsets are evaluated repeatedly through cross-validation with a certain learning machine of interest. One can criticize wrapper methods for being brute-force methods that need great amount of computation in order to cover all of the search space. Conceptually this is true, but greedy search strategies have been devised that are computationally wise and robust against overfitting. For instance, two popular greedy search strategies are *forward selection*, which incrementally incorporates features into larger subsets, and *backward elimination* that starts with the set of all features and iteratively eliminates the least promising ones. Actual instantiations of these two strategies have been proposed in the related literature. For instance, the Gram-Schmidt orthogonolization procedure provides the basis for forward feature selection by allowing the addition of the feature that reduced the mean-squared error the most at each step [6].

3. *Embedded methods* try to build a prediction model that attempts to maximize the goodness of fit of the developed model and minimize the number of input features of the model. Such methods are reliant on the specifics of the utilized learning machine used in the prediction model. Embedded methods that incorporate feature selection as a part of their training process possess some interesting advantages such as reaching a solution faster by avoiding the retrain of the model for each feature subset and also making better use of the available data by not needing to split the data into training and validation subsets. Decision tree learning algorithms such as CART inherently include an embedded feature selection method [1].

Finally, feature extraction methods are a subclass of the general dimensionality reduction algorithms. These methods attempt to construct some form of combination of all or a subset of the initial features in order to develop a reduced-size feature space that represents the initial data with sufficient amount of accuracy. emphPrinciple Component Analysis (PCA), emphLatent Semantic Analysis (LSA), and non-linear dimensionality reduction methods are some of the representatives of these methods [9]. One of the drawbacks of this approach is that the developed features of the new feature space are not easily interpretable since they may not have an obvious or straightforward human understandable interpretation.

3   Proposed Method

In practice, wrapper methods employ the prediction performance of the learning machine to assess the relative usefulness of the selected features. Therefore, a wrapper method needs to provide three main decisions with regards to the following concerns: First, a search strategy, such as the greedy forward selection and backward elimination strategies introduced earlier, needs to be created that would guide the process of searching all of the possible feature space efficiently. Second, methods for the assessment of the prediction performance of the learning machine need to be defined in order to guide the search strategy, and Third, the choice for the appropriate learning machine needs to be made. As it can be seen wrapper methods rely on the prediction performance of the learning machine, which is not a viable strategy since they require the class labels that are not present in document clustering. However, filter methods use other unsupervised measures such as DF, TS and others to evaluate the usefulness of the selected features and hence rank the available features accordingly and select the most competent, which makes them suitable for the task of document clustering.

As was discussed earlier, filter methods have been criticized for selecting redundant or locally optimum features from the feature space [15]. This is due to the fact that they tend to select the top best features based on a given measure without considering the different possibilities of feature composition available in the feature space. It seems enticing to create a hybrid strategy based on filter methods and wrapper methods to overcome their limitations and reap their individual capabilities. An efficient hybrid strategy would provide two main benefits for document clustering: 1) it decreases the chance of being trapped by a local optimum feature sets through the use of an iterative greedy search strategy; 2) both supervised and unsupervised measures can be used to evaluate the discriminative ability of the selected features in each iteration of the process. Here, we propose such a hybrid strategy.

Our proposed method for feature selection and document clustering employs a greedy algorithm, which iteratively chooses the most suitable subset of features locally in each round. The chosen features are used to create corresponding document clusters using the k-Means algorithm. The resulting document clusters are then employed as local contexts for the greedy algorithm to choose a new subset of features from each cluster. The representative features of each cluster are chosen such that they could maximally discriminate the documents within that cluster. The union set of all local features of clusters is developed, which would serve as the newly selected feature subset. Formally said, our approach is a combination of the expectation maximization algorithm accompanied by a greedy search algorithm for traversing the feature space, and an unsupervised feature ranking technique. Expectation maximization algorithms are employed for maximum likelihood estimation in domains with incomplete information. A typical EM algorithm estimates the expectation of the missing information based on the current observable features in its *E-step*. In the *M-step*, the missing information are replaced by the expected value estimates computed in the *E-step* in order to develop a new estimation that can maximize the complete data likelihood function. These steps are iterated until a certain stopping criterion is satisfied.

Liu et al. [11] have proposed a general framework for the employment of EM for text clustering and feature selection. We employ their formulation of the problem statement within their framework and provide our own instantiation of the provided skeleton. Here, the basic assumption is that a document is created by a finite mixture model between whose components and the clusters there exists a one-to-one correspondence. Therefore, the probability of all documents given the model parameters can be formulated as follows:

$$p(D|\theta) = \prod_{i=1}^{N} \sum_{j=1}^{|C|} p(c_j|\theta) p(d_i|c_j, \theta) \tag{1}$$

where D denotes the document set, N the number of documents in the data set, c_j the jth cluster, $|C|$ the number of clusters, $p(c_j|\theta)$ the prior distribution of cluster c_j, and $p(d_i|c_j, \theta)$ the distribution of document d_i in cluster c_j. Further, since we use the bag-of-words representation, we can assume that document features (terms) are independent of each other given the document class label. Hence, the likelihood function developed in the above equation can be re-written as

$$p(D|\theta) = \prod_{i=1}^{N} \sum_{j=1}^{|C|} p(c_j|\theta) \prod_{t \in d_i} p(t|c_j, \theta) \tag{2}$$

where t represents the terms in d_i, and $p(t|c_j, \theta)$ the distribution of term t in cluster c_j. Since not all of the terms in a document are equally relevant to the main concept of that document, $p(t|c_j, \theta)$ can be regarded as treated as the sum of relevant and irrelevant distributions:

$$p(t|c_j, \theta) = z(t)p(t \text{ is relevant}|c_j, \theta) + (1 - z(t))p(t \text{ is irrelevant}|\theta) \tag{3}$$

where $z(t) = p(t \text{ is relevant})$, which is the probability that term t is relevant. Therefore, the likelihood function can be reformulated as below:

$$p(D|\theta) = \prod_{i=1}^{N} \sum_{j=1}^{|C|} p(c_j|\theta) \prod_{t \in d_i} \left[z(t)p(t \text{ is relevant}|c_j, \theta) \right.$$

$$\left. + (1 - z(t))p(t \text{ is irrelevant}|\theta) \right] \tag{4}$$

Now, the expectation maximization algorithm can be used to maximize the likelihood function by iterating over the following two step:

$$\text{E-step: } \hat{z}^{(k+1)} = E(z|D, \hat{\theta}^{(k)}) \tag{5}$$

$$\text{M-step: } \hat{\theta}^{(k+1)} = argmax_\theta \, p(D|\theta, \hat{z}^{(k)}) \tag{6}$$

In the following, we provide details of each of the two steps of the customized EM algorithm for our hybrid feature selection and document clustering method.

It should be noted that in our proposed method we assume that the number of correct document classes is known *a priori*, denoted **k**.

Lets assume that the vector **Y** represents the class labels for each of the N documents; therefore, $|\mathbf{Y}| = N$, and $\mathbf{Y}_i$ would denote the class label of the ith document. In the first iteration of the process, the values for **Y** are unknown for which we assign randomly picked values to the class labels; hence, each document is randomly classified into one of the **k** clusters. Now, we would like to suppose that the clusters developed based on random label assignments are the best classification representatives of the available documents. Therefore, it is desirable to find the set of features locally within each cluster that provide the best approximation of the available data in that cluster. For this purpose, let us proceed with some definitions.

Definition 1. *Let D_i be the set of documents in cluster i, D_i^j be the jth document in D_i, and t be a given term in the feature space. Local Document Frequency of t (in D_i), denoted $\mathcal{LDF}(D_i, t)$, is defined as follows:*

$$\mathcal{LDF}(D_i, t) = \sum_{j=1}^{|D_i|}(t \in D_i^j ? 1 : 0) \tag{7}$$

which is the number of documents in which the term t has appeared.

Definition 2. *Let C_i be the ith cluster, and D_i be the set of documents in C_i. A feature such as t is assumed to be a competent feature of C_i iff:*

$$\forall j \neq i \in \mathbf{k} : \quad \mathcal{LDF}(D_i, t) > \mathcal{LDF}(D_j, t) \tag{8}$$

Informally stated, a feature is only a competent feature of a given cluster if it's local document frequency is highest in that given cluster compared to all other clusters.

Definition 3. *Let C_i be the ith cluster. A competent set for C_i, denoted $\mathcal{COMP}(C_i)$, is defined as follows:*

$$\mathcal{COMP}(C_i) = \{t \mid t \text{ is a competent feature of } C_i\} \tag{9}$$

The competent set of features for each cluster possess the highest occurrence rate locally over all of the available clusters; therefore, they have a high chance of being a discriminant feature. This is because their local document frequency measure behavior is quite distinct from the same feature in the other clusters.

With the above definitions, we are able to locally identify those features that are competent. The competent set of features for cluster is hence identified. The union of all these features over all of the clusters is generated, which would represent the new feature space. Once the new feature space is developed, they are employed to cluster all of the documents once more using the k-Means algorithm. The k-Means algorithm would provide new values for **Y**. The values of this vector are employed in order to identify the new competent set of features for each cluster, which would consequently be used to re-cluster the documents. This process is iteratively repeated until a relatively stable cluster setting is reached.

The stopping criterion of the iterative process is based on the distance of the clusterings in the last two iterations. In other words, when the distance of these clustering is less than a threshold τ. The distance between two clusterings is computed by considering one of these clustering as the natural class labels and calculating the accuracy the other clustering.

4 Performance Evaluation

In the following, configurations of the running environment, data sets, and measuring methods for the experiments is explained. Results and corresponding analyzes are presented afterwards.

4.1 Experimental Settings

The proposed algorithm and the rivals have been implemented in Java. All experiments has been performed on an Intel Xeon 1.83GHz with 4GB of RAM.

Four data sets has been used to conduct the experiments. Table 1 shows the properties of the data sets. In this table, n_d, n_w, k, and $\hat{n}_c$ represent the total number of documents, the total number of terms, the number of natural classes, and the average number of documents per class respectively. Balance, in the last column, is the ratio of the number of documents in the smallest class to the number of documents in the largest one. The distance metric used for the k-means algorithm is Cosine similarity.

The *k1b* data set is prepared by the WebACE project [7]. In this data set, each document is a web page from the subject hierarchy of Yahoo!. *NG17-19* is a subset of a collection of messages obtained from 20 different newsgroups known as *NG20*. All three classes of the *NG17-19* data set are related to political subjects; hence, difficult to separate the documents by the clustering algorithms. The *hitech* and *reviews* data sets contain San Jose Mercury newspaper articles.

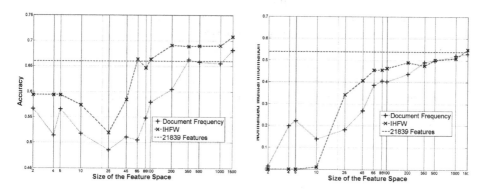

Fig. 1. Results for the k1b Data Set

Table 1. Data Sets

Data	Source	n_d	n_w	k	$\hat{n}_c$	Balance
k1b	WebACE	2340	21839	6	390	0.043
NG17-19	3 overlapping groups from NG20	2998	15810	3	999	0.998
hitech	San Jose Mercury(TREC)	2301	10080	6	384	0.192
reviews	San Jose Mercury(TREC)	4063	18483	5	814	0.098

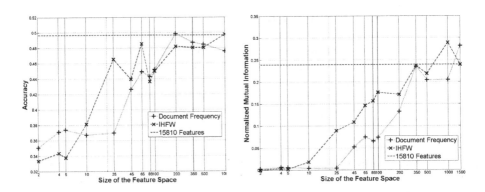

Fig. 2. Results for the NG17-19 Data Set

The former is about computers, electronics, health, medicine, research, and technology. Topics of the latter are food, movies, music, radio, and restaurants.

Two preprocessing steps including stemming and removal of the words appearing in less than three documents has been applied to all data sets.

In order to evaluate the efficiency of the proposed algorithm, the results has been compared to the DF method [14,11] which is a straightforward but efficient feature selection algorithm. In addition, the original k-Means algorithm has been applied to the data sets with all words in the feature space. The TS algorithm [14,11] is another possible rival method. However, the computing time of this algorithm is not comparable neither with IHFW[2] nor with DF. Since feature weighting schemes such as $tf - idf$ obscures/improves the efficiency of any dimensionality reduction method, comparison with TC or any other method which takes advantage of these schemes is unfair.

In addition to the accuracy of the clusterings (according to the class labels), the NMI [12] is reported due to its growing popularity.

NMI formula is presented here in Equation 10, where l is a cluster and h is a class of documents, n_h and n_l are the number of their corresponding documents, $n_{h,l}$ is the number of documents in class h as well as cluster l, and n is the size of the data set. Higher NMI values indicate high similarity between the clustering and the class labels.

---

[2] We refer to our proposed method as IHFW, hereafter.

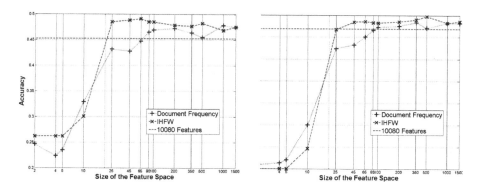

Fig. 3. Results for the hitech Data Set

$$NMI = \frac{\sum_{h,l} n_{h,l} \log \frac{n \cdot n_{h,l}}{n_h n_l}}{\sqrt{(\sum_h n_h \log \frac{n_h}{n})(\sum_l n_l \log \frac{n_l}{n})}} \tag{10}$$

Since k-Means is a randomized algorithm, the methods have been performed 10 times for each size of the feature space and the average is reported. The threshold used for the stopping criterion (maximum distance of the last two clusterings) is equal to 0.1. Smaller values may increase both the clustering performance and computation time.

4.2   Results and Discussions

Figures 1, 2, 3, and 4 illustrate the experimental results for all data sets. The x-axis is logarithmically scaled in all figures. The two line charts for each data set depict the evaluation metrics described in Section 4.1.

In almost all cases, the NMI value obtained by the proposed algorithm outperforms the DF method. There are only rare cases (feature sizes of 2, 4, 5, and 10 in Figure 1) that DF outperformed the proposed algorithm. The distance between DF and the proposed algorithm NMI values is higher when the number of features is reduced aggressively in particular. The difference between the NMI values is decreased as the number of selected features increases.

The NMI values for the proposed algorithm tend to reach that of k-Means as the number of selected features grows to 10% of the total number of words. For the *hitech* data set, the NMI value given by our algorithm outperforms the original k-Means even for very small feature sizes as illustrated in Figure 3.

Table 2. Comparing the Running Time (in Seconds) of IHFW with k-Means

a	k1b	NG17-19	hitech	reviews
$IHFW_{max}$	542.27	619.73	619.75	619.72
$IHFW_{average}$	**246.69**	383.16	288.87	**295.77**
k-Means	383.96	**173.93**	**233.0**	443.51

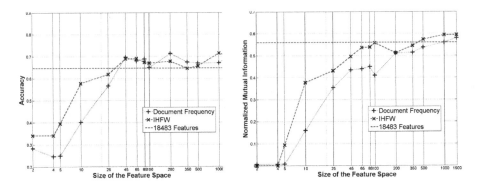

Fig. 4. Results for the reviews Data Set

The accuracy of the proposed algorithm also exceeds that of DF in most cases. For all feature sizes greater than 65 over all data sets except the *NG17-19*, the accuracy of the proposed algorithm is either higher than the accuracy of k-Means or there is at most 1% loss of accuracy. With regard to the *NG17-19* data set, at most 6% loss of accuracy for feature sizes greater than 65 is observed. It is notable that 65 features is 0.29%, 0.41%, 0.64%, 0.34% for the *k1b*, *NG17-19*, *hitech*, and *reviews* data sets respectively. Therefore, one can state that this algorithm is capable of aggressively reducing the size of the feature space with either negligible loss of accuracy or boosted accuracy.

The computation time of the DF algorithm is obviously dramatically lower than IHFW due to its simplicity. Table 2 shows the average and maximum running time for IHFW and k-Means. The average values have been computed over all feature sizes.

5   Concluding Remarks

In this paper, we have proposed a feature selection method that benefits from the advantages of both filter and wrapper methods. The method is conceptually based on the expectation maximization algorithm. It uses a greedy strategy for locally selecting the most competent set of features from the feature space. The method is advantageous over filter methods since it uses an iterative EM based feature selection strategy, which is *more likely* to reach the globally optimum feature set. In addition, it augments the capability of wrapper methods by allowing them to be used in the document clustering field where class labels are not available. The proposed method has been evaluated on various data sets whose results show promising improvement in terms of accuracy and normalized mutual information compared with several existing methods.

For the future, other clustering algorithms such as Bisecting k-Means and Spectral Clustering can be incorporated to the algorithm and experiments conducted. An interesting study regarding clustering algorithms is the study of dynamic clustering algorithms such as X-means in which the number of clusters is not determined beforehand. Moreover, other feature ranking measures can be localized (as it was the case for DF in this paper) and applied to see how the algorithm can improve the performance of clustering.

References

1. Breiman, L.: Classification and Regression Trees. Chapman & Hall/CRC, Boca Raton (1998)
2. Chua, S., Kulathuramaiyer, N.: Semantic feature selection using wordnet. In: WI 2004: Proceedings of the IEEE/WIC/ACM International Conference on Web Intelligence, Washington, DC, USA, pp. 166–172. IEEE Computer Society, Los Alamitos (2004)
3. Dash, M., Choi, K., Scheuermann, P., Liu, H.: Feature selection for clustering - a filter solution. In: ICDM, pp. 115–122 (2002)
4. Dhillon, I., Kogan, J., Nicholas, C.: Feature Selection and Document Clustering, Survey of Text Mining: Clustering, Classification, and Retrieval (2004)
5. Forman, G.: An extensive empirical study of feature selection metrics for text classification. The Journal of Machine Learning Research 3, 1289–1305 (2003)
6. Guyon, I., Elisseeff, A.: An introduction to variable and feature selection. The Journal of Machine Learning Research 3, 1157–1182 (2003)
7. sam Han, E.h., Boley, D., Gini, M., Gross, R., Hastings, K., Karypis, G., Kumar, V., Mobasher, B., Moore, J.: Webace: a web agent for document categorization and exploration. In: Proc. of the 2nd International Conference on Autonomous Agents, pp. 408–415. ACM Press, New York (1998)
8. Jain, A., Zongker, D.: Feature selection: evaluation, application, and small samplepperformance. IEEE Transactions on Pattern Analysis and Machine Intelligence 19(2), 153–158 (1997)
9. Jolliffe, I.T.: Principal component analysis. Springer, New York (2002)
10. Kohavi, R., John, G.H.: Wrappers for feature subset selection. Artificial Intelligence 97(1-2), 273–324 (1997)
11. Liu, T., Liu, S., Chen, Z., Ma, W.-Y.: An evaluation on feature selection for text clustering. In: ICML, pp. 488–495 (2003)
12. Strehl, A., Ghosh, J.: Cluster Ensembles-A Knowledge Reuse Framework for Combining Partitionings. In: Proceedings of the National Conference on Artificial Intelligence, pp. 93–99. AAAI Press, MIT Press, Menlo Park, Cambridge (1999) (2002)
13. Wolf, L., Shashua, A.: Feature selection for unsupervised and supervised inference: The emergence of sparsity in a weight-based approach. J. Mach. Learn. Res. 6, 1855–1887 (2005)
14. Yang, Y.: Noise reduction in a statistical approach to text categorization. In: Proceedings of the 18th Ann Int ACM SIGIR Conference on Research and Development in Information Retrieval (SIGIR 1995), pp. 256–263. ACM Press, New York (1995)
15. Zhao, Z., Liu, H.: Spectral feature selection for supervised and unsupervised learning. In: ICML, pp. 1151–1157 (2007)

Cost-Based Sampling of Individual Instances

William Klement[1], Peter Flach[2], Nathalie Japkowicz[1], and Stan Matwin[1,3]

[1] School of Information Technology and Engineering
University of Ottawa, K1N 6N5 Ottawa, Ontario, Canada
{klement,nat,stan}@site.uottawa.ca
[2] Department of Computer Science, University of Bristol
Bristol BS8 1UB, United Kingdom
Peter.Flach@bristol.ac.uk
[3] Institute of Computer Science, Polish Academy of Sciences, Poland

Abstract. In many practical domains, misclassification costs can differ greatly and may be represented by class ratios, however, most learning algorithms struggle with skewed class distributions. The difficulty is attributed to designing classifiers to maximize the accuracy. Researchers call for using several techniques to address this problem including; under-sampling the majority class, employing a probabilistic algorithm, and adjusting the classification threshold. In this paper, we propose a general sampling approach that assigns weights to individual instances according to the cost function. This approach helps reveal the relationship between classification performance and class ratios and allows the identification of an appropriate class distribution for which, the learning method achieves a reasonable performance on the data. Our results show that combining an ensemble of Naive Bayes classifiers with threshold selection and under-sampling techniques works well for imbalanced data.

Keywords: Class Imbalance, Sampling, Cost-Based Learning.

1 Introduction

In many practical domains, machine learning methods encounter difficulties when the class distribution is highly skewed. In fact, when misclassification costs differ greatly among classes, most machine learning methods face a demanding task to achieve acceptable performance. For instance, classifying patients with head injuries, whether they require a CT scan or not, results in highly skewed data. Often, the size of the positive class is significantly smaller than that of the negative class. This may be attributed to a low frequency of serious head injuries, or, may be due to physicians being over cautious in treating head injuries. Regardless of the reason, this is a real-life example of the natural existence of the skewed class distribution problem. The same problem presents itself in many other domains. Examples of such include: fraud detection [9], anomaly detection, information retrieval [14], thyroid disease [1], and detection of oil spills [13]. To make matters worse, data insufficiency, a very small size of the minority class, is another problem that commonly co-occurs with skewed class distributions.

Y. Gao and N. Japkowicz (Eds.): Canadian AI 2009, LNAI 5549, pp. 86–97, 2009.

Both can be viewed as related to the cost of obtaining examples in the minority class. Machine learning researchers emphasize the need to distinguish between the problem of skewed class distribution and that of the small minority class [3]. In dealing with the presence of both problems, most machine learning methods struggle to achieve less-than-desired levels of performance. In recent years, machine learning researchers made several recommendations to deal with such data [3]. In this paper we base our implementations on several such recommendations including; using a probabilistic, rather than tree-based, classifier [17,3], adjusting the decision threshold [17,3], and adopting an under-sampling, rather than over-sampling, strategy to balance the training data [6]. In the quest of an appropriate sampling strategy, our investigation focuses on the question; *what is the correct distribution for a learning algorithm?* [17].

In this paper, we propose a cost-based sampling technique that assigns weights to individual training instances based on their individual misclassification costs. This novel approach allows us to answer the questions: (1)*what is the cost associated with the performance of a learning algorithm?*, and (2)*what is an appropriate class distribution for a learning method to perform well on the data set?*. Our sampling approach reveals the relationship between the minority class ratio in the training data and the performance of the learning method. We measure the performance by the average sensitivity (true positive rate), the average specificity (true negative rate), and the area under the ROC curve (AUC). Our experiments compare the performance of two classification models. The first is a standard Naive Bayes classifier trained on several data samples containing various class distributions including the original distribution. The second classification model is an ensemble of Naive Bayes classifiers, each of which is trained on a balanced data sample. The balanced data is obtained by means of randomly under-sampling the majority class without replacement. Further, each classifier in the ensemble employs classification threshold selection for a maximized value of the F-measure. Our results show that combining random under-sampling and decision threshold selection into an ensemble of Naive Bayes learners is an effective approach to achieve a balanced classification performance on both classes simultaneously. Finally, our results indicate that our sampling approach does not affect the performance measured by the AUC metric.

In his paper, we review classification with a skewed class distribution, describe the random under-sampling strategy, and present our novel cost-based sampling. Further, we present our experiment design and results followed by a discussion and future work.

2   Classification with a Skewed Class Distribution

Machine learning researchers have taken interest in the problem of skewed class distributions and unequal misclassification costs [7,3,17,19]. Provost [17] suggests that machine learning classifiers struggle when dealing with imbalanced data due to making two assumptions, the first is that the goal is to maximize accuracy, and the second is that the classifier will operate on data drawn from the same distribution as the training set. Consequently, common machine learning algorithms appear to predict the majority class [7]. Intuitively, we want the classifier to perform better on the minority class. To this extent, many researchers have considered taking separate error rates for positive and negative classes [2], raised concerns about maximizing the accuracy [18], and studied the relationship between accuracy and AUC, the latter is based on the trade off between the rates of true positives and false positives [15].

Proposed solutions follow two approaches. The first is based on modifying the distribution in the data and, the second, on adjusting the learning algorithm to adapt to the skewness in the class distribution. Sampling methods modify the class distribution by increasing or decreasing the frequency of one class. These are known as over-sampling and under-sampling respectively. Sampling is to artificially balance the training set by duplicating instances in the minority class (over-sampling the minority class) or by removing instances in the majority class (under-sampling the majority class). It has been shown that under-sampling outperforms over-sampling [7] and over-sampling can lead to over-fitting [3]. The use of sampling techniques gives rise to the question: *what is the correct distribution for a learning algorithm?*. Furthermore, *what is the appropriate sampling strategy for a given data set?* [17]. Alternatively, the learning algorithm can be adjusted to adapt for class imbalance. Cost-based learning is another such technique where instances in the minority class are assigned higher misclassification costs than those in the majority class. Cost-based learning can be divided into three categories; making a specific learning algorithm sensitive to cost [5,8], assigning examples to their lowest risk class [4,16,21], and converting an arbitrary learning algorithm to become cost sensitive [4,22]. Adjusting the probability estimation or adjusting the classification threshold can also help counter the problem [17]. Finally, building classifiers that learn each class separately can also be used to counter the problem of imbalance. Most of these solutions have been discussed in [12,3].

3   Random Under-Sampling

Our random under-sampling keeps the entire minority class and randomly samples, without replacement, an equal proportion of the majority class. This results in a balanced data set which we use for training. The testing set, however, retains the original class distribution of the data. Our choice of under-sampling, as opposed to over-sampling, is based on the finding that under-sampling performs better [7]. The main disadvantage of under-sampling is the loss of potentially useful instances in the majority class [19]. To limit this possible loss, we combine an ensemble of Naive Bayes classifiers trained on multiple randomly under-sampled balanced training sets, then, we average their predicted probabilities. The next section describes a cost-based sampling to help identify the appropriate class distribution for which the learning method performs well, therefore, identifies a suitable sampling strategy.

4   Cost-Based Sampling

This section proposes a cost-based sampling strategy that modifies the class distribution according to a misclassification cost function. Altering the class distribution in the training data is formally introduced in [11] and is designed to assist the learning with highly skewed data by imposing non-uniform class misclassification costs [19]. It has been shown that altering the class distribution in the training data and altering misclassification costs are equivalent [19,11].

Let X be a set of n instances where the ith instance is a vector x_i of values for attributes $A_1, \cdots, A_n$. Let $C \in \{+, -\}$ be the class labels for a binary classification problem. It is important to point out that this model is not limited to a binary class representation, however, we assume binary classification for simplicity. Let f be a misclassification cost function whose values are $s_i \in \{0, 1, 2, \cdots, S\}$, i.e. f is an ordinal function between 0 and S. In general, we represent costs as weights assigned to individual instances and modify the ratio of positive to negative instances according to these weights. In particular, we duplicate each instance as positive or negative according to the ratio of its corresponding weight. The algorithm of duplicating instances follows that, for each instance x_i; if x_i is positive, then, we duplicate s_i instances of x_i and label them as positives, further, we duplicate $S - s_i$ instances of x_i and label them as negatives. If x_i is negative, we duplicate $S - s_i$ instances of x_i and label them as positives, as well as, we duplicate s_i instances of x_i and label them as negatives.

To illustrate this process, consider examples (a) and (b) listed in table 1. The tables show instance number i, its corresponding label, its misclassification cost s_i, the final number of positive instances n^+, and the final number of negative instances n^- in the data. For these examples, let $s_1 \in \{1, 2, 3, \cdots, 10\}$ with a maximum value $S = 10$. In table 1.a, when $s_1 = s_2 = 8$, our approach produces the same ratio of positives to negatives of 10 instances each. Instance x_1 is duplicated as $s_1 = 8$ positive instances and as $S - s_1 = 10 - 8 = 2$ negative instances. Similarly, x_2 is duplicated as $S - s_1 = 10 - 8 = 2$ positive instances and as $s_1 = 8$ negative instances. The resulting class ratio remains equal because $s_1 = s_2 = 8$. In example 1.b, $s_1 = 3$ and $s_2 = 6$ and instance x_1 is duplicated three times as a positive instance and seven times as a negative instance. Similarly, x_2 is duplicated as four negative instances and as six positive instances. In this case, the resulting class distribution is $7 : 13 = 53.9\%$ proportional to the specified cost weights of $3 : 6 = 50\%$. The representation of the function f used in the

Table 1. Examples of cost-based sampling

(a)						(b)				
i	Label	s_i	n^+	n^-		i	Label	s_i	n^+	n^-
1	+	8	8	2		1	+	3	3	7
2	-	8	2	8		2	-	6	4	6
Total:			10	10		Total:			7	13

above examples, naturally, corresponds to assigning weights of importance to each instance in the data set. In particular, a weight assignment of 10 indicates *"most important"* assignment, a weight of 1 is *"least important"*, and a weight of 0 is *"not important"*. Instances with 0 weight (*not important*) are removed from the data set. If we let n^+ and n^- be the number of positives and negatives in the original data respectively, then, the number of positives n'^+ and negatives n'^- duplicates, respectively, can be calculated by:

$$n'^+ = \sum_{i=1}^{n^-}(S - s_i) + \sum_{j=1}^{n^+} s_j \quad \text{and} \quad n'^- = \sum_{j=1}^{n^-} s_j + \sum_{i=1}^{n^+}(S - s_i)$$

5   Experiment Design

In this work, we use our cost-based sampling method to determine an appropriate cost (class distribution) for which the performance of a single Naive Bayes classifier is improved on selected data sets. In order to obtain a benchmark performance, we use an additional classifier built using a combination of several techniques. Collectively, these techniques improve the performance of the Naive Bayes learning on data with highly skewed class distribution. This classifier combines an ensemble of ten Naive Bayes classifiers, each of which is trained on a balanced data by means of random under-sampling without replacement. Each of these 10 classifiers is tuned with classification threshold selection for a maximized F-measure. The F-measure is a combination of precision and recall. When maximized for the minority class, the F-measure maximizes both precision and recall. This process of optimizing the F-measure is performed by a 3-fold cross-validation on the training set. In the Weka software [20], this is an option listed for the ThresholdSelector meta-classifier. Finally, these 10 classifiers are combined by averaging their predicted probabilities.

5.1   The Data Sets

We select various data sets from the UCI Machine Learning Repository [1] as listed in table 2. These data sets have various degrees of imbalance. The table shows the number of instances n, n^+ of which are positives and n^- are negatives. The ratio of the minority class $\frac{n^+}{n}$ is a percentage and s is a sampling percentage by which we randomly under-sample the training set for balancing. For example, the discordant patient data from the thyroid database, contains 3772 instances, 58 of which are discordant and 3714 are negatives instances. The minority class in this set is 1.54% of the total size. When we under-sample the training set, we randomly select 3% of the training set without replacement to balance it. Effectively, we select 1.5% positive instances (i.e. all the positive instances) and 1.5% from the negatives. The minority class remains complete while the negative class is under-sampled to the same size of the minority class. The data sets in table 2 have minority class sizes of less than 25%. The most imbalanced is the discordant data with a mere 1.5% positives. The spect, the hepatitis, and the

Table 2. Details of the selected UCI data sets [1]

Data Set	n	n^+	n^-	$\frac{n^+}{n}$	s	Description
discordant	3772	58	3714	1.54	3.0	Thyroid disease records
ozone-onehr	2536	73	2463	2.88	5.7	Ozone Level Detection
hypothyroid	3163	151	3012	4.77	9.5	Thyroid disease
sick	3772	231	3541	6.12	12.2	Thyroid disease
ozone-eithhr	2534	160	2374	6.31	12.6	Ozone Level Detection
sick-euthyroid	3163	293	2870	9.26	18.5	Thyroid disease
spect	267	55	212	20.60	41.0	SPECT images.
hepatitis	155	32	123	20.65	41.2	Hepatitis Domain.
adult	40498	9640	30858	23.80	47.61	Adult income census

`adult` data have the highest proportion of positive instances of approximately 20%. The sampling ratio s is associated with the size of the minority class. The idea is that when we randomly sample the training set, we obtain $s = 2 \times \frac{n^+}{n}$ to balance the data. One half of the sampled instances is obtained from the minority class and the other half is randomly obtained form the majority class. This under-sampling strategy is only applied to the training set. The test set retains the original class distribution. Finally, these data sets vary in two ways, the percentage of the minority class and the size of the positive instances. The `discordant`, the `ozone-onehr`, the `spect`, and the `hepatitis` data have fewer positives than 100.

5.2   Testing and Evaluation

For the purpose of evaluation, we randomly select (without replacement) 66% of the data for training and 33% for testing. This random sampling preserves the original class distribution which is important, at least, for testing. This sampling is repeated 10 times, i.e. 10 different random seeds of sampling. The result is 10 pairs of training and testing sets for each data set. Our experiment proceeds with training a single Naive Bayes classifier on the original training samples, randomly under-sampling these training sets and training an ensemble of 10 Naive Bayes classifiers, and applying our cost-based sampling to these training sets, then training a single Naive Bayes classifier. For the latter, we repeat our sampling method for cost values of $1, 2, 3, \cdots, 10$ applied to the negative class only. The positive class (the minority class) always gets a fixed cost ratio of 10. The end result is 12 classifiers for each pair of training/testing sets (one Naive Bayes classifier trained on the original class distribution, one ensemble of classifiers trained on balanced under-samples, and 10 Naive Bayes classifier trained on our cost-based sampling, one for each cost value). Each of these 12 classifiers is tested on the corresponding holdout test set which contains the original class distribution. This process of constructing classifiers is repeated 10 times for each data set because we have 10 pairs of training/testing sets.

To measure the performance of our 12 classifiers, we average three metrics for all 10 pairs of training/testing data. Therefore, for each data set, we obtain the average sensitivity (the average true positive rate), the average specificity (the average true negative rate), and the average AUC (or the average area under the ROC curve). We omit the use of accuracy due to its well-known deficiency in measuring performance for highly skewed class distribution [18]. Instead, we measure the AUC. Furthermore, the idea of performance, in the case of highly skewed class distribution, is to obtain good performance on both classes simultaneously. The sensitivity and specificity depict just that. We plot the average values for the above three metrics against the size of the positive class to show how cost ratios affect the performance. These results are presented in the following section.

6   Results

We begin with results obtained on the `adult` data. Figure 1 shows the average sensitivity and the average specificity (recorded for testing data) plotted against the positive class ratio (in the training data) on the y-axis and on the x-axis respectively. All values are normalized between 0 and 1. The performance of the Naive Bayes classifier, trained with the original class distribution, is indicated by the □ and by the ○ symbols. In figure 1, this classifier achieves an average specificity (○) above 90% (on the y-axis) and an average sensitivity (□) just over 50% (also on the y-axis) for the original positive class ratio of just over 20%. The lower average sensitivity is consistent with the detrimental effect of the majority negative class in the training data. The performance of our ensemble of Naive Bayes classifiers, trained on the balanced under-sampled data, is indicated by the △ and by the ▽ symbols. The ensemble achieves an average sensitivity (△) of 89% and an average specificity (▽) of 72% (both are on the y-axis) for a positive class ratio of 50% (on the x-axis). The dotted line shows the average sensitivity obtained from testing a single Naive Bayes classifier trained on sampled data of positive class ratios shown on the x-axis. The solid line

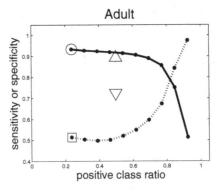

Fig. 1. Naive Bayes requires a positive class ratio of 80% to do well

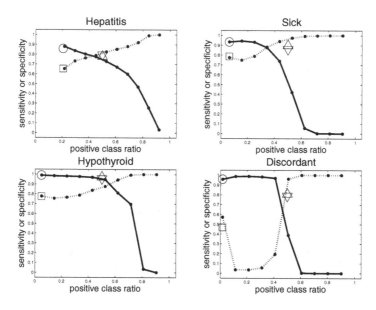

Fig. 2. Naive Bayes achieves higher specificity than sensitivity with the original class distribution. A positive class ratio between 40-50% produces best average performance.

represents the average specificity achieved by this classifier. The sampling, in this case, is based on our cost-based sampling approach presented in this paper. These two lines show that the single Naive Bayes classifier achieves a much higher average specificity than average sensitivity when trained on data of less than 80% positive instances. At the 80% positive class ratio, the two lines cross and the average sensitivity increases higher than the average specificity. This suggests that this classifier performs much better on the negative class for a positive class ratio less than 80%. Above that, this classifier performs better on the positive class with higher sensitivity than specificity. Thus, over-sampling the positive class to a ratio of 85%, in this case, produces a reasonable average sensitivity of 84% and an acceptable average specificity of 75%. This performance is balanced on both classes and is similar to that of our ensemble with under-sampling. Therefore, we can conclude that both strategies of balancing with under-sampling or over-sampling to 85% positives can offset the devastating effects of class imbalance.

The next set of observations are focused on data sets: hepatitis, sick, hypothyroid, and discordant. Their results are shown in figure 2. These results represent the classic expected performance on data with skewed class distributions. In these four data sets, the Naive Bayes classifier, trained on the original class distribution, achieves higher average specificity ($\bigcirc$) than average sensitivity ($\square$) with a significant gap between them. For the hepatitis and the sick data, the ensemble of classifiers ($\triangle$, $\triangledown$), using random under-sampling, performs reasonably well and comes close to achieving a balanced performance on both classes (a small gap between the two lines). For the hepatitis data, sampling

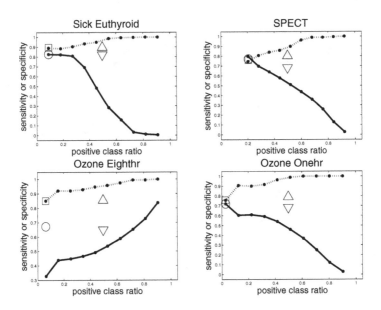

Fig. 3. Naive Bayes (with original class distribution or with under-sampling) performs well except on the `eighthr` where cost-based sampling performs significantly better

the positive instances to be 45% of the training data performs best. This value is very close to 50% where the positives have the same cost as the negatives. In this case, the ensemble performs really well in achieving an average sensitivity of 78% and an average specificity of 80%. The situation is very similar in the results for the `sick` data. However, the point at which the sensitivity curve (the dotted line) intersects with the specificity line (the solid line) occurs at the 35% positive class ratio. The ensemble also performs well and close to the point of intersection. In the case of `hypothyroid` and `discordant` data, the ensemble of classifiers produces slightly better performance than that of our sampling method. The sensitivity △ and the specificity ▽ are both vertically higher than the point where the two curves intersect. The results in figure 2 suggest that the random under-sampling is very appropriate for this data. Furthermore, in all these four data sets, the intersection point occurs close to the 50% positive class ratio indicating that a Naive Bayes performs best with equal costs on these data sets.

The final set of results are shown in figure 3. For sets `sick-euthroid`, SPECT, and `onehr` data, Training a Naive Bayes classifier on these data sets with the original class distribution or training an ensemble of Naive Bayes classifier on randomly under-sampled data produce equal results. Our cost-based sampling shows that the meeting point of sensitivity and specificity occur very close to the original class distribution, at a very low positive class ratio. For the `eighthr` data set, however, the situation is reversed. The best performance occurs closer to positive class ratio of 1. Sampling the positive class ratio to be over 90% results in the Naive Bayes achieving an average sensitivity of 100% and an

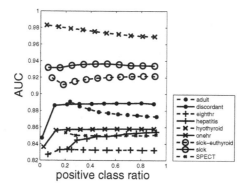

Fig. 4. Naive Bayes (with original class distribution or with under-sampling) performs well except on the `Eighthr` where cost-based sampling performs significantly better

average specificity of 84%. The ensemble of classifiers remains unable to achieve such performance. In fact, the ensemble performs no better than a Naive Bayes trained with the original class distribution ($\square$ and $\triangle$ are equal, $\bigcirc$ and $\triangledown$ are also almost level). In this case, our cost-based sampling results in the Naive Bayes classifier achieving a significantly better performance than other methods. The gap between the sensitivity and the specificity curves is large, however, the highest performance occurs for a positive class ratio of 90%.

Now, consider the AUC values shown in Figure 4. The figure shows that the AUC exhibits little to no change (less than 2%) when the class ratio changes. This is expected because changing the cost effectively selects a different point along the same ROC curve. When the ratio of the positive class is very small, the segment of interest lies on the bottom-left corner of the ROC curve. As the cost changes, this point of interest slides higher along the ROC curve towards the decreasing diagonal where the costs of positives and negatives are equal. The mostly flat lines of AUC values in figure 4 suggest that the ROC curve does not change dramatically. One can argue that our cost-based sampling strategy may produce noise by duplicating instances of positive and negative labels, however, the little to no change of the AUC suggests that the learning method, the Naive Bayes learning, remains unaffected by this noise. Hence, the added noise does not severely corrupt the data, however, this may be attributed to our choice of the learning method, the Naive Bayes learning which estimates probabilities based on the relative frequencies of attribute values, relative to the class prior. A different learning method may be more susceptible to the added noise.

7   Discussion and Future Work

This paper presents a novel cost-based sampling method that counters the detrimental effects of class imbalance to reveal the relationship between performance and class ratio. Understanding this relationship allows for the identification of an appropriate class distribution for a particular learning method applied to the given data, effectively, assisting the selection of a suitable sampling strategy.

When dealing with a skewed class distribution, we propose the use of an ensemble of classifiers, in this case, multiple Naive Bayes classifiers, and combine them with threshold selection and random under-sampling techniques. To our knowledge, combining these three techniques is a novel contribution to tackle imbalanced data. Our experiments show that this combination strategy is a reasonable approach that works well. Our results confirm that a Naive Bayes classifier suffers from the devastating effects of class imbalance, however, for some data sets, it can perform reasonably well despite the skewed class distribution. Our cost-based sampling method reveals that this performance is attributed to the original class distribution being the appropriate class ratio for Naive Bayes learning on these data sets.

An advantage of our cost-based sampling strategy lies in assigning cost weights to instances individually. Consequently, we can represent imbalance within a single class as well. In this work, we only use equal costs for instances in the same class. Future studies can address such issues. Furthermore, an anonymous reviewer suggests that our approach allows the representation of a variety of cost functions that may better reflect the problem domain, however, a disadvantage may arise. Our sampling approach is based on duplicating instances then setting their labels to positives or negatives according to their cost weights. One can argue that this may add noise to the data. In this paper, the AUC values of the classifier show little to no change when we change the class ratio suggesting that our method, in this case, is safe. This may be due to our selected learning method, the Naive Bayes method. Other learning methods, possibly less robust to noise, may be affected by this potential side-effect. To avoid such effects, our weighting strategy may be applied to the testing set to produce cost-sensitive evaluation. Finally, our approach may be extended to one-class learning (when the minority class is missing) and to multi-class learning. These are few of the many issues subject to further investigations.

References

1. Asuncion, A., Newman, D.J.: UCI Machine Learning Repository. University of California, School of Information and Computer Science (2007),
   http://www.ics.uci.edu/~mlearn/MLRepository.html
2. Cardie, C., Howe, N.: Improving minority class prediction using case-specific feature weights. In: Proc. of 14th Int. Conf. on Machine Learning, pp. 57–65 (1997)
3. Chawla, N.V., Japkowicz, N., Kolcz, A. (eds.): Proc. of ICML, Workshop on Learning from Imbalanced Data Sets (2003)
4. Domingos, P.: Metacost: A general method for making classifiers cost-sensitive. In: Proc. of 5th Int. Conf. on Knowledge Discovery and Data Mining, pp. 155–164 (1999)
5. Drummond, C., Holte, R.C.: Exploiting the cost (in)sensitivity of decision tree splitting criteria. In: Proc. of 17th Int. Conf. on Machine Learning, pp. 239–246 (2000)

6. Drummond, C., Holte, R.C.: C4.5, Class imbalance, and Cost Sensitivity: Why Under-Sampling beats Over-Sampling. In: Proc. of the ICML Workshop on Learning from Imbalanced Datasets II (2003)

7. Drummond, C., Holte, R.C.: Severe Class Imbalance: Why Better Algorithms Aren't the Answer. In: Gama, J., Camacho, R., Brazdil, P.B., Jorge, A.M., Torgo, L. (eds.) ECML 2005. LNCS (LNAI), vol. 3720, pp. 539–546. Springer, Heidelberg (2005)

8. Fan, W., Stolfo, S., Zhang, J., Chan, P.: AdaCost: misclassification cost-sensitive boosting. In: Proc. of 16th Int. Conf. on Machine Learning, pp. 97–105 (1999)

9. Fawcett, T., Provost, F.: Adaptive Fraud detection. Data Mining and Knowledge Discovery (1), 291–316 (1997)

10. Hettich, S., Blake, C.L., Merz, C.J.: UCI Repository of machine learning databases. University of California, Irvine, Dept. of Information and Computer Sciences (1998), http://www.ics.uci.edu/~mlearn/MLRepository.html

11. Elkan, C.: The foundations of cost-sensitive learning. In: Proc. of 17th Int. Joint Conf. on Artificial Intelligence (2001)

12. Japkowicz, N. (ed.): Proc. of AAAI 2000 Workshop on Learning from Imbalanced Data Sets, AAAI Tech Report WS-00-05 (2000)

13. Kubat, M., Holte, R.C., Matwin, S.: Machine learning for the detection of oil spills in satellite radar images. Machine Learning (30), 195–215 (1998)

14. Lewis, D.D., Catlett, J.: Heterogeneous uncertainty sampling for supervised learning. In: Proc. of 11th Int. Conf. on Machine Learning, pp. 179–186 (1994)

15. Ling, C.X., Huang, J., Zhang, H.: AUC: a statistically consistent and more discriminating measure than accuracy. In: Proc. of 18th Int. Conf. on Machine Learning, pp. 519–524 (2003)

16. Margineantu, D.: Class probability estimation and cost-sensitive classification decisions. In: Proc. of 13th European Conf. on Machine Learning, pp. 270–281 (2002)

17. Provost, F.: Learning with Imbalanced Data Sets 101. In: Invited paper for the AAAI 2000 Workshop on Imbalanced Data Sets (2000)

18. Provost, F., Fawcett, T., Kohavi, R.: The case against accuracy estimation for comparing induction algorithms. In: Proc. of 15th Int. Conf. on Machine Learning, pp. 43–48 (1998)

19. Weiss, G.M., McCarthy, K., Zabar, B.: Cost-Sensitive Learning vs. Sampling: Which is Best for Handling Unbalanced Classes with Unequal Error Costs? In: Proc. of the Int. Conf. on Data Mining, pp. 35–41 (2007)

20. Witten, I.H., Frank, E.: Data Mining: Practical machine learning tools and techniques, 2nd edn. Morgan Kaufmann, San Francisco (2005)

21. Zadrozny, B., Elkan, C.:: Learning and making decisions when costs are probabilities are both unknown. In: Proc. of 7th Int. Conf. on Knowledge Discovery and Data Mining, pp. 203–213 (2001)

22. Zadrozny, B., Langford, J., Abe, N.: Cost-Sensitive Learning by Cost-Proportionate Example Weighting. In: Proc. of IEEE Int. Conf. on Data Mining (2003)

Context Dependent Movie Recommendations Using a Hierarchical Bayesian Model

Daniel Pomerantz and Gregory Dudek

School of Computer Science
McGill University
Montreal, Quebec
{dpomeran,dudek}@cim.mcgill.ca

Abstract. We use a hierarchical Bayesian approach to model user preferences in different contexts or settings. Unlike many previous recommenders, our approach is content-based. We assume that for each context, a user has a different set of preference weights which are linked by a common, "generic context" set of weights. The approach uses Expectation Maximization (EM) to estimate both the generic context weights and the context specific weights. This improves upon many current recommender systems that do not incorporate context into the recommendations they provide. In this paper, we show that by considering contextual information, we can improve our recommendations, demonstrating that it is useful to consider context in giving ratings. Because the approach does not rely on connecting users via collaborative filtering, users are able to interpret contexts in different ways and invent their own contexts.

Keywords: recommender system, Expectation Maximization, hierarchical Bayesian model, context, content-based.

1 Introduction

Recommender systems are becoming more and more widespread with many websites such as Amazon.com™able to provide personalized recommendations as to what products a customer will like. If the customer likes the product that has been recommended, then she is more likely to both buy the specific product and continue shopping there in the future.

One problem with many current recommender systems is that they fail to take into account any *contextual* information. That is, they do not consider important questions such as when, where, and with whom you will use the item. For example, a couple looking to see a movie on a date is recommended movies such as *Finding Nemo* and *Shrek* because they previously watched similar movies with their kids and enjoyed them. Those systems that do incorporate contextual information do so in a way that does not allow users to define their own contexts.

Traditional recommender systems can only answer the question, "What item should I use?" In this paper, we focus on the movie domain, but the ideas can be generalized to other domains such as books or online blogs. We will demonstrate the usefulness of storing contextual information and adapt the hierarchical

Y. Gao and N. Japkowicz (Eds.): Canadian AI 2009, LNAI 5549, pp. 98–109, 2009.

Bayesian model described in [13] and [11] in order to answer two questions: "In setting X, what movie should I watch?" and "Given that I gave movie M a score of S in context C, what would I think about it in a different context?" Throughout this paper, we will refer to these two problems as "Standard Context Recommendation" and "Different Context Evaluation."

One way to model a user in different contexts is to create separate movie accounts for each context. This is not ideal, however, as it very often happens that *some* of the user's tastes do not change in different contexts. Without sharing information between contexts, users would have to rate many movies in *each* context in order to get a good prediction. However, if we can share the similarities between contexts, we will not require users to rate as many movies. The hierarchical Bayesian model is ideal for our purposes as it allows us to share information between contexts, but at the same time, it allows the preferences in different contexts to be entirely different. In this way, we avoid forcing each user to rate an excessive number of movies. These techniques can then be extended to other domains in addition to movies such as books or online blogs.

2   Background Information and Related Work

To make a recommendation, we need to predict a rating r_p or usefulness of an item or movie m for a user u. We can then select the products that have the highest usefulness to the user. This paper will discuss ways to improve the predicted rating of a product, since once we calculate this, we can easily make a recommendation by sorting.

The two most common techniques to predict the usefulness of a movie are *collaborative-filtering algorithms* and *content-based models*. In each case, we normally represent each movie as a vector with each entry in the vector representing the amount of one particular feature (e.g. humor, Brad Pitt, bad acting, etc). Collaborative-filtering based techniques determine a set of similar users. Once the system has determined similar neighbors, it can make a recommendation based on assuming that similar users will like similar movies.

Content-based approaches model a user by determining the important features to each user based on previous ratings that the same user has made. Using the model, they will recommend items that are similar to other items that the user has rated highly. This paper will focus on the content-based technique. There are several ways to model a user. One model is based on a linear approach. Alspector [2] suggests that the recommender system should store a set of weight vectors corresponding with the user's preference of each feature. Another commonly used approach is to predict the same rating as the nearest item or the average of the k nearest items (nearest neighbor) [10]. Here we briefly summarize these approaches.

2.1   Representing Users

One approach to modeling users is a linear model which proposes that every user can be modelled as a vector of real numbers. This vector relates to the movie vector in that each element represents how much the user likes the presence of the corresponding feature. Once we learn these weights, denoted by w_u, we can make a prediction r_p as to whether a user u will like a movie m based on:

$$r_p(u, m) = \overrightarrow{w_u} \overrightarrow{m} . \tag{1}$$

We can use machine learning algorithms to learn the weights given a set of training data. One method is to compute a least squares linear regression [12]. This corresponds with assuming that every actual rating r_a differs from the predicted rating by adding Gaussian noise. For space considerations, we refer the reader to [12] for more details of this method as well as other approaches to learning the weights such as Support Vector Machines and Naive Bayes models.

Another content-based approach that has been used is the nearest neighbor approach or nearest k-neighbors approach. In this non-linear method, we calculate the similarity of a previously rated item to the item in question and then take a weighted (by the similarity) average of the ratings in the training data. One measure used to find similarity is the cosine similarity measure [7]. Another possibility is to use the inverse Euclidean distance to calculate similarity. Once this is calculated for all rated movies, we select the k-most similar previously rated movies and calculate a weighted average of the k movies to make a prediction.

2.2   Dimensionality Reduction

One common problem with recommender systems is that the dimensionality of the feature space is very large. This often causes the problem to be ill-posed. The dimensionality can often be lowered because many features are redundant and others are useless. For example, there is a large correlation between the features *Keanu Reaves* and *bad acting*, meaning these features redundant. Other features appear in only a few movies, and can be dropped without much information loss. There are several ways to reduce the dimensionality of the space. Goldberg et al. [4] suggest using a gauge set of movies. The idea here is to find an ideal set of movies which all users should be asked about. Similarly, one could create a gauge set of features. Some other possibilities are to reduce the dimensionality based on approaches using information gain or mutual information or Independent Component Analysis (ICA). We use Principal Component Analysis (PCA), a technique that is geared towards storing only the most useful data. For a further comparison on dimensionality results, see [9].

2.3   Connecting Users

One of the downsides of looking at users separately is a user has to rate several movies before being able to be given a useful prediction on a new movie. This is referred to as the "new user" problem. Zhang and Koren propose a hierarchical Bayesian model to solve this by relating each user's preference weights to each other. The model assumes that, as in Section 2.1, for any given user, there

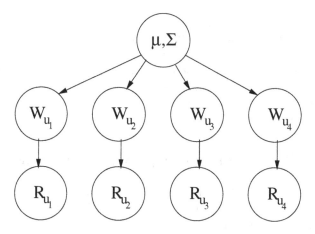

Fig. 1. Hierarchical model: First a mu and sigma are "chosen" for the entire "population." Then based on this, each user's weight vector is a Gaussian random vector. Finally, given the user's preference weights, the rating given to each movie can be determined, but it is not entirely deterministic due to noise. Note that R_i is observed.

is a linear relationship between the movie vector and the rating. The model relates each user's weights to each other. See Figure 1. It assumes that there is a mean vector μ and covariance matrix Σ^2 that exist for all users as the mean and covariance respectively of all user preferences. Each user's weight vector is then a Gaussian random vector with mean and covariance matrix μ and Σ^2 respectively. In other words, each user is merely a random sampling from the normal distribution. After determining the weight vector $\vec{w_u}$ for each user u, the rating for a movie with features $\vec{m}$ is a normal random variable with mean $\vec{w_u}\vec{m}$ and variance σ_u where σ_u is a noise factor that is calculated for each user. They use Expectation Maximization to estimate the unknown weights w_u. This solution is good for dealing with the new user problem because a new user is initially given a set of weights (that of the average user) and the model gradually adjusts these weights to fit the user. Since each user's weights are generated from a normal distribution, any particular set of weights is allowed so that after rating enough movies, the user's weights under this algorithm will converge with the weights from a standard linear regression.

2.4   Context-Based Recommendations

Adomavicius and Sankaranarayanan [1] explore using context to find similar users and similar items. They start with standard collaborative filtering algorithms, which use a similarity measure between two items, and extend the measure to include additional context dimensions. Ono et. al. [6] use a Bayesian network in which contexts, users, and items all combine together to form "impressions" (e.g. funny, depressing, etc.) of the movie, which in turn leads to ratings. They estimate the probability of a rating given the user, context, and item.

The downside of these approaches is they require sharing information about contexts between users. We would like to build a model for every user that is not dependent on other users. This allows each user to have his own definition of a context. For example, while the majority of users may like watching romantic movies on a date, there may be some users that prefer not. While it is theoretically possible to design an algorithm that determines which users treat contexts in one way, these are difficult parameters to estimate. Additionally, by designing a content-based model, we can easily allow users to add their own contexts.

3   Content-Based Context-Dependent Recommendations

Naively, one might think that modeling a user's preferences in different situations could be handled simply by considering each user as several different people. That is, for each user we maintain a different profile for every different possible context that they have rated movies in. However, the user's preferences in different contexts are possibly correlated even if they are not exactly the same. Since gathering enough data to accurately model every context separately is quite difficult, it is beneficial to use the ratings from one context to learn ratings in another. Otherwise every time a new context is added for a specific user, there would be no information about the ratings in that context. Thus you would suffer from a "new context problem."

The Hierarchical Bayesian model is well suited for our situation because it can give a prediction for a context without requiring as many movies to be rated. In order to give a context-dependent recommendation, we adapt the model proposed by Zhang and Koren in [13]. Rather than each branch of the tree corresponding to a *user*, we design one tree for every user and let each branch correspond to a specific *context*. By incorporating an average weight and variance into our model, we help avoid the problem of over-fitting or ill-posed problems that otherwise would occur frequently in context ratings. Often while there are more movies rated than the number of dimensions of the feature space, there are not more movies rated in a specific context than the number of dimensions.

3.1   Estimating the Weights

We need to estimate the weights for each context. If we know the generative μ and Σ^2, then we can estimate the weights W_c of each branch using a linear regression with a prior. If we know the weights W_c of each branch, then we can estimate μ and Σ^2 using maximum likelihood. These situations are typically solved by expectation maximization. After making an initial guess for μ and Σ^2, we estimate the weights. Then using these new weights, we adjust our guess of μ and Σ^2. We repeat this until the variables all stabilize. For spatial reasons, we do not present the derivation of the formulas here but merely present the resulting algorithm. See [13] and [11] for further information.

1. Make an initial guess for μ and Σ^2.
2. E step: For each context c, estimate $P(w_c|R, \mu, \Sigma^2)$ where R is the set of ratings given by the user.
3. M step: Reestimate μ and Σ^2 using the new user weights.
4. Repeat steps 2 and 3 until all variables stabilize.

In step 2, in order to estimate $P(w_u|R, \mu, \Sigma^2)$, for each context we keep track of the variance of the weights, denoted by Σ_c as well. By keeping track of the variance or certainty of our approximation to each weight vector, we can better estimate the covariance Σ^2 of the entire setup. The formulas for estimating $P(w_c|R, \mu, \Sigma^2)$ are:

$$w_c = \left((\Sigma^2)^{-1} + \frac{S_{xx,c}}{\sigma_\epsilon^2} \right)^{-1} \left(\frac{S_{xy,c}}{\sigma_\epsilon^2} + (\Sigma^2)^{-1}\mu \right) . \tag{2}$$

$$\Sigma_c^2 = \left((\Sigma^2)^{-1} + \frac{S_{xx,c}}{\sigma_\epsilon^2} \right)^{-1} . \tag{3}$$

where σ_ϵ^2 is the variance of the noise once the weights are determined (assumed to be known), $S_{xx,c}$ is the sample covariance for the specific user (i.e. Take the matrix composed of all the different feature vectors of movies that the user rated and multiply it by its transpose.) and $S_{xy,c}$ is the matrix created by taking the vector of movies rated and multiplying it by the actual ratings given.

In step 3, the mean and covariance matrices are estimated by:

$$\mu = \frac{1}{C} \sum_c w_c , \qquad \Sigma^2 = \frac{1}{C} \Sigma_c^2 + (w_c - \mu)(w_c - \mu)^T . \tag{4}$$

where C is the number of contexts for the user.

Looking at equation 2, we see that as the number of movies rated in a given context goes to infinite, the weights converge to the standard linear model because the overall mean and sigma become very small compared to the other terms.

3.2   Estimating the Noise Per User: σ_ϵ

In [13], it is assumed that σ_ϵ is given. In [11], they propose solving for σ_ϵ during the EM process as well during the M step by measuring the error on the training data. However, since we are assuming that the number of ratings in a given context is often less than the number of dimensions of the feature space, this spread is often quite low and is not a useful measurement. Since σ_ϵ represents the amount of noise added to the linear model, we estimate σ_ϵ heuristically by setting it equal to the variance of the error on the training data using *non-context* linear weights, which is the noise in the non-context linear model. This is done using the least-squares linear regression outlined in [12]. We then leave it constant throughout the EM algorithm.

3.3   Reducing the Dimensionality

By allowing users to have different weight vectors for each context we increase the dimensionality of the solution. If we normally had d weights to solve for, we now have to solve for cd weights where c is the number of contexts for the user. If a movie is represented as a vector of size n where n is the number of features, then we will now have a dimensionality of cn, which is in general much larger than the size of the training set. Thus it is often necessary to reduce the dimensionality of the space.

We chose to solve this problem using Principal Component Analysis (PCA) for two reasons. The first is that it is a relatively efficient model. We can pre-compute the eigenvalues and eigenvectors over the entire movie database thus allowing us to quickly give a rating at run time. The other benefit of the algorithm is we can judge the amount of precision lost by the reduction and adjust the number of eigenvalues used accordingly.

4   Experiments

We gathered data using the online website Recommendz. This site is accessible at http://www.recommendz.com and has been used in previous papers [3]. The site has over 3000 users and has been running for several years. Unfortunately, most of the ratings previously given are in context-independent settings and are not useful to these experiments. However, we did gather context-dependent data from fifteen users who rated on average about thirty movies. This is a small number of movies compared with other algorithms, but is useful for demonstrating the effectiveness of the algorithm on a small sample.

When a user rates a movie on the Recommendz system, they are required to give a numerical rating (on a scale of 1-10) along with a feature that they thought was important in the movie and the amount of it (on a scale of 1-10). These are used to estimate the movie feature vector. There are approximately 1500 movie features in the database. To reduce the dimensionality, we ran PCA on the movie features.

We compared our Hierarchical Bayesian algorithm to four different algorithms under the context ratings, each of which were tested on all fifteen users. The first algorithm ignores context dependency and predicts user ratings using the weight vector computed by a least squares regression on non-context dependent ratings as described in Section 2.1. This is what would happen if you do not consider contexts at all and share all the information between contexts.

The second algorithm we used involved separating the data completely from one context to another and then performing a linear regression. This would be the same as a user creating a different account for each different context. We expect that this approach will not work well in practice because given the small number of ratings given in each context, the model does not have enough data to learn the parameters. The third algorithm is the k-nearest neighbor algorithm described in Section 2.1. Finally, we ran a "hybrid" algorithm which simply averages the k-nearest neighbor algorithm and the new EM algorithm.

For each of these algorithms, we ran the data twice. In the first run, while performing cross-validation, we left out *all* the ratings from the movie we were leaving out, and not just the one from that context. That is, if a user rated the same movie several times but in different contexts, we left out each rating. This tests whether we can solve the "Standard Context Recommendation" problem. In the second run, we left out only that specific rating, potentially keeping ratings of the same movie in a different context. This tests the algorithm's ability to solve the "Different Context Evaluation" problem.

5 Experimental Results

We performed leave one out cross-validation on all data in the set. We first determined the ideal number of eigenvalues to use. The mean absolute error is presented in Figure 2. Based on the cross-validation results, we determined the optimal number of eigenvalues to use was eleven because all of the graphs other than the "separate users" graphs have smallest error with approximately eleven eigenvalues. When we use too few eigenvalues, too much data is lost. When we use too many eigenvalues, over-fitting occurs because the training set is small.

We evaluated the mean absolute error (MAE), the median absolute error, and the F-Score of each algorithm. While MAE is an intuitive error measure, F-Score, which depends on both the precision and recall of the algorithm, is considered a more accurate measure of error in recommender systems because the most important criterion for a recommender system is that it recommends the top movies [5]. In evaluating the F-Score, we divided our rankings into percentiles.

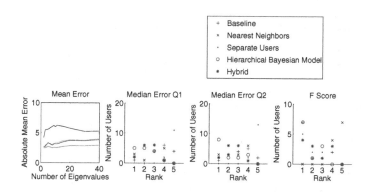

Fig. 2. From left to right: A graph showing the amount of error of the various algorithms as a function of how many eigenvalues were used. Note that each algorithm except for "separate users" (the top line) performs best with approximately eleven eigenvalues, which has very high error regardless. Next, the relative performance of the various algorithms under mean, median, and F-Score. The Hierarchical Bayesian and hybrid models perform best on the most users.

Table 1. The results of the recommender when using cross validation. The first set of results are in solving the "Standard Context Recommendation" (SCR) problem. The second set of results are in the "Different Context Evaluation" problem (DCE). For F-Score M, F-Score Q, and F-Score O, we calculate the success of the algorithms in selecting the top 50%, 25%, and 12.5% respectively. Bold items mean optimal relative performance. Note that both the Bayesian model and Hybrid model are original contributions of this paper.

Algorithm (SCR)	Mean Error	Median	F-Score M	F-Score Q	F-Score O
Linear Regression	2.9378	2.4045	**.676**	.449	.263
Separate Users	6.5767	4.9286	.545	.406	.259
Bayesian Model	2.8411	2.000	.668	**.469**	**.296**
Item Similarity	2.3444	2.1772	.383	.117	.018
Hybrid (Item + Bayes)	**2.2508**	**1.7958**	.591	.389	.182
Algorithm (DCE)	Mean Error	Median	F-Score M	F-Score Q	F-Score O
Linear Regression	2.0565	1.655	.696	.554	.418
Separate Users	6.5767	4.9286	.545	.406	.259
Bayesian Model	1.9109	**1.1569**	.681	**.557**	.351
Item Similarity	2.0847	1.8709	.643	.479	**.447**
Hybrid (Item + Bayes)	**1.8188**	1.4633	**.708**	.389	.408

We compared how successful each algorithm was at putting movies in various quantiles. For the numerical performance of the various algorithms, see Table 1.

When all contexts are left out of the training set for a specific movie, the hybrid algorithm, combining item similarity and the Bayesian model, has the smallest mean and median error. The Bayesian model performs better under this measure than the linear regression and separate users approach, but worse than the item similarity on its own. In measuring the F-Score, the linear regression model is best at determining which elements belong in the top half with the Bayesian model a close second. When considering the algorithms success at putting movies into the top *quarter*, the results are flipped, with the Bayesian model slightly outperforming the linear regression model. In the top *eighth*, the Bayesian model maintains the highest score again, this time with a larger gap.

When we leave only the one rating out, but leave all the others from different contexts in, as expected, the predicted ratings are closer to the actual ratings. This shows that ratings in one context are indeed correlated with ratings in another context. We aim to show that while they are correlated, they are not determinative. In this case the hybrid algorithm has the best results in mean error, and the Bayesian model has smallest error for median error. The hybrid algorithm has the strongest F-score for the median, but the Bayesian algorithm is strongest for the quarter F-score. Interestingly, the item similarity, which has a very low score in giving context ratings in the first run, has the highest score in the F-score when considering a successful match of the top 8th.

We wanted to consider the percentage of users the Bayesian algorithm worked best on. To do this, we broke our results down by user to rank the various algorithms. The data is shown in Figure 2. Along the x-axis is the relative rank

Table 2. Examples of predictions given by the hierarchical Bayesian network in different contexts. Some of the movies are the same, but the list varies.

Generic	Guys Night Out	Romantic Evening
The Net	Ace Ventura	A Simple Twist of Fate
The Transporter	Die Hard	Mona Lisa Smile
Star Trek	Highlander	The Terminal
Fantastic Four	Mortal Kombat	The Net
The Mask	Billy Madison	Mean Girls

of the algorithm (i.e. 1st, 2nd, 3rd, 4th, and 5th). On the y-axis is the number of users for which each algorithm has that rank. The hybrid and hierarchical Bayesian algorithm each have the largest number of users for which they rank first or second. An example of some of the movie recommendations given by the algorithm in different contexts is given in Table 2.

6   Discussion

An important factor in contrasting our results with others is that in our data set, the average user rated approximately 30 movies. Other studies have shown smaller errors using larger training sets (75-100 ratings per user) [13], but that was not the main goal of this work. Unfortunately, it is difficult to collect data as standard data sets such as the Netflix™ set do not keep track of context information and are thus not applicable to our work. Additionally, we wanted to demonstrate the effectiveness of the Hierarchical Bayesian algorithm on a small sample size points. Given enough ratings, it would even be reasonable to treat each context as a different user. However, our algorithm does not require a user to rate dozens of movies in each context. The algorithm gives a good approximation for each separate context even when only four or five movies have been rated in that context, thus dealing with the "new context" problem. With a smaller mean and median error than baseline linear algorithms, the Hierarchical Bayesian model accomplishes what we want: it shares information between contexts without requiring that all information be shared. Even on users for which the Bayesian model does not work best (i.e. those who the nearest neighbor model works well for), we can still improve the nearest neighbor recommendation by averaging it with the Bayesian prediction.

In answering the "Standard Context Recommendation" question, the mean and median error are lowest on the hybrid algorithm. The baseline linear algorithm works as well as the Bayesian model at selecting the top 50 % of the movies, but not as well at selecting the top 12 % of the movies. Thus we conclude that the baseline algorithm is good at giving a coarse guess of "good or bad" for a movie in a context, but the Hierarchical Bayesian model is best at distinguishing movies in a finer manner. This makes sense, since it is very often the case that in a different context, a user would still have the same general feeling for a movie (e.g. good vs. bad), but would have variations within this category (e.g. excellent vs. good). In this situation, the F-Score is more appropriate to use as a measure of error than mean or median because this question resembles more closely the traditional question of "What movie should I watch?"

The Bayesian model has the lowest median error in answering the "Different Context Evaluation" problem, showing that it is best able to use ratings from different contexts without automatically assigning the same value. While the results in the F-score are unclear, we consider the mean and median error a more appropriate measure to answering the second question. In the first question, it did not matter how poor movies are ranked as they are not presented to the user anyway. In this case, however, it does matter. The user will ask about a specific movie in a specific context, and we want to give as accurate an answer as possible. Additionally, looking at only the top movies can be very misleading. If a user rates a few movies very highly in all contexts (a very reasonable assumption since many users enjoy their favorite movie in almost every setting), then the item similarity algorithm is almost guaranteed to give the correct answer since the closest item to a movie is always itself. Since we only look at the top 12.5% of movies, many movies fit into this category, causing the item similarity algorithm to have an exceptionally strong score. The other algorithms do not have this benefit because they are parametric. In summary, the very top movies are often the same in various contexts, but after that there is diversity.

In answering both questions, the mean and median error were smallest in the Hierarchical Bayesian model or the hybrid algorithm. The hybrid model performs better than the item similarity algorithm, showing that even if we do not assume a linear model, the Hierarchical Bayesian model can be useful. The Hierarchical Bayesian model performs much better than other linear models. While the raw error is relatively high, the size of the training set is quite small and the results show that the Hierarchical Bayesian model (or in some cases a hybrid form of it) is better than baseline algorithms in making context-dependent recommendations. The Hierarchical Bayesian model is strongest at recommending the top movies when they are previously unrated. The Bayesian model is best at predicting a score of a movie when given a previous rating in a different context.

7   Conclusions

We designed a Hierarchical Bayesian model [13] to learn different weights in different contexts. This algorithm answers two questions: "What movie should I watch in context X?" and "Given that I gave a rating to movie M in context X, what would I think of movie M in context Y?" We compared our algorithm to several other techniques, some of which share *all* information between contexts and some of which share *none* of the information between contexts. We found that our algorithm or a hybrid algorithm performed at least as well in most forms of measurement. Because the approach is content-based, the algorithm does not assume that the preferences of every user change in the same way depending on the context. This allows users to have personal interpretations of contexts or even to add their own new contexts. This work demonstrates that it is useful to store contextual information. Naive approaches do not incorporate contexts as effectively as the Hierarchical Bayesian model. A potential area to explore is creating a hybrid of the content-based approach discussed here with a context-dependent collaborative filtering approach.

References

1. Adomavicius, G., Sankaranarayanan, R., Sen, S., Tuzhilin, A.: Incorporating contextual information in recommender systems using a multidimensional approach. ACM Trans. Inf. Syst. 23(1), 103–145 (2005)
2. Alspector, J., Kolcz, A., Karunanithi, N.: Comparing Feature-based and Clique-based User Models for Movie Selection. In: Proc. of the 3rd ACM Conf. on Digital Libraries, Pittsburgh, PA, pp. 11–18 (1998)
3. Garden, M., Dudek, G.: Semantic Feedback for Hybrid Recommendations in Recommendz. In: Proc. of the 2005 IEEE International Conf. on E-Technology, E-Commerce and E-Service (Eee 2005), Hong Kong, China (2005)
4. Goldberg, K., Roeder, T., Gupta, D., Perkins, C.: Eigentaste: A Constant Time Collaborative Filtering Algorithm. Information Retrieval Journal 4(2), 131–151 (2001)
5. Herlocker, J., Konstan, J., Terveen, L., Riedl, J.: Evaluating Collaborative Filtering Recommender Systems. ACM Transactions on Information Systems 22, 5–53 (2004)
6. Ono, C., Kurokawa, M., Motomura, Y., Asoh, H.: A Context-Aware Movie Preference Model Using a Bayesian Network for Recommendation and Promotion. User Modelling, 247–257 (2007)
7. Salton, G.: Automatic Text Processing. Addison-Wesley, Reading (1989)
8. Sarwar, B.M., Karypis, G., Konstan, J.A., Riedl, J.: Item-base Collaborative Filtering Recommendation Algorithms. In: Proc. of the 10th International World Wide Web Conf., WWW, vol. 10 (2001)
9. Vinay, V., Cox, I., Wood, K., Milic-Frayling, N.: A Comparison of Dimensionality Reduction Techniques for Text Retrieval. In: ICMLA, pp. 293–298 (2005)
10. Yang, Y.: An Evaluation of Statistical Approaches to Text Categorization. Information Retrieval 1(1), 67–88 (1999)
11. Yu, K., Tresp, V., Schwaighofer, A.: Learning Gaussian Processes from Multiple Tasks. In: ICML 2005: Proc. of the 22nd international Conf. on Machine Learning, New York, NY, USA, pp. 1012–1019 (2005)
12. Zhang, T., Iyengar, V.S., Kaelbling, P.: Recommender Systems Using Linear Classifiers. Journal of Machine Learning Research 2, 313–334 (2002)
13. Zhang, Y., Koren, J.: Efficient Bayesian Hierarchical User Modeling for Recommendation Systems. In: Proc. of the 30th Annual International ACM SIGIR Conf. on Research and Development in Information Retrieval (SIGIR 2007), New York, NY, USA (2007)

Automatic Frame Extraction from Sentences

Martin Scaiano and Diana Inkpen

School of Information Technology and Engineering,
University of Ottawa
mscai056@uottawa.ca, diana@site.uottawa.ca

Abstract. We present a method for automatic extraction of frames from a dependency graph. Our method uses machine learning applied to a dependency tree to identify frames and assign frame elements. The system is evaluated by cross-validation on FrameNet sentences, and also on the test data from the SemEval 2007 task 19. Our system is intended for use in natural language processing applications such as summarization, entailment, and novelty detection.

1 Introduction

Many natural language processing tasks could benefit from new algorithms which use enhanced semantic structures and relations. Researchers have been successfully applying semantic role labeling (SRL) to tasks such as question answering [1], machine translation [2], and summarization [3]. The next step is a deeper semantic representation; but before such as representation can be used, a method for automatically creating this representation is needed. The goal of this paper is to describe a new method for this task.

Our semantic representation is based on frames [4] which represent events, objects, and situations. The specific variations and roles of a frame are defined as frame elements. This has the benefit of representing a particular situation independent of how it is expressed or phrased. Consider purchasing something; some common parameters (frame elements) are the buyer, the seller, the object in question (theme), and the price. The following three sentences represent the same situation but use different words and structures:

1. Bob bought a red car from Mary.
2. Mary's red car was sold to Bob.
3. Bob purchased a car from Mary, that was red.

Our system relies on Berkeley FrameNet [5] for frame definitions. FrameNet is a resource that provides definitions of about 800 frames, spanning about 10000 lexical units (words that evoke a particular frame). FrameNet also provides annotated examples of the frames and lexical units being used in English. FrameNet is currently the most comprehensive resource for frame definitions and one of the few resources for semantic role labeling.

Y. Gao and N. Japkowicz (Eds.): Canadian AI 2009, LNAI 5549, pp. 110–120, 2009.

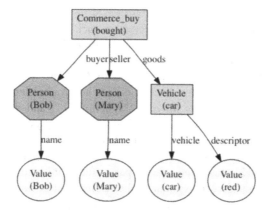

Fig. 1. Frame representation of the sentence "Bob bought a red car from Mary"

Figure 1[1] shows the result of our system for the first of the above sentences (the other sentences would have nearly identical representations)[2]. The rectangles represent frames from FrameNet, the octagons represent people, while the ovals represent words which are functioning as values for frame elements. Each edge (arrow) represents a frame element and it is labeled. The text in the top of each node represents the type (frame name, person, or value) while the text below indicates the word which the node was derived from.

1.1 Semantic Role Labeling

Semantic role labeling is the operation of identifying the semantic relation between words. Usually roles are labeled for predicates of verbs; some general-purpose roles are as follows: agent, patient, instrument, time, location, and frequency.

The set of semantic roles can differ from resource to resource. FrameNet uses role labels (called frame elements) that are specific to the frame being annotated, while resources like PropBank [6] and VerbNet [7] provide only a small set of general-purpose roles.

The first statistical SRL system was developed by Gildea and Jurafsky [8] using FrameNet; subsequent research lead to the development of VerbNet and PropBank.

1.2 SemEval 2007 Task 19: Frame Semantic Structure Extraction

At SemEval 2007, task 19 [9] was dedicated to frame extraction. The task relied on FrameNet but provided additional training data for systems. The competition involved identifying in each sentence, which frames from the training data and FrameNet were evoked by the given lexical units, and subsequently identifying and labeling frame elements.

---

[1] We used Graphviz to visualize the xml output of our system. http://www.graphviz.org/
[2] FrameNet represents buying and selling as different frames though they inherit from a common super frame.

The resulting labeled texts from each system were evaluated based on precision and recall of detecting all the expected frames in each of the three texts. Our system was not implemented at the time of the competition, but it can be compared, for evaluation purposes, to the two participating systems, using the same data.

The UTD system [10] applies Support Vector Machines (SVM) and Maximum Entropy (ME) to the tasks of frame disambiguation and frame element labeling. Their system uses established methods for disambiguation [11] and when an inadequate amount training data is available (less than 5 sentences) the system randomly chooses the sense. The system uses established features and methods for identifying grammatical functions, frame element boundaries, and frame element labels.

The LTH system [12] uses a dependency parser for its initial representation. Before doing frame disambiguation, the system applies a small set of rules to filter out words which tend to cause performance degradation (in particular prepositions). Frame disambiguation is done using a SVM classifier on the resulting words. The authors extended the collection of lexical units for various frames by using WordNet [13]. Frame element labeling is done in the dependency tree using features from the target word, and some immediate related words (parent and children of the node).

The CLR system participated only in the frame identification task. Its method of identifying the frames was based on heuristic rules [14].

Our system is very different from UTD and CLR, and it is similar to LTH in some aspects, including the fact that our initial data structure is a dependency parse, except that our dependency trees are augmented with shallow semantic information. Unique to our system is the idea of using the completely annotated texts from SemEval 2007 and FrameNet to train classifiers with negative frame identification (identifying when a word does not evoke a frame) instead of using special rules to remove unlikely frames or words. Our method's frame identification features differ from the LTH and UTD features, in particular the use of a boolean list of evokable[3] frames and the boolean list of child nodes. Also, we tested several machine-learning models for the frame element labeling task (one model per lexical unit, one model per frame, and one model per location per frame) which are described below.

2 Method Description

Like most SRL methods, our method relies on machine learning for most of the decisions[4]. Our method accomplishes the following sub-tasks, each of which is described in more detail in the following subsections:

1. Initial sentence parsing;
2. Frame assignment;
3. Identification of people, entities, locations and other special types;
4. Role recognition (identification of frame elements).

---

[3] Frames, which may possibly be evoked, based on matching lexical units from a given text.

[4] Our system is implemented in Java so that it can easily interface with Weka [15]. FrameNet provides no particular API, only xml files containing the data, so we implemented our own Java interface.

2.1 Initial Sentence Parsing

Typically SRL systems use a syntactic parser for the initial parsing; our method's initial parse is obtained with the Machinese Semantic parser by Connexor Oy[5] [16], which produces a dependency tree augmented with identification of named entities and their classes, and shallow semantic types for a variety of nouns.

The semantic parser provides about 30 dependency relations, which form a good foundation for our frame element labeling. Additionally, the parser does an effective job of identifying named entities, and their semantic classes (location, person, country, etc.).

2.2 Identification of Frames

Our frame identification method uses machine learning classifiers from Weka (Waikato Environment for Knowledge Analysis) [15] applied to features extracted from the parse tree. We tried Naïve Bayes (NB), Decision Trees (J48) and SVM (SMO) from Weka, using default parameter settings, and found all results to be very close for this task, as shown in Table 1. We chose these three classifiers in order to test a classifier that is known to work well with text (NB), a classifier whose output is understandable to human readers (J48), and one that is known to produce good classification results for many tasks (SVM).

The semantic parser lemmatizes each word, then a machine learning model specific to that lemma is used to classify what frame, if any, the word evokes.

The training data for the models is extracted from FrameNet's annotated examples and from SemEval 2007 task 19. Because the SemEval 2007 training was completely annotated, including words that were not frame evoking words or frame elements, we used it to extract negative examples (examples of lexical units that did not evoke any frames) for training.

Models were produced for each lemmatized head word of all lexical units. Each model determined if particular word should evoke a frame and if so which frame.

Some of the lexical units in FrameNet have very few or no training examples, thus we choose to train frame identification models on examples from all lexical units which could evoke the given frame. This approach showed no decline in precision and recall, and allows the system to operate on lexical units which would not have had enough training data.

Table 1. 10-fold cross-validation results for frame identification. A baseline classifier that always chooses the most frequent frame produces results of 74%; therefore our results are a significant improvement.

	Precision	Recall	F-Measure
Naïve Bayes	96%	99%	97.5%
Decision Trees	96%	99%	97.5%
SVM	97%	99%	98%

[5] http://www.connexor.eu

The features used for the frame identification are as follows:

1. A boolean list for each possibly relevant frame for this lemma. The value defaults to false, unless the lexical unit in its entirety (all lexemes) for a given frame matches in the given example. Models are selected by simply matching a single word (the head word) though any lexical units containing the given head word may not fully match, thus this feature should assist the classifier in filtering inappropriate lexical units and thus inappropriate frames. This feature increased both precision and recall.
2. The dependency relation (as assigned by Machinese Semantics) of the word in question. This feature alone provides good results for assigning frames.
3. A boolean list of all child dependency relations, with the value being true if that dependency relation is present in the given sentence, otherwise false. This feature is motivated by the idea that neighboring words could help disambiguate frames. This feature provided a 10% increase in recall.

2.3 Identification of Entities

Although there are various techniques and systems available for identifying named entities, our system relies on the semantic parser's ability to identify entities. When a word is tagged as an entity, special rules convert the words into a final representation. The rules are as follows:

- If a word is tagged as a named entity and specifically a location, then a special location object is made; the tagging will also indicates the type of location, i.e., city, country, continent.
- If a word is tagged as a named entity and specifically a person, male or female, then a special person object is created. The word that was tagged as named entity is labeled as a "name" associated with the person. Any immediately adjacent words which are also named entities are added to the person frame as "name" labels. Future work is needed to determine how to combine our representation of people and entities with those from FrameNet.
- If a word is tagged as named entity but is not of a human type, we create a special general-purpose entity object. This object may represent an organization, person, location or other named entity. Like with the person object, a "name" is associated with the object and any adjacent words that are named entities.
- If a word is tagged as a human (but not a named entity) we still create a person object for the word. This often occurs with words like, "he", "she", and "you". The parser can recognize a variety of words as being people without them being a named entity. This does not provide any special benefits during evaluation, though it is our belief that this will keep the representation uniform and be more useful during future processing.

2.4 Assigning Frame Elements

Assigning labels to frame elements is effectively the SRL task, a topic that has been extensively researched in recent years.

Table 2. Distribution of frame elements by location relative to the frame word. Positive class are actually frame elements, while negative class are not frame elements.

	Grandparents	Parents	Siblings	Children	Grandchildren	Total
Positive	1,763	16,323	50,484	146,405	10,555	225,529
Negative	49,633	77,863	224,606	148,929	251,620	752,651

Our system assigns frame elements using machine learning classifiers trained on the FrameNet annotated corpus. Gildea and Jurafsky were the first to implement semantic role labeling for FrameNet [8], and the task has since been studied frequently. Our system assigns frame elements from a dependency tree provided by the Machinese Semantic parser. Our method is significantly distinct from other researchers because our system starts with a dependency tree, which provides guidance for semantic role labeling.

To assign frame elements for a particular frame the system applies a classifier to each of the words surrounding the frame evoking word. Each word is classified with either a frame element or "none", indicating no frame element and no association. 50% of all frame elements are immediate children of the frame evoking word in the dependency tree. Of all child relations about 50% are themselves frame elements.

Our method considers any word within two edges in dependency graph; thus grandparents, parents, siblings, children and grandchildren are all considered possible frame element.

From the FrameNet annotated data and the SemEval 2007 task 19 data, 978,180 training examples were extracted; among them, 225,529 of the examples were positive (of actual frame elements), while the remainder were negative examples (not frame elements). Table 2 shows the distribution of frame elements and negative examples by location in the dependency tree relative to a frame word.

About 1000 examples of frame elements could not be found within 2 edges of the frame. All the missing frame elements were in incomplete parse trees (cases when the parser had problems parsing the sentences, which resulted in a fragmented parse tree). We believe that, had the parse been complete, then most frame elements in those examples would have been found within two edges of a frame word.

The classifier is configured so that one word may only act as a single frame element, though on rare occasions a single word may actually represent two frame elements.

A frame evoking word may also be a frame element of it's own frame. This is common of general-purpose frames, such as food, where the word may specify a specific sub-type of frame, such as spaghetti, or sandwich.

When training our classifier, we experimented with several machine-learning configurations, but our best results were found by training one model per frame, which is then applied to each word being classified. The other configurations that we tried are one model per lexical unit and one model per location per frame. One model per lexical unit was meant to address the possible reversal of subject and object in some frames, but may have suffered from less training data than one model per frame. One model per location per frame involved sub-dividing the task of labeling frame elements by their location relative to the frame evoking word in the dependency tree: grandparent, parent, sibling, child, grandchild. It was thought that locations carried consistent classification properties, but this configuration performed worse than one model per frame, which is more general.

During development, we tested several features including the lexical unit that evoked the frame, part of speech, voice of the verb, location of the word in question (relative to frame word in the parse tree), general purpose semantic role label of frame word and frame element word, leading prepositions, semantic classes of words, and if the frame element word is an animate object.

The features that provided the best balance between precision and recall were the dependency relation of the frame evoking word, the dependency relation of the classified word and the dependency relation of the intermediate word, if one existed. Intermediate words exist for locations such as grandchildren which have an intermediate word that is a child of the frame evoking word; grandparent which has an intermediate word that is the parent word; and siblings words which have an intermediate word that is the parent word.

We were surprised to find that using different classifiers did not provide large differences in our results. There are small trade-offs between precision and recall, as shown in Table 3. Perhaps small improvements can be obtained in future work by combining them in an ensemble, but our focus was of finding the best features and generic models for classification.

Table 3. 10-fold cross validation results for frame element classification

	Precision	Recall	F-measure
Naïve Bayes	74%	52%	61.0%
Decision Trees	82%	45%	58.1%
SVM	82%	43%	56.8%

3 Evaluation on Additional Test Data

The cross-validated results are very good, as is often the case when the validation data is similar in nature to the training data. Therefore, we decided that further evaluation was needed. The SemEval 2007 task 19 was a frame extraction task and could provide data for evaluation. Plain text was provided to the competing systems; each system was required to identify all frames, tag various entities, and label frame elements in a specific output format.

Three systems entered the task for the frame identification and entity-tagging component. Only two of the systems completed the entire task of labeling frame elements.

3.1 Evaluation Issues

Our system was not directly intended to support the required output for the SemEval 2007 task 19 and thus some adjustments to our system and to the evaluation script were required.

While developing our system we found that the FrameNet annotated data contained various inconsistencies in the frame elements names (capitalization, use of space and underscores, and even name mismatches). For consistency our system adjusts all frame element names to use spaces and lower case letters. The evaluation script used a case sensitive comparison of frame element names. Since the case of our frame elements did not necessarily match that of the gold standard, we matched all the frame elements in lower case.

The evaluation of results was founded on the assumption that tokenization would be consistent between the gold standard and the system. Our system's tokenization was significantly different from the expected tokenization and a method for conversion between tokens was required. This conversion is not exact, and occasionally leads to correct frames and frame elements being considered incorrect.

Unlike FrameNet and the expected output of SemEval 2007 task 19, our system does not select frame elements as sequences of text but selects an object (frame, entity or word representing a value) from the final representation as frame elements; this is best show in Figure 1. When the results of our system are evaluated on the task data, some of the correct frame elements are considered incorrect for having differing boundaries.

The training texts contained new frames and frame elements that were not available in FrameNet. This was intended to test a systems ability to include new frames from annotated examples. Our system easily added the new frames from the training data into the frame identification models, but the frame element labeling models were not updated to work for these new frames, therefore our system could not get correct labels in these cases.

All the difficulties above are estimated to cause no more than 3% discrepancy in precision and recall.

Table 4. Frame identification results on the SemEval 2007 Task 19 test data, which consisted in three texts, entitled Dublin, China and Work. The results of the three participating systems are shown, plus our results for the method that we present in this paper. Our system did not participate in the SemEval task because it was developed afterwards. Our system missed some frames due to low-level tokenization and format issues, since our system was not designed specifically for the SemEval task.

Text	System	Precision	Recall	F-measure
Dublin				
	* Our system	0.7070	0.3162	0.4370
	CLR	0.6469	0.3984	0.4931
	LTH	0.7156	0.5184	0.6012
	UTD	0.7716	0.4188	0.5430
China				
	* Our system	0.6401	0.4261	0.5116
	CLR	0.6302	0.4621	0.5332
	LTH	0.7731	0.6261	0.6918
	UTD	0.8009	0.5498	0.6457
Work				
	* Our system	0.7336	0.4132	0.5286
	CLR	0.7452	0.5054	0.6023
	LTH	0.8642	0.6606	0.7488
	UTD	0.8382	0.5251	0.6457

Table 5. Frame element labeling results on the SemEval 2007 Task 19 test data. Only two of the participating systems worked on this task (CLR worked only on the previous task, frame identification). Our system had the highest precision on the task of frame element labeling.

Text	System	Precision	Recall	F-measure
Dublin				
	* Our system	0.63507	0.22027	0.32710
	LTH	0.54857	0.36345	0.43722
	UTD	0.53432	0.26238	0.35194
China				
	* Our system	0.56323	0.26245	0.35806
	LTH	0.57410	0.40995	0.47833
	UTD	0.53145	0.31489	0.39546
Work				
	* Our system	0.71053	0.28000	0.40170
	LTH	0.67352	0.30641	0.54644
	UTD	0.61842	0.45970	0.40978

3.2 Results

The SemEval task was evaluated on three texts, each having a different topic and different frequencies of frames. The task proved to be much more difficult than the task of extracting frames from FrameNet examples, which contain annotations only for one frame for each example.

Our final results for frame identification compared to the other systems are shown in Table 4. Our system's recall and F-measure are significantly lower than that of the other two systems, but our precision is comparable.

The combined results of frame identification and frame element labeling are shown in Table 5. Our system's recall and F-measure still tend to be the weakest, but our precision is the best in two of the three evaluations.

This evaluation has shown that our system is a good foundation for our future work. Since our final goal is not simply to extract frames, but to apply the representation to other tasks, we can now focus on optimizing and improving our system for those tasks.

4   Conclusions and Future Work

We have developed a method for frame element extraction that has good precision and should be usable for future research. We plan many improvements in future work.

Cross-validation was an inadequate quality measure for the machine learning models. For future evaluations we are considering using more held-out texts, such as the 3 texts from SemEval 2007 and any completely annotated texts from FrameNet. Using hold-out texts seems to provide a better evaluation of how the system will work on new domains and texts with varying distributional patterns of frames and frame elements.

The use of dependency relations as foundation for SRL (frame element labeling in our system) has produced excellent precision. Although numerous feature combinations were tested the most important features were the dependency relations provided by the initial parser, when used in conjunction with the features of the neighboring words.

Future work on the system includes evaluating new features for frame element labeling, such as a frame evoked by a frame element word (if one exists) and considerations for any intermediate frame elements. Also semantic type checking using WordNet is a possibility.

Extending FrameNet with information from other sources, to increase its coverage [17], is another direction for future work. We are particularly concerned with adding more training data for the frames that have too few manually annotated examples.

We plan to focus on how to use the extracted frame-based representations in tasks such as summarization, novelty detection, and entailment, exploiting the fact that our representations have high precision. When recall is insufficient, we can fall back of the shallow methods based on syntactic dependencies that we used before for the three mentioned tasks.

References

1. Narayanan, S., Harabagiu, S.: Question and Answering based on semantic structures. In: Proceeding of the 20th International Conference on Computational Linguistics (COLING), pp. 693–701 (2004)
2. Boas, H.C.: Bilingual FrameNet dictionaries from machine translation. In: the proceedings of the Third International Conference on Language Resources and Evaluation (LREC), pp. 1364–1371 (2002)
3. Melli, G., Wang, Y., Liu, Y., Kashani, M.M., Shi, Z., Glu, B., Sarkar, A., Popowich, F.: Description of SQUASH, the SFU question and answering summary handler for the DUC-2005 Summarization Task. In: Proceedings of the HLT/EMNLP Document Understanding Conference (DUC) (2005)
4. Fillmore, C.J.: Frame semantics and the nature of language. In: Annals of the New York Academy of Sciences: Conference on the Origin and Development of Language and Speech, vol. 280, pp. 20–32 (1976)
5. Fillmore, C.J., Ruppenhofer, J., Baker, C.: FrameNet and representing the ink between semantic and syntactic relations. In: Huan, C., Lenders, W. (eds.) Frontiers in Linguistics. Language and Linguistics Monograph Series B, vol. I, pp. 19–59 (2004)
6. Palmer, M., Kingsbury, P., Gildea, D.: The Proposition Bank: An Annotated Corpus of Semantic Roles. Computational Linguistics 31, 71–106 (2005)
7. Kipper, K., Dang, H.T., Palmer, M.: Class based construction of a verb lexicon. In: Proceedings of the 17th National Conference on Artificial Intelligence (AAAI 2000). (2000)
8. Gildea, D., Jurafsky, D.: Automatic Labeling of Semantic Roles. Computational Linguistics 28, 245–288 (2002)
9. Baker, C., Ellsworth, M., Erk, K.: SemEval-2007 Task 19: Frame Semantic Structure Extraction. In: Proceedings of the Fourth International Workshop on Semantic Evaluations (SemEval 2007), pp. 99–104 (2007)

10. Bejan, C.A., Hathaway, C.: UTD-SRL: A Pipeline Architecture for Extracting Frame Semantic Structures. In: Proceedings of the Fourth International Workshop on Semantic Evaluations (SemEval 2007) (2007)
11. Florian, R., Cucerzan, S., Schafer, C., Yarowsky, D.: Combining classifiers for word sense disambiguation. In: Natural Language Engineering (2002)
12. Johansson, R., Nugues, P.: LTH: Semantic Structure extraction using nonprojective dependency trees. In: Proceeding of the 17th International Workshop on Semantic Evaluations (SemEval 2007), pp. 227–230 (2007)
13. Fellbaum, C. (ed.): WordNet: An electronic lexical database. MIT Press, Cambridge (1998)
14. Litkowski, K.: CLR: Integration of FrameNet in a Text Representation System. In: Proceedings of the Fourth International Workshop on Semantic Evaluations (SemEval 2007), pp. 113–116 (2007)
15. Witten, I.H., Frank, E.: Data Mining: Practical machine learning tools and techniques, 2nd edn. Morgan Kaufmann, San Francisco (2005)
16. Järvinen, T.: Multi-layered annotation scheme for treebank annotation. In: Nivre, J., Hinrichs, E. (eds.) TLT 2003. Proceedings of the Second Workshop on Treebanks and Linguistic Theories, pp. 93–104 (2003)
17. Shi, L., Mihalcea, R.: Semantic Parsing Using FrameNet and WordNet. In: Proceedings of the Human Language Technology Conference (HLT/NAACL 2004) (2004)

Control of Constraint Weights for a 2D Autonomous Camera

Md. Shafiul Alam and Scott D. Goodwin

School of Computer Science, University of Windsor, Windsor, Canada

Abstract. This paper addresses the problem of deducing and adjusting constraint weights at run time to guide the movement of the camera in an informed and controlled way in response to the requirements of the shot. This enables the control of weights at the frame level. We analyze the mathematical representation of the cost structure of the search domain so that the constraint solver can search the domain efficiently. Here we consider a simple tracking shot of a single target without occlusion or other environment elements. In this paper we consider only the distance, orientation, frame coherence distance and frame coherence rotation constraints in 2D. The cost structure for 2D suggests the use of a binary search to find the solution camera position.

Keywords: camera control, virtual camera, virtual environment, graphical environment, computer game, video game, animation, constraint satisfaction.

1   Introduction

A general approach to camera control is to use a constraint satisfaction based technique. Complex properties can be easily represented as constraints. But conflicting requirements make it an over-constrained problem. Often constraint weighting is used to give each constraint a priority order and the weighted sum of the costs of violation for all the constraints is used as the objective function. Constraint satisfaction optimization is used to solve it.

Bares et al. [2, 3] and Bourne and Sattar [5] use user specified constraint weights based on the relative importance of the constraints in a particular type of shot. Bourne and Sattar [5, 6] and Bourne [7] use an example animation trace to deduce the desired values and weights for all the constraints. The resulting constraint weights and desired values are coupled together, and they are appropriate for a particular type of shot that is equivalent to the example animation trace.

Offline generation of constraint values and weights is not appropriate for interactive applications. Since there are numerous variations in types of shot, their visual effects, staging of players, motion of camera, etc., it is quite impossible to generate desired values and weights of constraints for all of them. Moreover, if there is a particular set of constraint values and weights for particular situation, there must be a relation between them. For that we need to consider the physical

Y. Gao and N. Japkowicz (Eds.): Canadian AI 2009, LNAI 5549, pp. 121–132, 2009.

significance of adding any weighted constraint cost to the objective function for the problem. For an autonomous camera it is necessary to find a relation between the requirements of a shot and its weighted constraint representation so that the constraints and their weights can be deduced automatically for any situation. This will make the camera control system modular which can be used in any environment without modification. Our approach is a starting point toward that goal. In our approach the physical significance of the weights are used to deduce them automatically from the requirements of the shot.

Bares et al. [2] and Pickering [13] find valid region of space before applying search in that volume. It is further extended by Christie and Normand [9] to give semantic meaning to each volume of valid region in terms of its corresponding cinematographic properties. It seems that Bourne [7] and Bourne and Sattar [8] are the first to identify the higher cost regions of the search domain for different weights. But they use only height, distance and orientation constraints, and their work is limited to either equal weights or only one constraint having a higher weight. To take advantage of this information about the search space they use a specialized constraint solver called a sliding octree solver to search the domain quickly. It prunes the regions with poor solutions quickly. But, until now mathematical modeling of the domain of search has not been studied. Consequently, the information about the structure of the domain could not be utilized to arrive at an exact solution or direct the search using that information. We use the cost structure of the domain to search it. This ensures an effective and informed way of pruning the domain, and no local minima problems. A complete description of the search can be found in Alam [1].

According to Halper et al. [10] one of the main challenges of the camera control problem is to find a balance between optimal camera positions and frame coherence. They do not use a constraint to enforce frame coherence. On the basis of the prediction about the future target positions they evaluate future camera positions and move the camera toward those positions. Bourne and Sattar [4] say that this does not ensure frame coherence and the method is fully dependent on the algorithm used to predict the movement of the target. Bourne and Sattar [5] and Bourne [7] introduce a frame coherence constraint in the weighted constraint representation of camera control problem. Since the properties of frame coherence constraints are different from those of visual constraints, we decouple the two types of constraints and use their relative priority to influence the search of the solution camera position.

We begin by describing our camera control system in Section 2. Section 3 describes how to use the constraints and their weights. Section 4 outlines the solution for the camera control system. This is followed by conclusions and directions for future research in Section 5.

2   The Camera Control System

In this paper, we shall consider a simple 2D tracking shot of a single target without any occluder. The most important visual properties are represented by the size of the subject's image in relation to the frame and viewpoint [Katz 11; Mascelli 12]. The viewpoint is determined by the orientation and the camera height or the angle of view. The image size is determined by the distance of the camera from the subject and the focal length of the camera. So, we can use a relatively larger range for the domain of distance and adjust the focal length after determining the most appropriate distance. The visual requirements of the distance and orientation can be expressed as constraints on the distance and azimuth angle of the polar coordinate system whose pole is at the position of the subject in the current frame and polar axis is along the horizontal projection of the direction of line of action (e.g., along the direction of facing or the direction of movement of the subject). These coordinates determine the desired position of the camera according to the visual requirements. Let the coordinates be (ρ_d, θ_d) with respect to the polar coordinate system (ρ, θ) mentioned before. The above mentioned visual requirements are applicable to static shots also. We shall call their respective constraints as visual constraints.

If the potential position of the camera is at (ρ_p, θ_p), the costs for the most important visual constraints are given by: $\rho_1 = |\rho_p - \rho_d|$, $\theta_1 = |\theta_p - \theta_d|$. If k_1 and l_1 are the corresponding weights, then the weighted cost for the visual constraints is: $k_1\rho_1 + l_1\theta_1$.

Similarly, since the frame coherence constraints of the camera are measured with respect to the past motion of the camera [Halper et al. 10; Bourne and Sattar 6, 8; Bourne 7] we specify them using the polar coordinate system (ρ', θ') with the pole at the position of the camera in the previous frame and horizontal projection of the direction of movement of the camera in the previous frame as the polar axis. Frame coherence constraints are related only to the smooth movement of the camera. So, we shall call them motion constraints. Let the desired position of the camera according to the motion constraints be (ρ'_d, θ'_d). If the potential position of the camera is at (ρ'_p, θ'_p) with respect to this coordinate system, then the costs for motion constraints are given by: $\rho_2 = |\rho'_p - \rho'_d|$, $\theta_2 = |\theta'_p - \theta'_d|$. If k_2 and l_2 are the corresponding weights, then the weighted cost for the motion constraints is: $k_2\rho_2 + l_2\theta_2$.

The total weighted cost for the problem is given by:

$$k_1\rho_1 + l_1\theta_1 + k_2\rho_2 + l_2\theta_2 \tag{1}$$

2.1 Determination of the Solution

The system of isocurves Γ_1 of visual constraints (Figure 1 shows a member γ_1 of the family Γ_1) are given by

$$k_1\rho_1 + l_1\theta_1 = c$$

where c is a constant. The cost on an isocurve is proportional to the distance cost when $\theta_1 = 0$. It is also proportional to the orientation cost when $\rho_1 = 0$. Let ρ_{10} and θ_{10} be the ranges of acceptable values for ρ_1 and θ_1 respectively. Since A and B are acceptable solutions lying at the edges of the two constraint

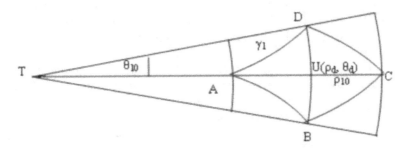

Fig. 1. An isocurve γ_1 of visual constraints

ranges keeping the other constraint's cost to zero, they must have the same total cost. Thus, the isocurve γ_1 passing through A will also pass through B, C, and D, and its equation is given by

$$\rho_1/\rho_{10} + \theta_1/\theta_{10} = 1$$

and the weights for distance and orientation are inversely proportional to ρ_{10} and θ_{10} respectively. Thus, the isocurves of total constant cost are given by

$$\rho_1/\rho_{10} + \theta_1/\theta_{10} = \text{constant}$$

If the total cost is greater than 1 the potential position for the camera is outside the domain of visual constraints and hence of less than acceptable quality. Here the constraint costs are normalized costs with equal weights, viz., 1.0.

Similarly, we can find the isocurves and the weights for the linear and angular frame coherence constraints from their desired values and the ranges of acceptable values. The results will be similar. The weights here will act as linear and angular acceleration / deceleration.

The following theorem reveals the cost structure of the domain for visual and motion constraints. We prove the theorem for visual distance and orientation constraints. It also holds for frame coherence distance and rotation constraints.

Theorem 1. *Isocurves of less cost are contained within isocurves of higher cost.*

Proof. Let γ_1 be an isocurve (Fig. 2) with total cost c given by: $\rho_1/\rho_{10}+\theta_1/\theta_{10} = c$. Let P be a point inside γ_1. Let TP meets with γ_1 at R $(\rho_d + \rho_1, \theta_d + \theta_1)$ and intersects with BD at Q. Then R has the total constraint cost of c. Since P is inside γ_1, $QP < QR = \rho_1$. Since the points P and R have the same angle RTC $= \theta_1$, the total cost at P is less than that at R. So, any point inside γ_1 has less cost than that on γ_1. Similarly, we can show that any point outside γ_1 has higher cost than that on γ_1. $\qquad\square$

Now, let the total weighted cost of visual and frame coherence constraints for a potential position of the camera be c. From (1) we have

$$k_1\rho_1 + l_1\theta_1 + k_2\rho_2 + l_2\theta_2 = c \qquad (2)$$

We have to find the point where c is the minimum.

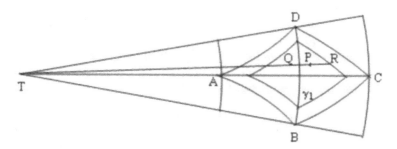

Fig. 2. Isocurves of less cost are contained within isocurves of higher cost

Frame coherence is necessary to keep a viewer's attention to the action in the image and to nothing else. Without it, the viewer's attention will be drawn to the camera [Thompson 14]. On the other hand, adequate coverage is the minimum requirement for the visual effect of a scene [Katz 11]. So, frame coherence must be given higher priority than the visual constraints. Frame coherence constraint cannot be relaxed beyond its acceptable range, but visual constraint can be relaxed as much as the situation demands provided there is adequate coverage. Consequently, frame coherence constraint will limit the search domain to promising regions of space. Within this region the frame coherence constraint can be relaxed more readily than other constraints. So, once we identify this region we can ignore frame coherence constraints. Also the nature of their effect on the cost potential is different from that of the visual constraints. So, we need to decouple the cost potential for frame coherence constraints from that for visual constraints to analyze the cost structure of the search space and take advantage of that to control the camera in an informed way.

To that end we decompose the above problem (2) into two - one for the frame coherence constraints and the other for the visual constraints:

$$k_1\rho_1 + l_1\theta_1 = c_1$$

$$k_2\rho_2 + l_2\theta_2 = c_2$$

where $c_1 + c_2 = c$. We have to find c_1 and c_2 such that c is the minimum. These two equations define two systems of isocurves for their respective constraints. All the points on a particular isocurve have the same cost with respect to the cost potential of the isocurve given on the left hand side of their respective equation.

Now, we can specify the visual and the motion weights separately by considering the acceptable domains of visual and motion constraints respectively. The weights are determined automatically once we identify the respective acceptable domains. Moreover, the solution will always be within their common domain if they have an acceptable common region. Thus, one need not consider the weights. Only the most appropriate desired positions and range of acceptable positions of camera with respect to visual constraints and frame coherence constraints need to be determined.

The points of intersection of the two families of isocurves will have the cost that is the total of the costs of the two isocurves. So, if we find the point of intersection of the two systems of isocurves that has the total minimum cost, that point will be the solution for the total problem. Obviously the locus of the point of contact of the two families of isocurves will contain the minimum cost point. The following theorem helps us find this point.

Theorem 2. *If within a certain region of space one of the visual or motion constraints has higher priority, then the total least cost for all of the visual and motion constraints will be at the end point of the locus of the point of contact within that region of the two systems of isocurves for visual and motion constraints that has the lowest cost for the higher priority constraints.*

Proof. Let the total weighted cost be given by (1). The points with constant weighted cost are given by

$$m_1(\rho_1 + n_1\theta_1) + \rho_2 + n_2\theta_2 = c$$

where c is a constant for the particular locus of the point of total constant cost. The systems of isocurves Γ_1 and Γ_2 with constant cost for visual and motion constraints respectively are given by:

$$m_1(\rho_1 + n_1\theta_1) = c_1 \tag{3}$$

$$\rho_2 + n_2\theta_2 = c_2 \tag{4}$$

where $c_1 + c_2 = c$.

Let P be the point of contact of the isocurve γ_2^1 of motion constraints with isocurve γ_1^1 of visual constraints and UPV be the locus of the point of contact (Fig. 3). The other curves from Γ_1 intersecting with γ_2^1 will contain P and hence by Theorem 1 they will have more visual cost than that on γ_1^1. Since all the points of γ_2^1 have the same motion cost, those other intersecting points will have more total cost for combined visual and motion constraints than that at P.

Let the isocurves γ_1^1 and γ_2^1 intersect with their respective polar axes at radial distances ρ_1' and ρ_2' respectively from their origins. Then, from (3) and (4) we see that the cost for their respective isocurves will be $m_1\rho_1'$ and ρ_2' respectively. So, the total cost at the point of contact will be

$$m_1\rho_1' + \rho_2' \tag{5}$$

Successive interior curves of one family will be in contact with the successive exterior curves of the other family. So, if the costs $m_1\rho_1'$ of successive interior curves of Γ_1 decrease, the costs ρ_2' of the corresponding tangential successive exterior curves of Γ_2 increase, and similarly the other way around.

So, in the total cost expression given by (5), if ρ_1' increases then ρ_2' decreases, and vice versa. Since (5) is linear in ρ_1' and ρ_2', and since ρ_1' and ρ_2' are non-negative and bounded, we can select m_1 sufficiently large within a certain region bounded by the isocurve γ_2^0 of motion constraints to make the visual constraints

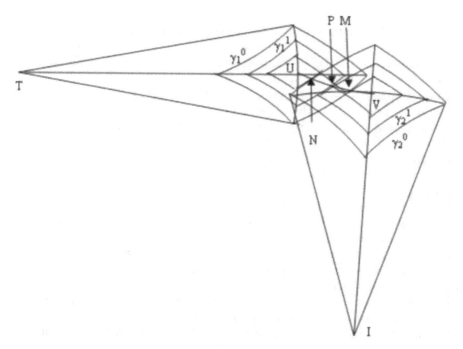

Fig. 3. Points of contact of the isocurves of visual and motion constraints

higher priority than the motion constraints within that region which will make the total cost in (5) minimum when ρ'_1 is the minimum. So, the minimum cost point will be at the end point N of the locus VMPN of the point of contact within that region and N has the lowest cost for the higher priority visual constraints.

Similarly, by making m_1 sufficiently small we can make the motion constraints higher priority than the visual constraints within a certain region bounded by the isocurve γ_1^0 of visual constraints, for which case the minimum cost solution will be at the end point M of the locus UNPM of the point of contact within that region where M has the lowest cost for the motion constraints within that region. Similarly, it can be shown that the theorem also holds for the other alignment of the isocurves of visual and motion constraints.   □

2.2   Effect of Weights

Depending upon the requirements of the shot there are many choices for the control of constraint weights. In Fig. 3 suppose we can relax the motion constraints readily within a certain region bounded by the isocurve γ_2^1 of motion constraints. Then Theorem 2 shows that the minimum cost solution will be at P. We can use this property of the constraints to better control the motion or the visual effect of the camera. For example, we can move the camera most vigorously (of course, frame coherently) by giving lower priority to the motion constraints for the whole domain of frame coherent motion. For this case the visual quality will be the maximum for the problem. To further increase the visual quality we can even further increase the region of lower priority of the motion constraints. But that will make the camera motion unsmooth. Or, we can give lower priority to the motion constraints for only a part of their maximum acceptable ranges of relaxation to move the camera very smoothly.

In the extreme cases, if the motion constraints cannot be relaxed at all (i.e., if it has higher priority than the visual constraints throughout the whole domain) then the solution will be at the desired position of the camera according to the motion constraints and the camera will be moving at the same speed in the same direction. On the other hand, if the visual constraint cannot be relaxed at all then the camera solution point will be at the desired position of the camera according to the visual constraints, and the camera will be moving erratically - always going to the best possible viewing position.

If only one of the motion constraints, say linear motion of the camera, can be relaxed and the other can not be relaxed at all then the camera will be moving in the same direction with variable speed. The solution camera position will be at any point on the straight line along the motion direction within the range of acceptable values for linear motion. Similar will be the case if only the rotation speed of the camera can be relaxed but its linear speed can not be relaxed.

If the visual constraints have higher priority than the motion constraints within a region bounded by an isocurve of motion constraints the weights for visual distance and orientation constraints that have isocurves with equal length axes will attract the camera solution point toward the desired position according to the visual constraints in a uniform manner from the desired position of camera according to the motion constraints. A higher weight for the visual distance constraint attracts the camera more rapidly towards the positions of desired distance than the positions of desired orientation, and a higher weight for the visual orientation constraint attracts the camera more rapidly towards the positions of desired orientation than the positions of desired distance.

The case will be similar for giving higher priority to the motion constraints within a region bounded by an isocurve of visual constraints. In this case the role of those two groups of constraints will be interchanged.

3   Using Constraints and Their Weights

3.1   Control of Weights at Frame Level

In Bares et al. [2, 3], Bourne and Sattar [5, 8] and Bourne [7] constraint weights are determined and applied at the level of a simple shot (i.e., for the whole length of a simple shot). They cannot be determined or specified at the frame level. During the transition the camera profiles are interpolated by Bourne and Sattar [5, 8] and Bourne [7]. In our approach, weights can be determined and applied at all levels including simple shot level and frame level. Application of weights at frame level is necessary if some constraints are affecting only portions of a simple shot.

Hierarchical control of weights is necessary to have finer control on the visual effect or camera movement. Depending on the situation they are controlled down to the frame level. Frame level weights will have the highest precedence, developing shot level weights will have the lowest precedence, and the simple shot level weights will have the precedence in between. For example, at the simple shot level if we can afford to have less control for visual quality within a region, we can apply stricter frame coherence weight within that region and the camera motion becomes very smooth. For portions of the simple shot it may be necessary to increase the frame coherence domain to its maximum possible extent to have a common domain to position the camera, thus reducing the weights for frame coherence constraints at the frame level.

Once new desired values and weights are assigned to the constraints, the camera will automatically be guided toward the desired position smoothly. No interpolation is necessary. The camera may be accelerated or decelerated radially or angularly at the frame level by higher level adjustment of the weights of camera motion and the relative priority of the motion constraints with respect to the visual constraints.

3.2   Strategy for Using the Constraints

The effect of all the requirements and hence the constraints of a shot on camera parameters and its motion are not similar. To control the camera intelligently, we need to have prior information about the effect of each constraint and the resultant effect of all the constraints before their application. Since decomposing the problem helps us to know and precisely control the effect of each constraint in the overall problem involving all the constraints of the camera, it is desirable to classify the constraints as visual constraints and motion constraints. Their combined effect will determine the position of the camera. The centre of view and view up vector determine the other three parameters of the camera. Finally, determining the field of view fixes the focal length. Thus all the seven parameters of the camera are determined.

For that we need to group the constraints in relation to the type of camera parameters or camera motion they are interacting with. The classification is given below:

1. **Camera Motion Constraints:** These are related to the frame coherent motion of the camera. They include frame coherent constraints, and all other constraints related to the movement of the camera, viz., slow or fast moving camera, jerky camera, ascending or descending camera, tracking camera, etc. This group will also include the constraints that will guide the camera to move to a desired region of space in the future frames by using prediction to avoid collision of the camera with the environment elements, or to avoid occlusion of, say, dramatic circle of interest by environment elements, or to transition to another shot within the same developing shot. Each of them will have the most appropriate value and a domain of acceptable values, the range of which determines the weight for it.

2. **Visual Constraints:** This group includes all other constraints. All these constraints are related to the quality of the image. They include, for example, distance, orientation, camera height, depth order, etc.
3. **Centre of View Constraints:** They include location of subject / subjects of the shot on the image, etc.
4. **Field of View Constraints:** These include such constraints as dramatic circle of interest, shot size on the image, object inclusion constraint, etc.

Some of the constraints may be hard constraints, e.g., frame coherence constraints and avoiding occlusion of eyes in extreme close up. These must be satisfied, however, most of the constraints are soft constraints. Each of the soft constraints in each group are satisfied in the best possible way with respect to other constraints in their respective groups using the weighted constraint method or any other method appropriate for that particular constraint. In this way, camera motion constraints and visual constraints will produce two acceptable domains with the most appropriate desired position in their respective centres. The position having the least total cost for these two groups of constraints will be the position of the camera.

Once the camera is positioned there, the centre of view is determined by considering the related constraints. Finally, the field of view is adjusted by using its related constraints. These two are also adjusted smoothly. The strategy described here is very similar to that of the real cameraman as he moves the camera smoothly to the best possible position and adjusts the centre of view and field of view accordingly.

3.3   Strategy for Using the Weights

Knowing the behaviour of the camera position (corresponding to visual constraints) and motion (corresponding to motion constraints) in relation to their respective weights, the camera module can control the camera in an informed way. Different strategies can be used for different types of shots to determine the appropriate weights for that type of shot. An example strategy would be to use a very restrictive domain (maybe a one point domain in the extreme case) for the camera motion constraints in the first attempt. More weight is given to visual constraints so that the solution lies on the outer isocurve of the restrictive domain of camera motion.

If there is no common region in the first attempt, the camera motion domain can be relaxed to its maximum acceptable range, but more weight can be given to it so that the solution lies within its domain. This can be achieved by searching along the bounding isocurve of visual constraints that bounds the acceptable region of visual constraints. If the second attempt fails, the domain for the visual constraints can be relaxed and the visual constraints can be given higher priority than the motion constraints. If the third attempt fails but still then we need a solution, the domain for the visual constraints can be relaxed to its maximum extent and the higher weight is maintained. The visual quality of the image may be very poor. This option can be used in such cases as during the computation of the next shot for which it is not possible to cut in the current frame.

This hierarchy of decision is based on the relative importance of frame coherent movement of the camera versus the visual effect of the scene. The above
hierarchy may be modified if the shot demands differently.

4   Solving for Camera Position

We consider only one scenario where the visual constraints have higher preference than the motion constraints within the domain of motion constraints and
lower preference outside that. We also assume that all the weights are constant.
So, both the systems of isocurves will be either convex or concave in each quad-
rant. According to Theorem 2, if the desired position of camera according to
the visual constraints is inside the domain of the motion constraints then the
solution camera position is at that desired position, otherwise the solution camera position will be at one of the points of contact of the outermost isocurve for
motion constraints with an isocurve for visual constraints where the cost of the
visual constraints is the minimum.

But there is no known exact solution for it. So, we need to search the outermost
isocurve for motion constraints to find the point where the total weighted cost of
the visual constraints is the minimum. We use a binary search in which the search
domain is successively refined into one half of its previous size and the search continues in the region which is known to contain the point with the minimum total weighted cost for the visual constraints. The search is described extensively in
Alam [1]. It finds the solution in real-time.

5   Conclusion

We have described our approach to automatic camera control in a virtual environment using a simple tracking shot of a single target in 2D without any
occluder. We use a weighted constraint representation for the camera control
problem. Here we consider only four constraints (visual distance, visual orientation, motion distance and motion rotation), and apply them purely reactively
to enable their use in a dynamic environment. Each of these constraints has an
optimal value and a range of acceptable values. They give rise to the weights for
the constraints. This relieves the user of specifying the weights. The linear and
angular speed of the camera can be controlled by increasing or decreasing the
range of acceptable values of their respective constraints. Also the relative priority between the groups of visual and motion constraints can be used to guide
the motion of the camera. For this we do not need any specific value. There are
only three cases such as higher, lower or equal priority.

The camera control system has been implemented in a simple tracking shot
without occlusion. The result is satisfactory for different values of constraints
and their weights. In all the cases the camera motion is found to be very smooth
and there is no jerking in the camera.

We have already extended the method to 3D, and included occlusion and
collision avoidance constraints [Alam 1]. Other constraints can be investigated

for inclusion in this framework. Determination of the maximum limits of the radial and angular acceleration and deceleration of the camera in relation to its radial and angular speed and visual requirements can be investigated.

References

Alam, M.S.: Control of Constraint Weights for an Autonomous Camera. Master's Thesis, School of Computer Science, University of Windsor, Windsor, Canada (2008)

Bares, W.H., McDermott, S., Boudreaux, C., Thainimit, S.: Virtual 3D camera composition from frame constraints. In: MULTIMEDIA 2000: Proceedings of the Eighth ACM International Conference on Multimedia, pp. 177–186. ACM Press, New York (2000a)

Bares, W.H., Thainimit, S., McDermott, S.: A model for constraint-based camera planning. In: AAAI 2000 Spring Symposium Series on Smart Graphics, Stanford, California, USA, March 2000, pp. 84–91 (2000b)

Bourne, O., Sattar, A.: Applying constraint satisfaction techniques to 3D camera control. In: Webb, G.I., Yu, X. (eds.) AI 2004. LNCS, vol. 3339, pp. 658–669. Springer, Heidelberg (2004)

Bourne, O., Sattar, A.: Applying constraint weighting to autonomous camera control. In: Artificial Intelligence and Interactive Digital Entertainment, Marina Del Ray, CA, USA, pp. 3–8. AAAI Press, Menlo Park (2005a)

Bourne, O., Sattar, A.: Evolving behaviours for a real-time autonomous camera. In: Proceedings of the Second Australasian Conference on Interactive Entertainment, Sydney, Australia, pp. 27–33 (2005b) ISBN 0-9751533-2-3/05/11

Bourne, O.: Constraint-Based Intelligent Camera Control for Interactive Digital Entertainment. PhD Thesis, Institute of Integrated and Intelligent Systems, Griffith University, Queensland, Australia (2006)

Bourne, O., Sattar, A.: Autonomous camera control with constraint satisfaction methods. In: Robin, S. (ed.) AI Game Programming Wisdom, March 2006, vol. 3, pp. 173–187. Charles River Media (2006)

Christie, M., Normand, J.-M.: A semantic space partitioning approach to virtual camera composition. In: Proceedings of the Annual Eurographics Conference, vol. 24, pp. 247–256 (2005)

Halper, N., Helbing, R., Strothotte, T.: A camera engine for computer games: managing the trade-off between constraint satisfaction and frame coherence. In: Chalmers, A., Rhyne, T. (eds.) Proceedings of the Eurographics 2001 Conference, Manchester, UK, September 2001, vol. 20(3), pp. 174–183 (2001)

Katz, S.: Film Directing Shot by Shot: Visualizing from Concept to Screen. Michael Wiese Productions, Studio City, CA 91604, USA (1991)

Mascelli, J.: The Five C's of Cinematography: Motion Picture Filming Techniques. Silman-James Press, USA (1998)

Pickering, J.H.: Intelligent Camera Planning for Computer Graphics. PhD Thesis, Department of Computer Science, University of York (2002)

Thompson, R.: Grammar of the Shot. Focal Press (1998) ISBN 0-240-51398-3

Training Global Linear Models for Chinese Word Segmentation[*]

Dong Song and Anoop Sarkar

School of Computing Science, Simon Fraser University
Burnaby, BC, Canada V5A1S6
dsong@alumni.sfu.ca, anoop@cs.sfu.ca

Abstract. This paper examines how one can obtain state of the art Chinese word segmentation using global linear models. We provide experimental comparisons that give a detailed road-map for obtaining state of the art accuracy on various datasets. In particular, we compare the use of reranking with full beam search; we compare various methods for learning weights for features that are full sentence features, such as language model features; and, we compare an Averaged Perceptron global linear model with the Exponentiated Gradient max-margin algorithm.

1 Introduction

The written form of many languages, including Chinese, do not have marks identifying words. Given the Chinese text "北京大学生比赛", a plausible segmentation would be "北京(Beijing)/大学生(university students)/比赛(competition)" (Competition among university students in Beijing). However, if "北京大学" is taken to mean Beijing University, the segmentation for the above character sequence might become "北京大学(Beijing University)/生(give birth to)/比赛(competition)" (Beijing University gives birth to competition), which is less plausible. Chinese word segmentation has a large community of researchers, and has resulted in three shared tasks: the SIGHAN bakeoffs [1,2,3]. Word segmentation can be treated as a supervised sequence learning (or tagging) task. As in other tagging tasks, the most accurate models are discriminative models such as conditional random fields (CRFs) [4], perceptron [5], or various max-margin sequence models, such as [6,7]. [5] provides a common framework collectively called *global linear models* to describe these approaches.

In this paper we show that using features that have been commonly used for Chinese word segmentation, plus adding a few additional global features, such as language model features, we can provide state of the art accuracy on several standard datasets using global linear models. In particular, the accuracy numbers obtained by our approach do not use any post-processing heuristics.

[*] This research was partially supported by NSERC, Canada (RGPIN: 264905) and by an IBM Faculty Award. Thanks to Michael Collins and Terry Koo for help with the EG implementation (any errors are our own), to the anonymous reviewers, and to the SIGHAN bakeoff organizers and participants.

Y. Gao and N. Japkowicz (Eds.): Canadian AI 2009, LNAI 5549, pp. 133–145, 2009.

Several types of ad-hoc post-processing heuristics are commonly used by other systems to obtain high accuracy on certain data sets but not others. The main contribution of this paper is to motivate the various choices that need to be made while training global linear models. We provide experimental evidence for choices made that provide state of the art accuracy for Chinese word segmentation.

2   Global Linear Models

Michael Collins [5] provides a common framework called *global linear models* for the *sequence learning* task (also called *tagging*): Let $\mathbf{x}$ be a set of inputs, and $\mathbf{y}$ be a set of possible outputs. In our experiments, $\mathbf{x}$ are unsegmented Chinese sentences, and $\mathbf{y}$ are the possible word segmentations for $\mathbf{x}$.

- Each $x \in \mathbf{x}$ and $y \in \mathbf{y}$ is mapped to a d-dimensional feature vector $\Phi(x,y)$, with each dimension being a real number, summarizing partial information contained in (x,y).
- A weight parameter vector $\mathbf{w} \in \Re^d$ assigns a weight to each feature in $\Phi(x,y)$, representing the importance of that feature. The value of $\Phi(x,y) \cdot \mathbf{w}$ is the score of (x,y). The higher the score, the more plausible it is that y is the output for x.
- The function $GEN(x)$ generates the set of possible outputs y for a given x.

Having $\Phi(x,y)$, $\mathbf{w}$, and $GEN(x)$ specified, we would like to choose the highest scoring candidate y^* from $GEN(x)$ as the most plausible output. That is,

$$F(x) = \operatorname*{argmax}_{y \in GEN(x)} p(y \mid x, \mathbf{w})$$

where $F(x)$ returns the highest scoring output y^* from $GEN(x)$. A *conditional random field* (CRF) [4] defines the conditional probability as a linear score for each candidate y and a *global* normalization term:

$$\log p(y \mid x, \mathbf{w}) = \Phi(x, y) \cdot \mathbf{w} - \log \sum_{y' \in GEN(x)} \exp(\Phi(x, y') \cdot \mathbf{w})$$

In our experiments we find that a simpler global linear model that ignores the normalization term is faster to train and provides comparable accuracy.

$$F(x) = \operatorname*{argmax}_{y \in GEN(x)} \Phi(x, y) \cdot \mathbf{w}$$

For this model, we learn the weight vector from labeled data using the perceptron algorithm [5]. A global linear model is global is two ways: it uses features that are defined over the entire sequence, *and* the parameter estimation methods are explicitly related to errors over the entire sequence.

Table 1. Feature templates for (a) local features and (b) global features

1 word w
2 word bigram $w_1 w_2$
3 single character word w
4 space-separated characters c_1 and c_2
5 character bigram $c_1 c_2$ in any word
6 word starting with character c with length l
7 word ending with character c with length l 15 sentence confidence score
8 first and last characters c_1 and c_2 of any word   16 sentence language model score
9 word w immediately before character c (b)
10 character c immediately before word w
11 starting chars c_1, c_2 for 2 consecutive words
12 ending chars c_1, c_2 for 2 consecutive words
13 a word of length l and the previous word w
14 a word of length l and the next word w

(a)

3 Feature Templates and Experimental Setup

In this section, we look at the choices to be made in defining the feature vector $\Phi(x, y)$. In our experiments the feature vector is defined using local feature templates and global feature templates. For local features, the 14 feature types from [8] are used, shown in Table 1a. The local features for the entire sequence are summed up to provide global features.

In our experiments we also use global features previously used by [9,10] that are not simply a sum of local features over the sequence. These features cannot be decomposed into a sequence of local features, which we will henceforth refer to as *global features* (the italics are important!), are listed in Table 1b.

Sentence confidence scores are calculated by a model that is also used as the *GEN* function for the global linear model (for instance, in our experiments the sentence confidence score is provided by a baseline character-based CRF tagger and *GEN* is the n-best list it produces). Sentence language model scores are produced using the *SRILM* [11] toolkit[1]. They indicate how likely a sentence can be generated given the training data, and they help capture the usefulness of features extracted from the training data. We use a trigram language model trained on the entire training corpus. We normalize the sentence LM probability: $P^{1/L}$, where P is the probability-based language model score and L is the length of the sentence in words (not in characters). Using logs the value is $| \log(P)/L |$. We explore different methods for learning the weights for these *global features*[2].

---

[1] http://www.speech.sri.com/projects/srilm/
[2] Our global features are different from commonly used "global" features in the literature, which either enforce consistency in a sequence (e.g. ensuring that the same word type is labeled consistently in the token sequence) or examine the use of a feature in the entire training or testing corpus.

We build a baseline system which is a character based tagger using only the character features from Table 1a. We use the 'IOB' tagset where each character is tagged as 'B' (first character of multi-character word), or 'I' (character inside a multi-character word), or 'O' (indicating a single character word). The baseline system is built using the *CRF++* toolkit by Taku Kudo[3]. It is also used in our reranking experiments as the source of possible segmentations for the *GEN* function in our global linear models.

The experimental results are reported on datasets from the first and third SIGHAN bakeoff shared task datasets [1,3]. From the 1st SIGHAN bakeoff we use the Peking University (PU) dataset. From the 3rd SIGHAN bakeoff we use the CityU (City University of Hong Kong), the MSRA (Microsoft Research Asia) and UPUC (University of Pennsylvania and University of Colorado) datasets. We strictly follow the closed track rules, where no external knowledge is used[4].

4   Reranking vs. Beam Search

There are two choices for the definition of *GEN* in a global linear model:

- *GEN(x)* enumerates all possible segmentations of the input x. In this case, search is organized either using dynamic programming or using beam search.
- *GEN(x)* is the n-best output of another auxiliary model and the global linear model is used as a *reranking* model.

In this section we compare beam search with reranking across many different corpora to test the strength and weakness of both methods.

Reranking. To produce a reranking system, we produce a 10-fold split of the training data: in each fold, 90% of the corpus is used for training and 10% is used to produce an n-best list of candidates. The n-best list is produced using the character-based CRF tagger described earlier. The true segmentation can now be compared with the n-best list in order to train using an averaged perceptron algorithm [5] shown in Figure 1. This system is then used to predict the best word segmentation from an n-best list for each sentence in the test data.

We used the development set of the UPUC corpus to find a suitable value for the parameter n, the maximum number of n-best candidates. This oracle procedure proceeds as follows: 80% of the training corpus is used to train the CRF model, which is used to produce the n-best outputs for each sentence on the remaining 20% of the corpus. Then, these n candidates are compared with the true segmentation, and for each training sentence, the candidate closest to the truth is chosen as the final output. As we increase the value of n, for some sentences, its n-best candidate list is more likely to contain a segmentation that will improve the overall F-score (Figure 2). To balance accuracy and speed, we choose $n = 20$ in all our reranking experiments.

---

[3] http://crfpp.sourceforge.net/

[4] We do not even use the encoding of the dataset (dates and non-Chinese characters are used in encoding-specific heuristics to improve performance, we do not do this).

Inputs: Training Data $\langle(x_1, y_1), \ldots, (x_m, y_m)\rangle$; number of iterations T
Initialization: Set $\mathbf{w} = \mathbf{0}$, $\gamma = \mathbf{0}$, $\sigma = \mathbf{0}$
Algorithm:

> **for** $t = 1, \ldots, T$ **do**
>> **for** $i = 1, \ldots, m$ **do**
>>> $y_i' = \underset{y \in n\text{-}best\ list}{\mathrm{argmax}}\ \Phi(x_i, y) \cdot \mathbf{w}$
>>>
>>> y^b is closest to y_i in terms of f-score and $y^b \in n\text{-}best\ list$
>>> **if** $y_i' \neq y^b$ **then**
>>>> $\mathbf{w} = \mathbf{w} + \Phi(x_i, y^b) - \Phi(x_i, y_i')$
>>>
>>> **end if**
>>> $\sigma = \sigma + \mathbf{w}$
>>
>> **end for**
>
> **end for**

Output: Avg. weight parameter vector $\gamma = \sigma/(\mathrm{mT})$

Fig. 1. Averaged perceptron learning algorithm using an n-best list

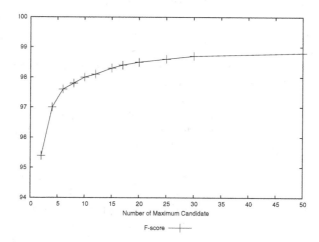

Fig. 2. F-score on the UPUC development set with different n

We do not use the algorithm in Figure 1 to train the weights for the *global features* defined in Table 1. The sentence confidence score feature weight and the language model feature weight is chosen to be 15 for the CityU corpus, to be 15 for the MSRA corpus, and to be 20 for the UPUC corpus. The reason for this choice is provided in Section 5.

Beam Search Decoding. In [8], instead of applying the n-best reranking method, their word segmentation system uses beam search decoding [12], where the global features are only those that are the sum of the local features.

In beam search, the decoder generates segmentation candidates incrementally. It reads one character at a time from the input sentence, and combines it with each existing candidate in two ways, either appending this new character to the last word, or considering it as the beginning of a new word. This combination process generates segmentations exhaustively; that is, for a sentence with k characters, all 2^{k-1} possible segmentations are generated. In global linear models which contain a normalization term it is common to use dynamic programming. However, for mistake-driven training such as the perceptron, beam search is more effective. We implemented the decoding algorithm following the pseudo-code described in [8] which is based on the algorithm in [12]. The beam size B is used to limit the number of candidates preserved after processing each character. The performance of the beam search system is compared with that of the n-best reranking system on the PU corpus from the first SIGHAN bakeoff, and on the CityU, MSRA, UPUC corpora from the third SIGHAN bakeoff (closed track). In the n-best reranking system, 20 is chosen to be the maximum number of n-best candidates. Using the approach described in Section 5 the weight for sentence confidence score and that for language model score are determined to be 15 for the CityU and MSRA corpora, 20 for the UPUC corpus, and 40 for the PU corpus. Similarly, using the dev set, the training iterations were set to 7 for the CityU and MSRA corpora, 9 for the UPUC corpus, and 6 for the PU corpus. In the beam search method, the beam size was set to be 16 for all corpora, and the number of iterations was set to be 7, 7 and 9 for the CityU, MSRA and UPUC corpora, respectively, corresponding to the iteration values we applied on each corpus in the reranking system. Table 2 shows the comparison between the averaged perceptron training using the beam search method v.s. the reranking method. For each corpus, the bold number represents the highest F-score. From the result, we see that on the CityU, MSRA and UPUC corpora, the beam search decoding based system outperforms the reranking using only local features. However, reranking based with *global features* is more accurate than the beam search decoding based system, except on the PU corpus.

For the PU corpus from the first SIGHAN bakeoff, the reranking does worse than beam search (and no better than the baseline). To see why we examine how many sentences in the gold standard also appear within the 20-best candidate list. For each corpus test set, the results are: CityU (88.2%), MSRA (88.3%), UPUC (68.4%), and PU (54.8%). For the PU test set, almost half of the true segmentations are not seen in the 20-best list, which seriously affects the reranking approach. While for the CityU and MSRA corpora, nearly 90% of the gold standard segmentations appear in the 20-best candidate list. Beam search has the advantage of not requiring a separate model to produce n-best candidates, but training and testing are much slower than reranking[5] and further research is required to make it competitive with reranking for Chinese word segmentation[6].

---

[5] We added the language model (LM) *global feature* as part of beam search, but could not use it in our experiments as training was prohibitively slow. Rescoring the final output using the LM probability led to lower accuracy.

[6] In general, in this paper we are not making general claims about algorithms, but rather what works and does not work for typical Chinese word segmentation datasets.

Table 2. Performance (in percentage) comparing averaged perceptron with beam search with reranking. F is F-score, P is precision, R is recall, R_{IV} is in-vocabulary (words in training) recall, and R_{OOV} is out of vocabulary recall. CRF with subword tagging is our implementation of [13]. Boldface is statistically significant improvement over all other methods (see [14] for detailed results).

Corpus	Setting	F	P	R	R_{IV}	R_{OOV}
	Avg. perc. with beam search	**94.1**	94.5	93.6	69.3	95.1
PU	Avg. perc. with reranking, global & local features	93.1	93.9	92.3	94.2	61.8
	Avg. perc. with reranking, local features	92.2	92.8	91.7	93.4	62.3
	Baseline character based CRF	93.1	94.0	92.3	94.1	61.5
	CRF with subword tagging	91.9	91.7	92.2	94.5	53.5
	Avg. perc. with beam search	96.8	96.8	96.8	97.6	77.8
	Avg. perc. with reranking, global & local features	**97.1**	97.1	97.1	97.9	78.3
CityU	Avg. perc. with reranking, local features	96.7	96.7	96.6	97.5	77.4
	Baseline character based CRF	95.7	95.7	95.7	96.5	78.3
	CRF with subword tagging	95.9	95.8	96.0	96.9	75.2
	Avg. perc. with beam search	95.8	96.0	95.6	96.6	66.2
	Avg. perc. with reranking, global & local features	95.8	95.9	95.7	96.9	62.0
MSRA	Avg. perc. with reranking, local features	95.5	95.6	95.3	96.3	65.4
	Baseline character based CRF	94.7	95.2	94.3	95.3	66.9
	CRF with subword tagging	94.8	94.9	94.6	95.7	64.9
	Avg. perc. with beam search	92.6	92.0	93.3	95.8	67.3
	Avg. perc. with reranking, global & local features	**93.1**	92.5	93.8	96.1	69.4
UPUC	Avg. perc. with reranking, local features	92.5	91.8	93.1	95.5	68.8
	Baseline character based CRF	92.8	92.2	93.3	95.5	70.9
	CRF with subword tagging	91.8	91.0	92.7	95.2	66.6

5 Learning Global Feature Weights

In this section we explore how to learn the weights for those features that are not simply the sum of local features. These so-called *global features* have an important property: they are real numbers that correspond to the quality of the entire segmentation, and cannot be identified with any portion of it. Algorithm 1 updates features for a segmentation but is restricted to local features collected over the entire segmentation (see line 6 of Algorithm 1 where the weight vector **w** is updated). For this reason, alternative strategies to obtain weights for global features need to be explored.

Learning Weights from Development Data. We use development data to determine the weight for the sentence confidence score S_{crf} and for the language model score S_{lm}.[7] In this step, each training corpus is separated into a training set, which contains 80% of the training corpus, and a development set containing the remaining 20% of the training corpus. Then, the perceptron algorithm is applied on the training set with different S_{crf} and S_{lm} weight values, and for various number of iterations. The weight values we test include 2, 4, 6, 8, 10,

---

[7] This process is the same for all datasets. Heuristics are tuned per dataset.

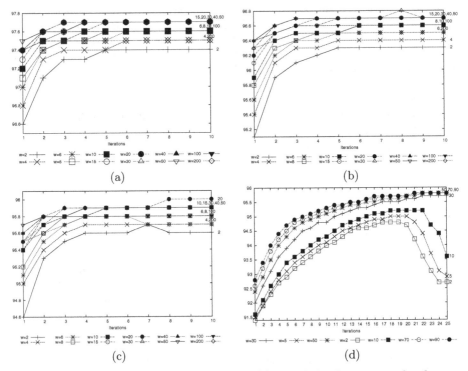

Fig. 3. F-scores on the (a) CityU, (b) MSRA, (c) UPUC development set for the avg. perceptron algorithm, and (d) UPUC development set for the EG algorithm

15, 20, 30, 40, 50, 100 and 200, across a wide range of scales. The reason for these discrete values is because we are simply looking for a confidence threshold over which the sum of local features can override global features such as the confidence score (cf. Figure 3). As we can see, there will be a significant number of testing scenarios (i.e. $12 \times 12 = 144$ testing scenarios) in order to pick the most suitable weight values for each corpus. To simplify the process, we assume that the weights for both S_{crf} and S_{lm} are equal – this assumption is based on the fact that weights for these global features simply provide an importance factor so only a threshold value is needed rather than a finely tuned value that interacts with other feature weights for features based on local feature templates. Figure 3 shows the F-scores on each of the three corpora using different S_{crf} and S_{lm} weight values with different number of iterations t. From the tables, we observe that when the weight for S_{crf} and S_{lm} increases, F-score improves; however, if the weight for S_{crf} and S_{lm} becomes too large to overrule the effect of weight learning on local features, F-score suffers. For our experiments, the weight for S_{crf} and S_{lm} is chosen to be 15 for the CityU corpus, 15 for the MSRA corpus, and 20 for the UPUC corpus. Iterations over the development set also allow us to find the optimal number of iterations of training for the perceptron algorithm which is then used on the test set.

Table 3. F-scores (in percentage) obtained by using various ways to transform global feature weights and by updating their weights in averaged perceptron learning. The experiments are done on the UPUC and CityU corpora.

Method	F-score (UPUC corpus)		F-score (CityU corpus)	
	held-out set	test set	held-out set	test set
Without global features	95.5	92.5	97.3	96.7
Fixed global feature weights	**96.0**	**93.1**	**97.7**	**97.1**
Threshold at mean to 0,1	95.0	92.0	96.7	96.0
Threshold at mean to -1,1	95.0	92.0	96.6	95.9
Normalize to [0,1]	95.2	92.1	96.8	96.0
Normalize to [-1,1]	95.1	92.0	96.8	95.9
Normalize to [-3,3]	95.1	92.1	96.8	96.0
Z-score	95.4	92.5	97.1	96.3

Learning Weights from Training Data. The word segmentation system, designed by Liang in [15], incorporated and learned the weights for real-valued mutual information (MI) features by transforming them into alternative forms:

- Scale the values from $[0, \infty)$ into some fixed range [a, b], where the smallest value observed maps to a, and the largest value observed maps to b.
- Apply z-scores instead of the original values. The *z-score* of value x from $[0, \infty)$ is defined as $\frac{x-\mu}{\sigma}$ where μ and σ represent the mean and standard deviation of the distribution of x values.
- Map any value x to a if $x < \mu$, the mean value from the distribution of x values, or to b if $x \geq \mu$.

We use Liang's methods to learn weights for our two global features during perceptron training, instead of manually fixing their weight using the development set. We experiment with the transformations on the two global features defined previously with the UPUC and CityU corpora[8]. Table 3 provides the performance on their development and test sets. Z-scores perform well but do not out-perform fixing global feature weights using the development set. The likely reason is that the two global features have different properties than the MI feature. They do not have shared components across different training sentences and they describe the entire sentence unlike the MI features.

6   Exponentiated Gradient

In this section, we explore the use of max-margin methods for global linear models. In many tasks, the use of large margin or max-margin methods provides better generalization error over unseen data. We would like to know if Chinese word segmentation can benefit from a max-margin approach. We implement the

---

[8] Due to the large number of experimental settings, we do not test on the CityU and PU corpora due to their size.

batch exponentiated gradient (EG) algorithm [6,7] with the same feature set as the perceptron experiments, including the two global features defined previously, and compare the performance on the UPUC corpus in the reranking setting.[9] In EG, a set of dual variables $\alpha_{i,y}$ is assigned to data points $\mathbf{x}$. Specifically, to every point $x_i \in \mathbf{x}$, there corresponds a distribution $\alpha_{i,y}$ such that $\alpha_{i,y} \geq 0$ and $\sum_y \alpha_{i,y} = 1$. The algorithm attempts to optimize these dual variables $\alpha_{i,y}$ for each i separately. In the word segmentation case, x_i is a training example, and $\alpha_{i,y}$ is the dual variable corresponding to each possible segmented output y for x_i. EG is also expressed as a global linear model:

$$F(x) = \underset{y \in GEN(x)}{\operatorname{argmax}} \ \Phi(x, y) \cdot \mathbf{w}$$

The weight parameter vector $\mathbf{w}$ is expressed in terms of the dual variables $\alpha_{i,y}$:

$$\mathbf{w} = \sum_{i,y} \alpha_{i,y} \left[\Phi(x_i, y_i) - \Phi(x_i, y) \right]$$

Given a training set $\{(x_i, y_i)\}_{i=1}^{m}$ and the weight parameter vector $\mathbf{w}$, the margin on the segmentation candidate y for the ith training example is defined as the difference in score between the true segmentation and the candidate y. That is,

$$M_{i,y} = \Phi(x_i, y_i) \cdot \mathbf{w} - \Phi(x_i, y) \cdot \mathbf{w}$$

For each dual variable $\alpha_{i,y}$, a new $\alpha'_{i,y}$ is obtained as

$$\alpha'_{i,y} \leftarrow \frac{\alpha_{i,y} e^{\eta \nabla_{i,y}}}{\sum_y \alpha_{i,y} e^{\eta \nabla_{i,y}}} \quad \text{where } \nabla_{i,y} = \begin{cases} 0 & \text{for } y = y_i \\ 1 - M_{i,y} & \text{for } y \neq y_i \end{cases}$$

and η is the learning rate which is positive and controls the magnitude of the update. In implementing the batch EG algorithm, during the initialization phase, the initial values of $\alpha_{i,y}$ are set to be $1/$(number of n-best candidates for x_i). In order to get $\alpha'_{i,y}$, we need to calculate $e^{\eta \nabla_{i,y}}$. When each ∇ in the n-best list is positively or negatively too large, numerical underflow occurs. To avoid this, ∇ is normalized:

$$\alpha'_{i,y} \leftarrow \frac{\alpha_{i,y} e^{\eta \nabla_{i,y} / \sum_y |\nabla_{i,y}|}}{\sum_y \alpha_{i,y} e^{\eta \nabla_{i,y} / \sum_y |\nabla_{i,y}|}}$$

As before, the weight for global features is pre-determined using the development set and is fixed during the learning process. Considering the difference in training time between online update for perceptron learning and batch update for EG method, the maximum number of iterations is set to be larger (T $= 25$) in the latter case during parameter pruning. The weight for the global features are

---

[9] Because EG is computationally expensive we test only on UPUC. We obtain F-score of 93.1% on UPUC (lower than other corpora) so there is room for improvement using max-margin methods, however the baseline CRF model performs quite well on UPUC at F-score of 92.8%. This section is about comparing perceptron and EG.

Table 4. Performance (percentage) of the EG algorithms, compared to the perceptron learning methods. All are in the reranking setting. Cf. UPUC results in Table 2.

Setting	F	P	R	R_{IV}	R_{OOV}
EG algorithm, global & local features	93.0	92.3	93.7	96.1	68.2
EG algorithm, local features	90.4	90.6	90.2	92.2	69.7
EG algorithm, global & local features, 9 iterations	92.4	91.7	93.1	95.5	67.6
Avg. perc. global & local features	**93.1**	92.5	93.8	96.1	69.4

tested with 2, 5, 10, 30, 50, 70, and 90. Figure 3(d) shows the performance on the UPUC held-out set with various parameters. We select the number of iterations to be 22 and the weight for global features to be 90, and apply these parameters on the UPUC test set. Table 4 lists the resulting performance. Performance of the EG method with 22 iterations and with the same number of iterations (9 iterations) as the averaged perceptron method is provided, along with the use of different feature sets. The bold number represents the highest F-score.

From Table 4, we see that the averaged perceptron with global features provides the highest F-score. Continuing to run the EG algorithm for more iterations (T = 120) with the weight of global features being fixed at 90, Figure 4 shows the convergence in terms of the primal and dual objective functions. From the figure, we can see that the algorithm does in fact converge to the maximum margin solution on this data set. However, at iteration 120, the F-score remains

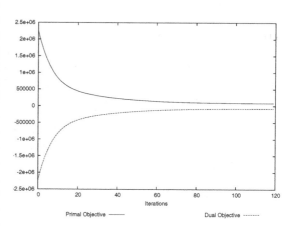

Fig. 4. EG algorithm convergence on UPUC

0.930, which is the same as the F-score produced in the 22nd iteration.

7   Accuracy

In this section, we compare the accuracy we obtain with the best known accuracy (to our knowledge) on each dataset from the first and third SIGHAN bakeoff [1,3]. On the PU corpus, we replicate the result in [8] and obtain $F = 0.941$ which is the best accuracy on that dataset. On the UPUC corpus the best result is obtained by Site 20[10]. They use date and non-Chinese character information in their features (this depends on the encoding). Plus one local feature they use

---

[10] Due to lack of space, we omit full references and ask that the reader refer to [1,3].

is the tone of the character which we do not use. They do not report results without the date and non-Chinese character features (they created these features for the training data using clustering). Their overall result using a log-linear tagger is $F = 0.933$ which is better than our result of $F = 0.931$. On the MSRA corpus the best result is obtained by Site 32. They use three different kinds of post-processing: dictionary based heuristics, date specific heuristics, and knowledge about named entities that is added to the training data. Without any post-processing, using a log-linear tagger with n-gram features the result obtained is $F = 0.958$ which matches our score of $F = 0.958$. On the CityU corpus, the best result is obtained by Site 15. They build a specialized tagger for non-Chinese characters using clustering, plus template-based post-processing. Without these steps the accuracy is $F = 0.966$ compared to our score of $F = 0.971$. Our system (avg. perceptron with global & local features) is the only one that consistently obtains good performance across all these datasets except for the Peking University (PU) dataset[11].

8 Summary

In this paper, we explore several choices in building a Chinese word segmentation system. We explore the choice between using global features or not, and the choices involved in training their feature weights. In our experiments we find that using a development dataset to fix these global feature weights is better than learning them from data directly. We compare reranking versus the use of full beam search decoding, and find that further research is required to make beam search competitive in all datasets. We explore the choice between max-margin methods and an averaged perceptron, and find that the averaged perceptron is typically faster and as accurate for our datasets. We show that our methods lead to state of the art accuracy and provide a transparent, easy to replicate design for highly accurate Chinese word segmentation.

References

1. Sproat, R., Emerson, T.: The 1st international chinese word segmentation bakeoff. In: Proceedings of the 2nd SIGHAN Workshop on Chinese Language Processing, Sapporo, Japan, ACL, pp. 123–133 (July 2003)
2. Emerson, T.: The 2nd international chinese word segmentation bakeoff. In: Proceedings of the 4th SIGHAN Workshop on Chinese Language Processing, Jeju Island, Korea, pp. 123–133 (October 2005)
3. Levow, G.A.: The 3rd international chinese language processing bakeoff. In: Proceedings of the 5th SIGHAN Workshop on Chinese Language Processing, Sydney, Australia, ACL, pp. 108–117 (July 2006)

---

[11] Testing on all available datasets would result in huge tables, so we report on the most recent dataset from the 5th SIGHAN, and we report on the 2nd SIGHAN data only for comparison with [8].

4. Lafferty, J., McCallum, A., Pereira, F.: Conditional random fields: Probabilistic models for segmenting and labeling sequence data. In: Proceedings of the 18th International Conf. on Machine Learning (ICML), pp. 282–289 (2001)
5. Collins, M.: Discriminative training methods for hidden markov models: Theory and experiments with perceptron algorithms. In: Proceedings of the Empirical Methods in Natural Language Processing (EMNLP), Philadelphia, PA, USA, ACL, pp. 1–8 (July 2002)
6. Kivinen, J., Warmuth, M.: Exponentiated gradient versus gradient descent for linear predictors. Technical Report UCSC-CRL-94-16, UC Santa Cruz (1994)
7. Globerson, A., Koo, T., Carreras, X., Collins, M.: Exponentiated gradient algorithms for log-linear structured prediction. In: ICML, pp. 305–312 (2007)
8. Zhang, Y., Clark, S.: Chinese segmentation with a word-based perceptron algorithm. In: Proceedings of the 45th Annual Meeting of the Association of Computational Linguistics, Prague, Czech Republic, ACL, pp. 840–847 (June 2007)
9. Sproat, R., Gale, W., Shih, C., Chang, N.: A stochastic finite-state word-segmentation algorithm for chinese. Comput. Linguist. 22(3), 377–404 (1996)
10. Song, D., Sarkar, A.: Training a perceptron with global and local features for chinese word segmentation. In: Proceedings of the 6th SIGHAN Workshop on Chinese Language Processing, pp. 143–146 (2008)
11. Stolcke, A.: SRILM – an extensible language modeling toolkit. In: Proceedings of the ICSLP, Denver, Colorado, vol. 2, pp. 901–904 (2002)
12. Collins, M., Roark, B.: Incremental parsing with the perceptron algorithm. In: Proceedings of the 42nd Meeting of the Association for Computational Linguistics (ACL 2004), Barcelona, Spain, pp. 111–118 (July 2004)
13. Zhang, R., Kikui, G., Sumita, E.: Subword-based tagging by conditional random fields for chinese word segmentation. In: Proceedings of the Human Language Technology Conference of the NAACL, New York City, USA, ACL, pp. 193–196 (June 2006)
14. Song, D.: Experimental comparison of discriminative learning approaches for chinese word segmentation. Master's thesis, Simon Fraser University (2008)
15. Liang, P.: Semi-supervised learning for natural language. Master's thesis, Massachusetts Institute of Technology (2005)

A Concurrent Dynamic Logic of Knowledge, Belief and Certainty for Multi-agent Systems[*]

Lijun Wu[1], Jinshu Su[2], Xiangyu Luo[3,4], Zhihua Yang[5], and Qingliang Chen[6]

[1] School of Computer Science and Engineering, University of Electronic Science and
Technology, Chengdu 410073, China
[2] School of Computer Science, National University of Defense and Technology,
Changsha 610054, China
[3] Department of Computer Science, Guilin University of Electronic Technology,
Guilin 541004, China
[4] School of Software, Tsinghua University, Beijing 10084, China
[5] Information school, Guangdong university of business stuides, Guangzhou 510320, China
[6] Jinan University, Guangzhou 510632, China
wljuestc@yahoo.com

Abstract. This paper extends the logic of Knowledge, Belief and Certainty
from one agent to multi-agent systems, and gives a good combination between
logic of knowledge, belief, certainty in multi-agent systems and actions that
have concurrent and dynamic properties. Based on it, we present a concurrent
dynamic logic of knowledge, belief and certainty for MAS, which is called
CDKBC logic. Furthermore, a *CDKBC* model is given for interpreting this
logic. We construct a *CDKBC* proof system for the logic and show the proof
system is sound and complete, and prove the validity problem for the system is
EXPTIME-complete.

1 Introduction

Modal logic has been proved to be a good tool for the formal representation of properties in systems or programs, such as knowledge, belief and many other mental atti-
tudes. Since 1962, a special kind of modal logics called epistemic logic (or the logic
of knowledge) and belief logic were introduced [1] and have been investigated widely
in philosophy [2], computer science [3], artificial intelligence [4], and game theory
[5]. However, these logics can not describe the change of knowledge and of beliefs
although their static counterparts provide wonderful ways to formalize systems.

In 1979, Fisher and Ladner proposed the idea of dynamic logic [6]. Following this
idea, researchers proposed and investigated the propositional dynamic logic, the first-
order dynamic logic [7], dynamic epistemic logic [8-12], concurrent dynamic logic
[13,14], and concurrent dynamic epistemic logic [15]. Though these logics can

---

[*] This work was supported by Postdoctor Foundation of China (No. 20070410978) ,the Major
State Basic Research Development Program of China (973 Program) (No.2009cb320503),
the National Natural Science Foundation of China (No.60763004), and the Young Science
Foundation of Guangxi Province of China (No.GuiKeQing0728090).

Y. Gao and N. Japkowicz (Eds.): Canadian AI 2009, LNAI 5549, pp. 146–157, 2009.
© Springer-Verlag Berlin Heidelberg 2009

express the dynamic properties of knowledge, belief and certainty, they just consider the combination of dynamic logic and one-dimension modal logics instead of multi-dimension modal ones, which has more expressive power.

The notion of knowledge, belief and certainty was first introduced by Lenzen [2]. Lenzen listed many syntactic properties of the notion for knowledge, belief and certainty, but he did not give any of their semantics. Kaile Su presented a computationally grounded logic of knowledge, belief and certainty [16]. Lamarre and Shoham provided a model theory of knowledge, belief and certainty, but they rejected S_5 [17]. However the above logics are just for one agent and did not consider the concurrent and dynamic properties of knowledge, belief and certainty.

The aim of this paper is to develop a concurrent dynamic logic of knowledge, belief and certainty for MAS (Multi-Agent System). And we will also present the proof system and then show its soundness and completeness. Our logic has the following significance. Firstly, it extends the logic of knowledge, belief, and certainty with concurrent dynamic modality [α]. Secondly, the knowledge operator and common knowledge operator satisfy the $S5_n{}^C$ axioms, but the belief operators B_i and certainty operator C_i need not satisfy the $S5_n$. Thirdly, the modality [α] has concurrent properties and we consider also the loop execution of action in order to express some statements such as "φ is true in every state that is reached after executing α an even number of times", which can not be expressed in the logic of [15].

2   A Concurrent Dynamic CDKBC Model for MAS

In order to present a concurrent dynamic model, we see first an example.

Example 1. Let us consider the scenario depicted in Figure 1: There are four squares in a 2×2 square, the bottom right one is dirty. Suppose that there are two robots and robot 1 stands at the top left one and robot 2 stands at the top right one. The robots can move up or down or left or right and can move from one square to another neighboring square. But the robots' moving-direction sensors are broken. And so they may get incorrect perception about its location. For instance, if a robot moves down from square (0,0) into square (0,1), it may think it moved right into square (1,0); thus the robot may confuse square (0,1) with (1,0). In the same way, the robot cannot correctly distinguish square (0,0) from (1,1). Therefore, what a robot perceive of the environment maybe is different from the environment.

An action is robot's moving from one square into another square. For example, robot 1 moving left is an action. Thus, the action set ACT1 of the first robot consists of four actions, and robot 1 moving left and right and up and down are denoted respectively by a_{11} and a_{12} and a_{13} and a_{14}. The action set ACT2 of the second robot consists of four actions too, and robot 2 moving left and right and up and down are denoted respectively by a_{21} and a_{22} and a_{23} and a_{24}. So the action set ACT consists of eight actions. We assume the goal of system (or say the task of two robots) is that two robots (one with a broom and one with a dustpan) move into the same square where is dirty so that they can clean the square. To say in detail, in every step, two robots concurrently excute the actions chosen nondeterministicly from four actions respectively, and after some steps like this, they stand in the same dirty square and all feel certainty that they are standing in the same square and believe the square is dirty.

(0,0), clean	(1,0), clean
(0,1), clean	(1,1), dirty

Fig. 1. The scenario with four squares where three squares are clean and the other is dirty

The example is related to "certainty' , 'belief' and concurrent actions. In order to solve situations like example above, we present a model, called Concurrent Dynamic model of knowledge, certainty, and belief, which consists of an environment and n agents.

We denote these n agents by the set $\{1,2,...,n\}$. Assume that every agent may not perceive of the environment completely and correctly, because of the device, network and etc. So what an agent perceives of environment maybe is different from environment. We divide a system state into two parts which one is environment state part (external state) and the other is agent state part (internal state) .An environment state is divided into n parts, which the i-th part is visible to agent i and is denoted as s_{en}^i. The environment state is denoted by s_{en}. It is obvious that $s_{en}=(s_{en}^1, s_{en}^2,..., s_{en}^n)$. Notice that s_{en}^i and s_{en}^j are overlapping partly, where $i,j =1,2,...n$. After perceiving the i-th part of environment, agent i gets its local state which is denoted as s_{in}^i $(i=1,2,...,n)$. We call s_{in} as an internal state and s_{en} as an external state or environment state. A global state s is a pair of an environment state and an internal state. Namely $s=(s_{en}, s_{in})=(s_{en}^1, s_{en}^2,..., s_{en}^n, s_{in}^1, s_{in}^2,..., s_{in}^n)$. We denote the set of all global states of system by S.

Given a global state $s=(s_{en}, s_{in})= (s_{en}^1, s_{en}^2,..., s_{en}^n, s_{in}^1, s_{in}^2,..., s_{in}^n)$, we denote s_{en}^i and s_{in}^i by $env^i(s)$, $int^i(s)$ $(i=1,2,...,n)$. Next we define three kinds of relations k_i, B_i and C_i .

Let s and t be two global states. For every agent i, if $int^i(s) = int^i(t)$, then we say that s and t are indistinguishable to agent i from the view point of knowledge and write $s\sim_k^i t$. Define relation $k_i=\{(s, t)| s\in S, t\in S, s\sim_k^i t \}(i=1,2,...,n)$. We denote the set of all relations k_i ($i=1,2,...,n$) by k.

For every agent i, if $int^i(t)=env^i(t)=env^i(s)$, then we say that s and t are indistinguishable to agent i from the view point of belief and write $s\sim_B^i t$. Define relation $B_i=\{(s, t)| s\in S, t\in S, s\sim_B^i t \}(i=1,2,...,n)$. We denote the set of all relations B_i ($i=1,2,...,n$) by B.

For every agent i, if $int^i(t)=env^i(s)$, then we say that s and t are indistinguishable to agent i from the view point of certainty and write $s\square_C^i t$. Define relation $C_i=\{(s, t)| s\in S, t\in S, s\sim_C^i t \}(i=1,2,...,n)$. We denote the set of all relations C_i ($i=1,2,...,n$) by C.

In our model, we assume that knowledge, belief and certainty can change. Namely, our model is dynamic. The change in knowledge, belief and certainty results from actions performed by agents. We denote the set of all actions of all agents by Act. For every action $\alpha\in$ Act, we introduce a binary relation $[\alpha]$ between a state and a set of states, which is associated with action α. $(s,W)\in[\alpha]$ iff executing the action α in state s leads to the set W of states. We denote the set of all binary relations $[\alpha]$ ($\alpha\in$ Act) by Λ.

Now we define concurrent dynamic model. Let Φ is a set of atomic propositions, used to describe basic facts about the system. Formally, a concurrent dynamic model

CDKBC of knowledge, belief and certainty for n agents over Φ and Act is a tuple $M=(S, V, K, B, C, \Lambda)$ such that:

S is the set of all global states of system. Every element of S can be denoted by $s=(s_{en}, s_{in})= (s_{en}^1, s_{en}^2,\ldots, s_{en}^n, s_{in}^1, s_{in}^2,\ldots, s_{in}^n)$.

$V: \Phi \rightarrow 2^S$ is a function, which to every atomic proposition, gives the set of all states in which the atomic proposition is true. We define a valuation function π, which assigns truth values to atomic propositions at every environment state. Thus, for every atomic proposition p and a global state s, we have $\pi(s)(p)= \pi(s_{en})(p) \in \{\text{true, false}\}$, which $s_{en}= (s_{en}^1, s_{en}^2,\ldots, s_{en}^n)$ is the environment state. Therefore, two different states that have the same environment states have the same truth valuation. Obviously, we can get V from π, and also get π from V. Thus, sometimes, we write $M=(S, V, K, B, C, \Lambda)$ as $M=(S, \pi, K, B, C, \Lambda)$.

K is the set of all relations K_i associated with agent i

B is the set of all relations B_i associated with agent i

C is the set of all relations C_i associated with agent i

Λ is the set of all binary relations $[\alpha]$ associated with the action α ($\alpha \in$ Act)

Let $M_n(\Phi,$ Act) be the class of all *CDKBC* models for n agents over Φ and Act. For notational convenience, we write M_n instead of $M_n(\Phi,$ Act).

Proposition 1. The relations K_i is an equivalent relation.
It is easy to prove the proposition by their definitions.
Let $R_K^i(s)=\{t \mid (s,t) \in K_i\}$, $R_B^i(s) = \{t \mid (s,t) \in B_i\}$ and $R_C^i(s)= \{t \mid (s,t) \in C_i\}$.

Proposition 2. For any model $M \in M_n$ and any state $s \in S$, we have $R_B^i(s) \subseteq R_C^i(s)$.
It is not difficult to prove the proposition by the definitions of $R_C^i(s)$ and $R_B^i(s)$.

3 Concurrent Dynamic Logic of Knowledge, Belief and Certainty

In this section, we introduce a Concurrent Dynamic Logic of Knowledge, Belief and Certainty, which is called *CDKBC* logic. The semantics of *CDKBC* logic is given in terms of *CDKBC* model above.

3.1 Syntax

Given a set P of atomic propositions, the language L_n of *CDKBC* logic is defined by the following BNF grammar:

   <wff>::= any element of P | ¬<wff> | <wff> $\wedge$ <wff> | K_i<wff> | B_i<wff> | C_i<wff> | E_G<wff> | C_G<wff> | [α]<wff>

The actions are defined as follows:

$\alpha := ?\alpha \mid (\alpha;\beta) \mid (\alpha \| \beta) \mid (\alpha \cup \beta) \mid \alpha^*$

The modality K_i allows us to represent the information that is instantaneously knowable or perceivable about the environment. Thus, the formula $K_i\varphi$ means φ is knowable or perceivable about environment to agent i. $B_i\varphi$ means that the agent i believes the property φ. $C_i\varphi$ means that the agent i feels certainty of the fact φ. Here we assume that every agent may not perceive of the environment completely and correctly, because of the device, network and etc. So the agent who believes a proposition may be not aware that the proposition might be false. G be a nonempty subset of $\{1,2,...,n\}$. $E_G\varphi$ means that every agent in G knows φ, $C_G\varphi$ means that φ is common knowledge of all agents in G.. The formula $[\alpha]\varphi$, where φ is a formula and α is an action in Act, is interpreted to mean that φ is true after action α.

The action $?\alpha$ is a test, action $(\alpha;\beta)$ is a sequential execution, $(\alpha||\beta)$ is a concurrent execution, action $(\alpha\cup\beta)$ is a nondeterministic choice, and action α^* is loop execution (meaning repeating 0 or more times, nondeterministically).

3.2 Semantics

In order to define the semantics of *CDKBC* logic (on *CDKBC* model), we introduce two operators between relations. Assume that R , R_1 and R_2 are binary relations, we define the composition $R_1 \circ R_2$ of two relations R_1 and R_2 as a binary relation $\{(u, W) \mid \exists W_1$ such that $(u, W_1) \in R_1$ and $\forall v \in W_1, \exists W_2$ such that $(v, W_2) \in R_2, W = \cup_{v \in W1}\{ W_2 \mid (v, W_2) \in R_2\}$, and define $R_1 \bullet R_2$ as a binary relation $\{(u, W) \mid \exists W_1$ such that $(u, W_1) \in R_1, \exists W_2$ such that $(u, W_2) \in R_2, W = W_1 \cup W_2\}$[15]. R^n is defined inductively: $R^1 = R$ and $R^{n+1} = R^n \circ R$ [7].

According to the definitions above, the binary relation $[\alpha]$ can be extended to all actions as follows:

$[\alpha;\beta] = [\alpha] \circ [\beta]$
$[\alpha||\beta] = [\alpha] \bullet [\beta]$
$[\alpha\cup\beta] = [\alpha] \cup [\beta]$
$[\alpha^*] = \cup_{n\geq 0} [\alpha]^n$

Let G be a nonempty subset of $\{1,2,...,n\}$. Define a state to be G-reachable from state s in k steps ($k\geq1$) if there exist states $s_0,s_1,...,s_k$ such that $s_0=s$, $s_k= t$ and for all j with $0\leq j\leq k-1$, there exists $i\in G$ such that $s_j\sim^i_K s_{j+1}$. We say t is G-reachable from s if t is G-reachable from s in k steps for some $k\geq1$.

Now given $M=(S, \pi, K, B, C, \Lambda)$, we define the interpretation of *CDKBC* logic formula φ by induction on the structure of φ as follows:

$(M, s)\models p$ iff $\pi(s)(p)$=true
$(M, s)\models \neg \varphi$ iff $(M, s)\not\models \varphi$
$(M, s)\models \varphi\wedge \psi$ iff $(M, s)\models \varphi$ and $(M, s)\models \psi$
$(M, s)\models K_i\varphi$ iff $(M, t)\models \varphi$ for all t such that $s\sim^i_K t$
$(M, s)\models B_i\varphi$ iff $(M, t)\models\varphi$ for all t such that $s\sim^i_B t$
$(M, s)\models C_i\varphi$ iff $(M, t)\models \varphi$ for all t such that $s\sim^i_C t$
$(M, s)\models E_G\varphi$ iff $(M, s)\models K_i\varphi$ for all $i\in G$.
$(M, s)\models C_G\varphi$ iff $(M, t)\models \varphi$ for all t that are G-reachable from s.
$(M, s)\models [\alpha]\varphi$ iff for all W such that $(s, W)\in [\alpha]$: $\exists t\in W, (M, t)\models\varphi$
$(M, s)\models [?\psi]\varphi$ iff $(M, s)\models \psi\Rightarrow\varphi$

Now we explicate some intuitions about knowledge, certainty, and belief. The semantic of $K_i\varphi$ means that not only φ is true of the environment, but also agent i perceive correctly the environment. The semantic of $C_i\varphi$ captures the intuition behind 'certainty' that, to agent i, the facts of which it feels certain appear to knowledge. Thus 'John is certain that' is equivalent to 'John is certain that John knows that'. The intuition behind 'belief' is that, agent i believes property φ but φ may be false of environment. For example, a robot believes that there is an obstacle in front of it, but maybe there is none.

3.3 A Case Study

We take still the example above. In the example, a global state can be represented as $((x_{1e}, y_{1e}), z_1, (x_{2e}, y_{2e}), z_2, (x_1, y_1), Z_1, (x_2, y_2), Z_2)$, Where $x_{1e}, y_{1e}, z_1, x_{2e}, y_{2e}, z_2, x_1, y_1, Z_1, x_2, y_2, Z_2$ are Boolean value, (x_{1e}, y_{1e}) and (x_{2e}, y_{2e}) indicate the locations that the first robot and the second robot are at respectively. z_1 expresses whether the square of the robot 1 is dirty, z_2 expresses whether the square of the robot 2 is dirty, (x_1, y_1) and (x_2, y_2) are used for what the first robot and the second robot perceive about their locations respectively.

Z_1 is the first robot's conjecture about the invisible part z_1 of the environment. Z_2 is the second robot's conjecture about the invisible part z_2 of environment.

According to the discussion above, we have several constraints on these variables:

$x_{1e} \wedge y_{1e} \Leftrightarrow z_1$ and $x_{2e} \wedge y_{2e} \Leftrightarrow z_2$ hold, which indicates that only square (1,1) is dirty; $(x_{1e} \Leftrightarrow y_{1e}) \Leftrightarrow (x_1 \Leftrightarrow y_1)$ and $(x_{2e} \Leftrightarrow y_{2e}) \Leftrightarrow (x_2 \Leftrightarrow y_2)$ hold, which means that robots can distinguish two neighboring squares;

Assume that we have a set Φ of primitive propositions, which we can think of as describing basic facts about the system. We may here take $\{p, q\}$ as the set Φ of primitive propositions, which p represents if two robots stand in the same room and q represents if the square where the first robot is standing is dirty. So, for the environment state $s_{env}=((1,0), 0, (1,0), 0)$, we may naturally define $\pi(s_{env})(p)=$true and $\pi(s_{env})(q)=$false. We assume the goal of system is that two robots (one with a broom and one with a dustpan) move into the same square where is dirty so that they can clean the square. To say in detail, in every step, two robots execute concurrently the actions chose nondeterministicly from four actions respectively, after some steps like this, they stand in the same dirty square and two robots feel certainty that they are standing in the same square, and believe the square is dirty. Thus the goal can be represented as $((a_{11} \cup a_{12} \cup a_{13} \cup a_{14}) \| (a_{21} \cup a_{22} \cup a_{23} \cup a_{24}))^*(p \wedge q \wedge C_1(p \wedge q) \wedge C_2(p \wedge q) \wedge B_1(p \wedge q) \wedge B_2(p \wedge q))$.

4   The CDKBC Proof System

In this section, we discuss the *CDKBC* proof system. It is mainly based on the dynamic logic [7], the dynamic epistemic logic [8], the concurrent logic [15] and the logic of knowledge, belief and certainty [16].

The *CDKBC* proof system consists of following axioms and inference rules:

A1. All tautologies of propositional calculus
A2. $(K_i\varphi \wedge K_i(\varphi \Rightarrow \psi)) \Rightarrow K_i(\psi)$, $i=1,2,\ldots,n$
A3. $K_i\varphi \Rightarrow \varphi$, $i=1,2,\ldots,n$
A4. $K_i\varphi \Rightarrow K_iK_i\varphi$, $i=1,2,\ldots,n$
A5. $\neg K_i\varphi \Rightarrow K_i\neg K_i\varphi$, $i=1,2,\ldots,n$
A6. $E_G\varphi \Leftrightarrow \wedge_{i \in G}K_i\varphi$

A7. $C_G\varphi \Rightarrow (\varphi \wedge E_G C_G \varphi)$

A8. $(B_i\varphi \wedge B_i(\varphi \Rightarrow \psi)) \Rightarrow B_i(\psi)$, $i=1,2,\ldots,n$

A9. $(C_i\varphi \wedge C_i(\varphi \Rightarrow \psi)) \Rightarrow C_i(\psi)$, $i=1,2,\ldots,n$

A10. $C_i\varphi \Rightarrow C_i K_i\varphi$, $i=1,2,\ldots,n$

A11. $C_i\varphi \Rightarrow B_i\varphi$, $i=1,2,\ldots,n$

A12. $[\alpha](\varphi \Rightarrow \psi) \Rightarrow ([\alpha]\varphi \Rightarrow [\alpha]\psi)$

A13. $[\alpha](\varphi \wedge \psi) \Leftrightarrow ([\alpha]\varphi \wedge [\alpha]\psi)$

A14. $[?\varphi]\psi \Leftrightarrow (\varphi \Rightarrow \psi)$

A15. $[\alpha;\beta]\varphi \Leftrightarrow [\alpha][\beta]\varphi$

A16. $[\alpha|\beta]\varphi \Leftrightarrow ([\alpha]\varphi \vee [\beta]\varphi)$

A17. $[\alpha \cup \beta]\varphi \Leftrightarrow ([\alpha]\varphi \wedge [\beta]\varphi)$

A18. $\varphi \wedge [\alpha][\alpha^*]\varphi \Leftrightarrow [\alpha^*]\varphi$

A19. $\varphi \wedge [\alpha^*](\varphi \Rightarrow [\alpha]\varphi) \Rightarrow [\alpha^*]\varphi$

A20. $K_i[\alpha]\varphi \Rightarrow [\alpha]K_i\varphi$, $i=1,2,\ldots,n$

R1. From φ and $\varphi \Rightarrow \psi$ infer $\varphi \Rightarrow \psi$

R2. From φ infer $K_i\varphi \wedge C_i\varphi$

R3. Form $\varphi \Rightarrow E_G(\psi \wedge \varphi)$ infer $\varphi \Rightarrow C_G\psi$

R4. From φ infer $[\alpha]\varphi$

Note that because we have that $\vdash C_i\varphi \Rightarrow B_i\varphi$, the proof system doesn't need to include the inference rule: from φ infer $B_i\varphi$.

Proposition 3. For any φ, $\psi \in L_n$, any structure $M \in M_n$, and any agent i,

(a) $M \models C_i\varphi \Rightarrow B_i\varphi$.

(b) $M \models C_i\varphi \Rightarrow C_i K_i\varphi$.

(c) $M \models [\alpha](\varphi \wedge \psi) \Leftrightarrow ([\alpha]\varphi \wedge [\alpha]\psi)$.

(d) $M \models [\alpha|\beta]\varphi \Leftrightarrow ([\alpha]\varphi \vee [\beta]\varphi)$.

(e) $M \models \varphi \wedge [\alpha][\alpha^*]\varphi \Leftrightarrow [\alpha^*]\varphi$.

Proof.

(a) Assume that $(M,s) \models C_i\varphi$, then $\forall t \in R_B{}^i(s)$, by Proposition 2, we have $t \in R_C{}^i(s)$. According to the assumption, it follows that $(M,t) \models \varphi$. By the semantics of $B_i\varphi$, it follows that $(M,s) \models B_i\varphi$. Using the same way as above, we have that if $(M,s) \models K_i\varphi$ then $(M,s) \models C_i\varphi$.

(b). Suppose that $(M,s) \models C_i\varphi$, then for every state t with $int^i(t) = env^i(s)$, we have that $(M,t) \models \varphi$. To prove $(M,s) \models C_i K_i\varphi$, we must prove that ,for every state u with $int^i(u) = env^i(s)$, we have that $(M,u) \models K_i\varphi$. It suffices to prove that, for every state v with $int^i(v) = int^i(u)$, $(M,v) \models \varphi$. However, we can get $int^i(v) = env^i(s)$ from $int^i(u) = env^i(s)$ and $int^i(v) = int^i(u)$. Thus, from $(M,s) \models C_i\varphi$, we have that $(M,v) \models \varphi$. This proves that $M \models C_i\varphi \Rightarrow C_i K_i\varphi$.

(c) We first prove $M \models [\alpha](\varphi \wedge \psi) \Rightarrow ([\alpha]\varphi \wedge [\alpha]\psi)$. If $M \models [\alpha](\varphi \wedge \psi)$, then for all $s \in S$, $(M, s) \models [\alpha](\varphi \wedge \psi)$, and so $(M,t) \models \varphi$ and $(M,t) \models \psi$ for all t such that $(s, W) \in [\alpha]$ and $t \in W$. Hence, $(M, s) \models [\alpha]\varphi$ and $(M, s) \models [\alpha]\psi$. So $M \models ([\alpha]\varphi \wedge [\alpha]\psi)$, and it follows that $M \models [\alpha](\varphi \wedge \psi) \Rightarrow ([\alpha]\varphi \wedge [\alpha]\psi)$. We can similarly prove $M \models ([\alpha]\varphi \wedge [\alpha]\psi) \Rightarrow [\alpha](\varphi \wedge \psi)$. Therefore, $M \models [\alpha](\varphi \wedge \psi) \Leftrightarrow ([\alpha]\varphi \wedge [\alpha]\psi)$.

(d) Let $M \models [\alpha \| \beta] \varphi$, then for all $s \in S$, $(M, s) \models [\alpha \| \beta] \varphi$. Hence for all W such that $(s, W) \in [\alpha \| \beta]$, we have that $\exists t \in W$, $(M, t) \models \varphi$, where $(s, W_1) \in [\alpha]$ and $(s, W_2) \in [\beta]$ and $W = W_1 \cup W_2$. Hence, if $t \in W_1$, then $(M, s) \models [\alpha] \varphi$, and if $t \in W_2$, then $(M, s) \models [\beta] \varphi$. Namely, $(M, s) \models ([\alpha] \varphi \vee [\beta] \varphi)$. Therefore, $M \models [\alpha \| \beta] \varphi \Rightarrow ([\alpha] \varphi \vee [\beta] \varphi)$. In a similar way, by semantics of action, we can also prove $M \models ([\alpha] \varphi \vee [\beta] \varphi) \Rightarrow [\alpha \| \beta] \varphi$.

(e) Let $M \models \varphi \wedge [\alpha][\alpha^*] \varphi$, then for all $s \in S$, we have that $(M, s) \models \varphi \wedge [\alpha][\alpha^*] \varphi$. It follows that $(M, s) \models \varphi$ and $(M, s) \models [\alpha][\alpha^*] \varphi$. By the definition of $[\alpha^*]$, $(M, s) \models [\alpha^*] \varphi$. Therefore, $(M, s) \models \varphi \wedge [\alpha][\alpha^*] \varphi \Rightarrow [\alpha^*] \varphi$ and $M \models \varphi \wedge [\alpha][\alpha^*] \varphi \Rightarrow [\alpha^*] \varphi$. The proof of other direction is easy. $\square$

Theorem 1. The *CDKBC* proof system is sound with respect to M_n.

By the Proposition 3 and the theorem 3.3.1 in [3], the proof of the soundness is straightforward.

5   Completeness and Complexity

In this section, we prove some fundamental results about the CDKBC proof system.

5.1   Completeness

One of our important results is the completeness of the CDKBC proof system. The proof of completeness is based on [7,15,18]. We first extend Fischer-Ladner closure.
5.1.1 The extended Fischer-Ladner closure.
We start by defining two functions:
$EFL: \Phi \rightarrow 2^{\Phi}$
$EFL_1: \{[\alpha] \varphi | \alpha \in Act, \varphi \in \Phi\} \rightarrow 2^{\Phi}$
The function EFL is defined inductively as follows:
$EFL(p) = \{p\}$ p is an atomic proposition
$EFL(\neg \varphi) = \{\neg \varphi\} \cup EFL(\varphi)$
$EFL(\varphi \wedge \psi) = \{\varphi \wedge \psi\} \cup EFL(\varphi) \cup EFL(\psi)$
$EFL(K_i \varphi) = \{K_i \varphi\} \cup EFL(\varphi) \cup EFL(C_i \varphi)$
$EFL(B_i \varphi) = \{B_i \varphi\} \cup EFL(\varphi)$
$EFL(C_i \varphi) = \{C_i \varphi\} \cup EFL(\varphi) \cup EFL(B_i \varphi)$
$EFL(C_G \varphi) = \{C_G \varphi\} \cup EFL(\varphi) \cup \{K_i C_G \varphi . i \in G\}$
$EFL([\alpha] \varphi) = EFL_1([\alpha] \varphi) \cup EFL(\varphi)$
The function EFL_1 is defined inductively as follows:
(a) $EFL_1([\alpha] \varphi) = \{[\alpha] \varphi\}$ α is an atomic program (or action)
$EFL_1([\alpha \cup \beta] \varphi) = \{[\alpha \cup \beta] \varphi\} \cup EFL_1([\alpha] \varphi) \cup EFL_1([\beta] \varphi)$
$EFL_1([\alpha \cap \beta] \varphi) = \{[\alpha \cap \beta] \varphi\} \cup EFL_1([\alpha] \varphi) \cup EFL_1([\beta] \varphi)$
$EFL_1([\alpha; \beta] \varphi) = \{[\alpha; \beta] \varphi\} \cup EFL_1([\alpha][\beta] \varphi) \cup EFL_1([\beta] \varphi)$
$EFL_1([\alpha^*] \varphi) = \{[\alpha^*] \varphi\} \cup EFL_1([\alpha][\alpha^*] \varphi)$
$EFL_1([?\psi] \varphi) = \{[?\psi] \varphi\} \cup EFL_1(\psi)$
To any formula φ, we call $EFL(\varphi)$ as the extended Fischer-Ladner closure of φ.

Proposition 4. If $\varphi \in L_n$, then $\varphi \in EFL(\varphi)$.

5.1.2 Filtration

Let Kripke structure $M=(S, V, K, B, C, \Lambda)$, to prove the completeness, we first construct the structure $M'=(S', V', K', B', C', \Lambda')$ from M, where

$S'=\{\Gamma \mid \Gamma$ is maximal consistent set$\}$;

$V'(\varphi)=\{\Gamma \mid \varphi \in \Gamma\}$;

K_i'is defined as follows : $\Gamma \sim^i_K \Delta$ iff $\{ w \mid K_i w \in \Gamma\}=\{ w \mid K_i w \in \Delta\}$, B_i'and C_i'are defined similarly ;

$m'(\alpha)=\{(\Gamma, W) \mid [\alpha]\varphi \in \Gamma$, and $W=\{\Delta \mid \varphi \in \Delta\}\}$.

Given a formula and a Kripke structure M', according to the section 5.1.1, we define a binary relation $\equiv$ on states $M'$of by:

$u \equiv v$ iff $\forall \psi \in EFL(\varphi)(u \in V'(\psi) \Leftrightarrow v \in V'(\psi))$.

Let

$[u]= u/_{EFL(\varphi)}=\{v \mid v \equiv u \}$;

$E([u])=\{ v \mid v \in S', [v]=[u]\}$;

$[W]=\{[t] \mid t \in W\}$;

$E([W])=\{ v \mid v \in E([t]), t \in W \}$;

By $EFL(\varphi)$, we have a new structure $M'/_{EFL(\varphi)}=(S'/_{EFL(\varphi)}, V'/_{EFL(\varphi)}, K'_{EFL(\varphi)}, B'/_{EFL(\varphi)}, C'/_{EFL(\varphi)}, \Lambda'/_{EFL(\varphi)})$, where:

$S'/_{EFL(\varphi)}=\{[u] \mid u \in S'\}$;

$V'/_{EFL(\varphi)}(\psi)=\{[u] \mid u \in V'(\psi)\}$;

$K_i'/_{EFL(\varphi)}$ is defined : $[u]\sim^i_{K/EFL(\varphi)}[v]$ iff $u \sim^i_K v$;

$B_i'/_{EFL(\varphi)}, C_i'/_{EFL(\varphi)}$ are defined similarly ;

$m'/_{EFL(\varphi)}(\alpha)=\{([u], [W]) \mid (v, W)\in m'(\alpha)\}$.

To be similar to [7], we have following propositions :

Proposition 5. For all $\psi \in EFL(\varphi)$, $u \in V'(\psi)$ iff $[u]\in V'/_{EFL(\varphi)}(\psi)$.

Proposition 6. For all $[\alpha]\psi \in EFL(\varphi)$

If $(u, W)\in m'(\alpha)$ then $([u], [W])\in m'/_{EFL(\varphi)}(\alpha)$;

If $([u], [W])\in m'/_{EFL(\varphi)}(\alpha)$, and $u \in V'/_{EFL(\varphi)}([\alpha]\psi)$, then $W \subseteq V'/_{EFL(\varphi)}(\psi)$.

5.1.3 Completeness.

We can prove the following small model theorem easily.

Theorem 2 (Small Model Theorem). If φ is satisfied in M_n, then φ is satisfied in a CDKBC model with no more than $2^{|\varphi|}$ states.

By [18], in order to prove the completeness, we only need to prove the truth lemma for $M'/_{EFL(\varphi)}$, which difficulty is to prove the case $<\alpha>C_G\psi$ of φ. To ensure that the true lemma holds for sentences of the form $<\alpha>C_G\psi$, we need the definition of good path: a good path from $\Gamma \in M'/_{EFL(\varphi)}$ for $<\alpha>C_G\psi$ is a path $\Gamma=\Gamma_0\sim_{K1}\Gamma_1\sim_{K2}\Gamma_2...\sim_{Kn}\Gamma_n$ ($K_i'\in G$, $i=1,2...,n$) which is satisfied with three following conditions:

(1)   There are n actions $\alpha_i(i=1,2...,n)$ such that $\alpha=\alpha_0\sim_{K1}\alpha_1\sim_{K2}\alpha_2...\sim_{Kn}\alpha_n$ ($K_i'\in G$, $i=1,2...,n$);

(2) $<\alpha_i>T\in \Gamma_i(i=1,2...,n)$;

(3) $<\alpha_n>\psi \in \Gamma_n$.

For a good path, we have two similar propositions in [18]

Proposition 7. Suppose $[\alpha]C_G\psi \in EFL(\varphi)$, if there is a good path from $\Gamma \in M'/_{EFL(\varphi)}$ for $<\alpha>C_G\psi$ then $<\alpha>C_G\psi \in \Gamma_0$.

Proposition 8. If $\Gamma_0 \wedge <\alpha>CG\psi$ is a consistent formula then there is a good path from $\Gamma \in M'/EFL(\varphi)$ for $<\alpha>C_G\psi$.
  This ensures that the truth lemma holds.

Proposition 9. (Truth Lemma) If φ is a consistent formula of L_n, then for all $\psi \in EFL(\varphi)$ and $[\Gamma] \in M'/_{EFL(\varphi)}$ then it holds that $(M'/_{EFL(\varphi)}, [\Gamma]) \models \psi$ iff $\psi \in [\Gamma]$.

Theorem 3. For any $\varphi \in L_n$, if $\models \varphi$ then $\vdash \varphi$.
Proof. Suppose it does not hold that $\vdash \varphi$, then $\neg\varphi$ is a consistent formula and so must be in some maximal consistent formula of L_n. By proposition 4, it follows that $\neg\varphi \in EFL(\neg\varphi)$, so $\neg\varphi \in [\Gamma]$. By proposition 9, it holds that $(M'/_{EFL(\varphi)}, [\Gamma]) \models \neg\varphi$, this contradicts with $\models \varphi$. Therefore it holds that $\vdash \varphi$.

5.2 Complexity

Proposition 10. A formula φ is satisfiable with respect to CDKBC M_n iff φ is M_n-consistent.

Proof. Let φ be satisfiable with respect to M_n. If φ is not M_n-consistent, then $\neg\varphi$ is provable in M_n. By the Theorem 1, $\neg\varphi$ is valid with respect to M_n, and so φ is not satisfiable with respect to M_n. Hence φ is M_n-consistent. For the other direction, if φ is M_n-consistent, then using the same method in section 5.1, we can prove that φ is satisfiable with respect to $M_{n.}$.

Proposition 11. If φ is M_n-consistent, then φ is satisfiable in some CDKBC model M with at most $2^{|\varphi|}$ states.

Proof. Let $sub(\varphi)$ be the set of all subformulas of φ and $sub^+(\varphi)=sub(\varphi) \cup \{\neg\psi|\psi \in sub(\varphi)\}$. And let $Acon(\varphi)$ be the set of maximal CDKBC M_n-consistent subsets of $sub^+(\varphi)$. Being similar to Lemma 3.1.2 of [3], we can show that every CDKBC M_n-consistent subsets of $sub^+(\varphi)$ can be extend to a maximal CDKBC M_n-consistent subset of $sub^+(\varphi)$, which is an element of $Acon(\varphi)$. Because ψ is M_n-consistent, a member of $Acon(\varphi)$ contains only ψ or $\neg\psi$ for every formula $\psi \in sub(\varphi)$. So $|Acon(\varphi)|$ is at most $2^{|sub(\varphi)|}$. Because $|sub(\varphi)|<=|\varphi|$, $|Acon(\varphi)|<=2^{|\varphi|}$.
  We now construct a model $M_\varphi=(S_\varphi, V, K, B, C, \Lambda)$, which is almost the same as that of section 5.1, except that we take $S_\varphi=\{s_v|v \in Acon(\varphi)\}$. It is obvious that $|S_\varphi|=|Acon(\varphi)|<=2^{|\varphi|}$. Like the proposition 6, we can show that if $v \in Acon(\varphi)$, then for all $\psi \in sub^+(\varphi)$, we have $(M_\varphi, s_v) \models \psi$ iff $\psi \in v$. From proposition 6, we have φ is satisfiable in M_φ. The proposition has been proved.

Proposition 12. A formula φ is satisfiable with respect to CDKBC M_n iff φ is satisfiable in some *CDKBC* model with at most $2^{|\varphi|}$ states.

Proof. If φ is satisfiable with respect to *CDKBC* M_n, by proposition 10, then φ is M_n-consistent. And by proposition 11, φ is satisfiable in some *CDKBC* model M with at most $2^{|\varphi|}$ states. For the other direction, if φ is satisfiable in some *CDKBC* model M with at most $2^{|\varphi|}$ states, then obviously φ is also satisfiable with respect to *CDKBC* M_n.

Now we discuss the validity problem for M_n (or the provability problem for *CDKBC* proof system).

Theorem 4. The validity problem for *CDKBC* M_n and provability problem for *CDKBC* system are decidable.

Proof. To decide if φ is provable in *CDKBC* proof system needs only to check if $\neg\varphi$ is *CDKBC* consistent. We now discuss how to check if $\neg\varphi$ is CDKBC consistent.

By the Propositions 10, 11, 12, we have that $\neg\varphi$ is *CDKBC* consistent iff $\neg\varphi$ is satisfiable with respect to some *CDKBC* model M with at most $2^{|\varphi|}$ states. So we can simple construct every model with at most $2^{|\varphi|}$ states. Note that the number of these models is finite. Then we can check if $\neg\varphi$ is true at some state in one of these models.

Theorem 5. The complexity of provability problem for *CDKBC* system is EXPTIME-complete.

Proof. Because Propositional Dynamic Logic (PDL) is part of *CDKBC* logic, Fisher and Ladner has proved that provability problem of proof system for PDL is EXPTIME-complete [6], which implies that the complexity of provability problem for *CDKBC* system is EXPTIME-complete.

6 Conclusions

The paper extends logic of Knowledge, Belief and Certainty from one agent to multi-agent systems, and gives a good combination between logic of knowledge, belief and certainty in multi-agent system and actions that have concurrent and dynamic properties. Based on it, we present a concurrent dynamic logic of knowledge, belief and certainty for MAS, which is called *CDKBC* logic. A *CDKBC* model is given for interpreting this logic. The relations between Knowledge, Belief and Certainty are also discussed, which certainty appears to knowledge and certainty entails belief. We construct a *CDKBC* proof system for the logic and show the proof system is sound and complete. We also prove the validity problem for the system is EXPTIME-complete and give an application of the logic.

References

1. Hintikka, J.: Knowledge and Belief. Cornell University Press, Ithaca (1962)
2. Lenzen, W.: Recent work in epistemic logic. Acta Philosophica Fennica 30, 1–219 (1978)
3. Fagin, R., Halpern, J., Moss, Y., Vardi, M.: Reasoning about knowledge. MIT Press, Cambridge (1995)

4. Meyer, J., Van der Hoek, W.: Epistemic logic for AI and computer science. Cambridge Tracts in theoretical computer science, vol. 41. Cambridge University Press, Cambridge (1995)
5. Aumann, R., Brandenburger, A.: Epistemic conditions for nash equilibrium. Econometrica 63, 116–1180 (1995)
6. Fisher, M.J., Ladner, R.E.: Propositional dynamic logic of regular programs. Journal of computer and system science 18(2), 194–211 (1977)
7. Harel, D., Kozen, D., Tiuryn, J.: Dynamic logic. MIT Press, Cambridge (2000)
8. van Ditmarsch, H., van der Hoek, W., Kooi, B.: Dynamic epistemic logic with assignment. In: Dignum, F., Dignum, V., Koenig, S., Kraus, S., Singh, M.P., Wooldridge, M. (eds.) Proceedings of the Fourth International Joint Conference on Autonomous Agents and Multi-Agent Systems, vol. 1, pp. 141–148. ACM Inc., New York (2005)
9. Plaza, J.: Logics of public communications. In: Proceedings of the 4th international symposium on methodologics for intelligent system, pp. 201–216 (1989)
10. Gerbrandy, J.: Bisimulation on planet Kripke. PhD thesis, university of Amsterdam, ILLC dissertation series DS-1999-01 (1999)
11. Van Bentem, J.: Logics of information update. In: Proceedings of TARK VIII, Los Altos, pp. 51–88 (2001)
12. van Ditmarsch, H.: Description of game actions. Journal of logic, language and information 11, 349–365 (2002)
13. Peleg, David: Concurrent dynamic logic. Journal of the ACM 34(2), 450–479 (1987)
14. Peleg, David: Communication in concurrent dynamic logic. J. Comput. Syst. Sci. 35, 23–58
15. van Ditmarsch, H., van der Hoke, W., Kooi, B.: Concurrent dynamic epistemic logic for MAS. In: Knowledge contributors, Dordrecht, pp. 45–82 (2003)
16. Su, K., Sattar, A., et al.: A computational grounded logic of knowledge, belief and certainty. In: AAMAS 2005, pp. 149–156 (2005)
17. Lamarre, P., Shoham, Y.: Knowledge, certainty, belief, and conditionalisation. In: KR 1994, pp. 415–424. Morgan Kaufmann, San Francisco (1994)
18. Baltag, A., Moss, L., Solecki, S.: The logic of public announcements, common knowledge and private suspicions. CWI Technical Report SEN-R9922 (1999)

Enumerating Unlabeled and Root Labeled Trees for Causal Model Acquisition

Yang Xiang, Zoe Jingyu Zhu, and Yu Li

University of Guelph, Canada

Abstract. To specify a Bayes net (BN), a conditional probability table (CPT), often of an effect conditioned on its n causes, needs to be assessed for each node. It generally has the complexity exponential on n. The non-impeding noisy-AND (NIN-AND) tree is a recently developed causal model that reduces the complexity to linear, while modeling both reinforcing and undermining interactions among causes. Acquisition of an NIN-AND tree model involves elicitation of a linear number of probability parameters and a tree structure. Instead of asking the human expert to describe the structure from scratch, in this work, we develop a two-step menu selection technique that aids structure acquisition.

1 Introduction

To specify a BN, a CPT needs to be assessed for each non-root node. It is often advantageous to construct BNs along the causal direction, in which case a CPT is the distribution of an effect conditioned on its n causes. In general, assessment of a CPT has the complexity exponential on n. Noisy-OR [7] is the most well known causal model that reduces this complexity to linear. A number of extensions have also been proposed such as [4,3,5]. However, noisy-OR, as well as related causal models, can only represent reinforcing interactions among causes [9]. The NIN-AND tree [9] is a recently proposed causal model. As noisy-OR, the number of probability parameters to be elicited is linear on n. Furthermore, it allows modeling of both reinforcing and undermining interactions among causes. The structure of causal interactions is encoded as a tree of a linear number of nodes which must be elicited in addition.

An NIN-AND tree can be acquired by asking the expert to describe the tree structure from scratch. When the number of causes is more than 3, describing the target NIN-AND tree accurately may be challenging. In this work, we develop a menu selection technique that aids the structure acquisition. We propose a compact representation through which an NIN-AND tree structure is depicted as a partially labeled tree of multiple roots and a single leaf, called a *root-labeled tree*. As there are too many root-labeled trees for a given number of causes, we divide the menu selection into 2 steps. In the first step, the human expert is presented with an enumeration of unlabeled trees for the given number of causes, and is asked to select one. In the second step, the expert is presented with an enumeration of root-labeled trees that are isomorphic to the selected unlabeled tree. The

Y. Gao and N. Japkowicz (Eds.): Canadian AI 2009, LNAI 5549, pp. 158–170, 2009.

two-step menu selection reduces significantly the total number of alternatives to be presented. It lowers the overall cognitive load to the expert and is expected to improve the accuracy and efficiency of NIN-AND tree model acquisition.

To implement the two-step menu selection, a proper set of tree structures must be enumerated at each step. For the first step, we draw from a technique from phylogenetics [2] for counting *evolutionary tree shapes*, which are unlabeled trees of a single root and multiple leaves. We extend the technique for counting to an algorithm for enumeration (generation) of unlabeled trees of multiple roots and a single leaf. For the second step, we develop a new algorithm to enumerate root-labeled trees isomorphic to a given unlabeled tree.

2   Background

This section is mostly based on [9]. An *uncertain cause* is a cause that can produce an effect but does not always do so. Denote a set of binary cause variables as $X = \{c_1, ..., c_n\}$ and their effect variable (binary) as e. For each c_i, denote $c_i = true$ by c_i^+ and $c_i = false$ by c_i^-. Similarly, denote $e = true$ by e^+ and $e = false$ by e^-.

A *causal event* refers to an event that a cause c_i caused an effect e to occur successfully. Denote this causal event by $e^+ \leftarrow c_i^+$ and its probability by $P(e^+ \leftarrow c_i^+)$. The causal failure event, where e is false when c_i is true, is denoted as $e^+ \nleftarrow c_i^+$. Denote the causal event that a set $X = \{c_1, ..., c_n\}$ of causes caused e by $e^+ \leftarrow c_1^+, ..., c_n^+$ or $e^+ \leftarrow x^+$. Denote the set of *all causes* of e by C. The CPT $P(e|C)$ relates to probabilities of causal events as follows: If $C = \{c_1, c_2, c_3\}$, then $P(e^+|c_1^+, c_2^-, c_3^+) = P(e^+ \leftarrow c_1^+, c_3^+)$.

Causes reinforce each other if collectively they are at least as effective in causing the effect as some acting by themselves. If collectively they are less effective, then they undermine each other. The following defines the 2 types of causal interactions generally.

Definition 1. *Let $R = \{W_1, W_2, ...\}$ be a partition of a set X of causes, $R' \subset R$, and $Y = \cup_{W_i \in R'} W_i$. Sets of causes in R reinforce each other, iff $\forall R'\ P(e^+ \leftarrow y^+) \le P(e^+ \leftarrow x^+)$. Sets of causes in R undermine each other, iff $\forall R'\ P(e^+ \leftarrow y^+) > P(e^+ \leftarrow x^+)$.*

Disjoint sets of causes $W_1, ..., W_m$ satisfy *failure conjunction* iff

$$(e^+ \nleftarrow w_1^+, ..., w_m^+) = (e^+ \nleftarrow w_1^+) \wedge ... \wedge (e^+ \nleftarrow w_m^+).$$

That is, collective failure is attributed to individual failures. They also satisfy *failure independence* iff $P((e^+ \nleftarrow w_1^+) \wedge ... \wedge (e^+ \nleftarrow w_m^+)) = P(e^+ \nleftarrow w_1^+) ... P(e^+ \nleftarrow w_m^+)$. Disjoint sets of causes $W_1, ..., W_m$ satisfy *success conjunction* iff $e^+ \leftarrow w_1^+, ..., w_m^+ = (e^+ \leftarrow w_1^+) \wedge ... \wedge (e^+ \leftarrow w_m^+)$. That is, collective success requires individual effectiveness. They also satisfy *success independence* iff $P((e^+ \leftarrow w_1^+) \wedge ... \wedge (e^+ \leftarrow w_m^+)) = P(e^+ \leftarrow w_1^+) ... P(e^+ \leftarrow w_m^+)$. It has been shown that causes are reinforcing when they satisfy failure conjunction and

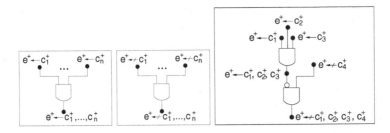

Fig. 1. (Left) Direct NIN-AND gate. (Middle) Dual NIN-AND gate. (Right) The structure of a NIN-AND tree causal model.

independence, and they are undermining when they satisfy success conjunction and independence. Hence, undermining can be modeled by a direct NIN-AND gate (Fig. 1, left), and reinforcement by a dual NIN-AND gate (middle).

As per Def. 1, a set of causes can be reinforcing (undermining), but the set is undermining (reinforcing) with another set. Such causal interaction can be modeled by a NIN-AND tree. As shown in Fig. 1 (right), causes c_1 through c_3 are undermining, and they are collectively reinforcing c_4. The following defines NIN-AND tree models in general:

Definition 2. *The structure of an NIN-AND tree is a directed tree for effect e and a set $X = \{c_1, ..., c_n\}$ of occurring causes.*

1. *There are 2 types of nodes. An* event *node (a black oval) has an in-degree ≤ 1 and an out-degree ≤ 1. A* gate *node (a NIN-AND gate) has an in-degree ≥ 2 and an out-degree 1.*
2. *There are 2 types of links, each connecting an event and a gate along input-to-output direction of gates. A* forward *link (a line) is implicitly directed. A* negation *link (with a white oval at one end) is explicitly directed.*
3. *Each terminal node is an event labeled by a causal event $e^+ \leftarrow y^+$ or $e^+ \nleftarrow y^+$. There is a single* leaf *(no child) with $y^+ = x^+$, and the gate it connects to is the* leaf gate. *For each* root *(no parent; indexed by i), $y_i^+ \subset x^+$, $y_j^+ \cap y_k^+ = \emptyset$ for $j \neq k$, and $\bigcup_i y_i^+ = x^+$.*
4. *Inputs to a gate g are in one of 2 cases:*
   (a) *Each is either connected by a forward link to a node labeled $e^+ \leftarrow y^+$, or by a negation link to a node labeled $e^+ \nleftarrow y^+$. The output of g is connected by a forward link to a node labeled $e^+ \leftarrow \cup_i y_i^+$.*
   (b) *Each is either connected by a forward link to a node labeled $e^+ \nleftarrow y^+$, or by a negation link to a node labeled $e^+ \leftarrow y^+$. The output of g is connected by a forward link to a node labeled $e^+ \nleftarrow \cup_i y_i^+$.*

An NIN-AND tree model for effect e and its causes C can be obtained by eliciting its structure (with $|C|$ roots) and $|C|$ single-cause probabilities $P(e^+ \leftarrow c_i^+)$ one for each root event in the structure. The CPT $P(e|C)$ can then be derived using the model.

By default, each root event in a NIN-AND tree is a single-cause event, and all causal interactions satisfy failure (or success) conjunction and independence. If a subset of causes do not satisfy these assumptions, suitable multi-cause probabilities $P(e^+ \leftarrow x^+)$, where $X \subset C$, can be directly elicited and incorporated into the NIN-AND tree model. The default is assumed in this paper.

Some additional notations used in the paper are introduced below. The number of combinations of n objects taken k at a time without repetition is denoted $C(n, k)$. We assume that the n objects are integers 0 through $n - 1$. Each combination is referred to as a k-combination of n objects. We assume $n < 10$ and k-combinations can be stored in an array, say, cb. Thus we refer to the i'th k-combination by $cb[i]$. We denote the number of k-combinations of n objects with repetition by $C'(n, k)$. Note $C'(n, k) = C(n + k - 1, k)$.

A *partition* of a positive integer n is a set of positive integers which sum to n. Each integer in the set is a *part*. A *base* of m *units* is a tuple of m positive integers $s = (s_{m-1}, ..., s_0)$. A *mixed base number* associated with a base s is a tuple $x = (x_{m-1}, ..., x_0)$ where $0 \leq x_i < s_i$. Each x_i ($i > 0$) has the *weight* $w_i = s_{i-1} * ... * s_0$ and the weight of x_0 is 1. Each integer k in the range 0 through $s_{m-1} * ... * s_0 - 1$ can be represented as a mixed base number x such that $k = \sum_{i=0}^{m-1} x_i * w_i$. For base $b = (3, 2)$, integers 0 through 5 can be represented in that order as $(0, 0), (0, 1), (1, 0), (1, 1), (2, 0), (2, 1)$. We denote an array z of k elements by $z[0..k - 1]$.

3   Compact Representation of NIN-AND Tree Structure

NIN-AND tree models allow a CPT of generally exponential complexity to be obtained by eliciting a tree structure and a linear number of probabilities of single-cause events. [9] relies on the human expert to describe the tree topology. When the number of causes is more than 3, accurate description may be cognitively demanding. As we will show, for 4 causes, there are 52 alternative NIN-AND tree structures.

A better alternative is to show the expert a menu of all possible structures from which one can be selected. To construct the menu, we need to enumerate alternative structures. To facilitate the enumeration, we seek a compact representation of NIN-AND tree structure.

First, we observe that the NIN-AND tree structure in Fig. 2 and that in Fig. 1 (right) correspond to the same causal model. In both structures, causes c_1 through c_3 are undermining, and they are collectively reinforcing c_4. We regard the structure in Fig. 2 as superfluous and that in Fig. 1 (right) as minimal, according to Def. 3 below. Our first step towards a compact structure representation is to adopt minimal structures.

Definition 3. *Let τ be an NIN-AND tree structure. If τ contains a gate t that outputs to another gate g of the same type (direct or dual), delete t and connect its inputs to g. If such deletion is possible, τ is* superfluous. *Apply such deletions until no longer possible. The resultant NIN-AND tree structure is* minimal.

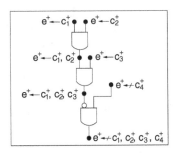

Fig. 2. An NIN-AND tree structure corresponding to the same causal model as that in Fig. 1 (right)

In a minimal NIN-AND tree structure, if the leaf gate g is a direct gate, then all gates outputting to g are dual, and their inputs are all from direct gates. That is, from the leaf towards root nodes, gates alternate in types.

This alternation implies that, for every minimal NIN-AND tree structure τ with a direct leaf gate, there exists a minimal NIN-AND tree τ' replacing each gate in τ with its opposite type, and vice versa. Therefore, if we know how to enumerate NIN-AND tree structures for a given number of causes and with a direct leaf gate, we also know how to enumerate structures with a dual leaf gate. Hence, our second step towards a compact structure representation is to focus only on minimal structures with direct leaf gates.

In a minimal structure with a direct leaf gate, types of all other gates are uniquely determined. If all root events are specified (i.e., root nodes labeled), then the causal event for every non-root node is uniquely determined. Note, however, specification of root events is partially constrained. For example, in Fig. 1 (right), since the leaf gate is dual, every root event connected to the top gate must be a causal success (rather than failure). Hence, our third step towards a compact structure representation is to omit labels for all non-root nodes.

In an NIN-AND tree structure, each gate node is connected to its unique output event. Hence, out final step towards a compact structure representation is to omit each gate node and connect its input event nodes to its output node.

As the result, our compact representation of the structure of each NIN-AND tree model is a minimal tree consisting of event nodes with only root nodes labeled. Its (implicit) leaf gate is a direct gate. Fig. 3 (a) shows the resultant representation for the NIN-AND tree in Fig. 1 (right). Following the convention in Def. 2, all links are implicitly directed (downwards away from labeled nodes). We refer to the graphical representation as *root-labeled tree*. Note that left-right order of parents makes no difference. For instance, Fig. 3 (b) is the same root-labeled tree as (a), whereas (c) is a different root-labeled tree from (a).

Theorem 1 establishes the relation between enumeration of root-labeled trees and enumeration of NIN-AND tree model structures.

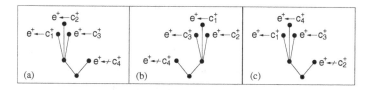

Fig. 3. Compact representations of NIN-AND tree structures

Theorem 1. *Let Ψ be the collection of NIN-AND tree models for n causes and Ψ' be the collection of root-labeled trees with n roots. The following hold:*

1. *$|\Psi| = 2\,|\Psi'|$.*
2. *For each NIN-AND tree model in Ψ, a unique tree in Ψ' can be obtained by minimizing the structure of the model, removing its gate nodes, and removing labels of non-root nodes.*
3. *For each tree in Ψ', minimal structures of 2 NIN-AND tree models in Ψ can be obtained by adding gate nodes to the tree, labeling the leaf gate as direct or dual, and labeling other non-root nodes accordingly.*

4   Enumeration of Unlabeled NIN-AND Tree Structures

Due to Theorem 1, we can enumerate NIN-AND tree model structures for n causes by enumerating root-labeled trees with n roots. The list of structures can then be presented to the expert for menu selection. However, when $n > 3$, there are too many root-labeled trees (and twice as many minimal model structures). With 4 roots, there are 26 root-labeled trees corresponding to 52 minimal NIN-AND tree model structures. With 5 roots, the numbers are 236 and 472.

To reduce the cognitive load to the expert, we divide the menu selection into 2 steps. In the first step, only unlabeled trees will be presented. With 4 roots, the menu size is 5. After the expert selects an unlabeled tree, either root-labeled trees or minimal NIN-AND tree structures corresponding to the choice will be presented for the second selection. For instance, if the unlabeled tree corresponding to Fig. 3 (a) is selected in the first step, a total of 4 root-labeled trees (or 8 NIN-AND tree structures) will be presented for second selection: at most $5 + 8 = 13$ items (rather than 52) presented in both steps. The two-step selection can be repeated until the expert is satisfied with the final selection. The advantage is the much reduced total number of menu items presented.

To realize the two-step menu selection, we first need to enumerate unlabeled trees (of a single leaf) given the number of roots. Many methods of tree enumeration in the literature, e.g., [1,8,6], do not address this problem. One exception is a technique from phylogenetics [2] for counting *evolutionary tree shapes*. The technique is closely related to our task but needs to be extended before being applicable:

First, [2] considers unlabeled directed trees with a single root and multiple leaves (called *tips*). Those trees of a given number of tips are counted. What we need to consider are unlabeled directed trees with a single leaf and multiple roots.

This difference can be easily dealt with, which amounts to reversal of directions for all links. Second, [2] represents these trees with a format incompatible with the standard notion of graph, where some link is connected to a single node instead of two (see, for example, Fig. 3.5 in [2]). Third, [2] considers only counting of these trees, while we need to enumerate (generate) them.

Nevertheless, the idea in [2] for counting so called *rooted multifurcating tree shapes* is an elegant one. Algorithm EnumerateUnlabeledTree(n) extends it to enumerate unlabeled trees with a single leaf and a given number n of roots.

Algorithm 1. *EnumerateUnlabeledTree(n)*
Input: the number of roots n.

1 *initialize list T_1 to include a single unlabeled tree of one leaf and one root;*
2 *for $i = 2$ to n,*
3   *enumerate partitions of i with at least 2 parts;*
4   *for each partition ptn of t distinct parts $(p_0, ..., p_{t-1})$,*
5   *create arrays $z[0..t-1]$, $s[0..t-1]$ and $cbr[0..t-1][][]$;*
6   *for $j = 0$ to $t - 1$,*
7   *$z[j] =$ number of occurrences of p_j in ptn;*
8   *$m = |T_{p_j}|$;*
9   *$s[j] = C'(m, z[j])$;*
10   *$cbr[j]$ stores $z[j]$-combinations of m objects with repetition;*
11   *count $= 1$;*
12   *for $j = 0$ to $t - 1$, count = count $* s[j]$;*
13   *for $q = 0$ to count $- 1$,*
14   *convert q to a mixed base number $b[0..t-1]$ using base $s[0..t-1]$;*
15   *subtree set $S = \emptyset$;*
16   *for $j = 0$ to $t - 1$,*
17   *if $z[j] = 1$, add tree $T_{p_j}[b[j]]$ to S;*
18   *else get combination $cb = cbr[j][b[j]]$;*
19   *for each number x in cb, add tree $T_{p_j}[x]$ to S;*
20   *$T' = MergeUnlabeledTree(S)$;*
21   *add T' to T_i;*
22 *return T_n;*

Line 1 creates $T_1 = (T_1[0])$ with $T_1[0]$ shown in Fig. 4 (a). The first iteration ($i = 2$) of *for* loop started at line 2 creates $T_2 = (T_2[0])$ with $T_2[0]$ shown in Fig. 4 (b). The next iteration ($i = 3$) creates $T_3 = (T_3[0], T_3[1])$ shown in (c) and

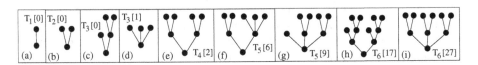

Fig. 4. (a), (b) The only unlabeled tree of one and 2 roots, respectively. (c), (d) The only trees of 3 roots. (e) A tree of 4 roots. (f), (g) Trees of 5 roots. (h), (i) Trees of 6 roots. Links are implicitly directed downwards.

(d). The loop continues to build each tree list of an increasing number of roots, until the list for $i = n$ roots is obtained.

In line 3, the set of partitions of i with 2 or more parts is obtained. For instance, for $i = 4$, the set is $\{\{3,1\}, \{2,2\}, \{2,1,1\}, \{1,1,1,1\}\}$. For $i = 5$, the set is $\{\{4,1\}, \{3,2\}, \{3,1,1\}, \{2,2,1\}, \{2,1,1,1\}, \{1,1,1,1,1\}\}$. Each partition signifies how a tree of 4 roots can be assembled from subtrees. For example, $\{3,2\}$ means that a tree of 5 roots can be assembled by merging a subtree of 3 roots (an element of T_3) with a subtree of 2 root (an element of T_2). The *for* loop started at line 4 iterates through each partition to enumerate the corresponding trees. Lines 5 through 12 count the number of trees from the given partition and specify indexes of subtrees to be merged in $cbr[]$. For each new tree, lines 13 through 21 retrieve relevant subtrees and merge them.

If a part p appears in a partition once (counted by $z[j]$ in line 7), then list T_p contributes one subtree to each new tree. This can be done in $m = |T_p|$ ways (line 8). It is counted by $s[j]$ in line 9, where $C'(m,1) = C(m,1) = m$. The $cbr[j]$ in line 10 will be $(cbr[j][0], ..., cbr[j][m-1])$ where each $cbr[j][k] = (k)$ is a 1-combination that indexes the m elements in T_p. For example, consider the iteration of the *for* loop started at line 4 with $ptn = \{3,2\}$ and hence $(p_0, p_1) = (3,2)$. After line 10, we have occurrence counting $(z[0], z[1]) = (1,1)$. It is used to produce $s[0] = C'(2,1) = 2$, $s[1] = C'(1,1) = 1$, and combinations $(cbr[0][0], cbr[0][1]) = ((0), (1))$ and $cbr[1][0] = (0)$.

The total number of distinct trees due to the partition is counted by the product of $s[j]$. For instance, line 12 produces $count = 2$, which says that there are 2 unlabeled trees in T_5 due to partition $\{3,2\}$.

If a part p appears in a partition $z[j] > 1$ times, then list T_p contributes $z[j]$ subtrees to each new tree. This can be done in $C'(m, z[j])$ ways (line 8). It is counted by $s[j]$ in line 9. For example, consider the iteration of the *for* loop started at line 4 with $ptn = \{3,3\}$ and hence $p_0 = 3$ and $z[0] = 2$. Now $s[0] = C'(2,2) = 3$ and $(cbr[0][0], cbr[0][1], cbr[0][2]) = ((0,0), (1,0), (1,1))$.

Each iteration of the *for* loop started at line 13 obtains a new tree in T_i. The relevant subtrees from earlier tree lists are retrieved and then merged into a new tree. Consider $ptn = \{3,2\}$, $(s[0], s[1]) = (2,1)$, $(cbr[0][0], cbr[0][1]) = ((0), (1))$ and $cbr[1][0] = (0)$ mentioned above. For $q = 1$, line 14 produces $(b[0], b[1]) = (1,0)$. Each iteration of the *for* loop started at line 16 adds one or more subtrees to S based on the mixed base number $b[]$. Since $(z[0], z[1]) = (1,1)$, the first iteration adds $T_3[1]$ to S and the next iteration adds $T_2[0]$. They are merged into unlabeled tree $T_5[6]$ shown in Fig. 4 (f).

Next, consider $ptn = \{3,3\}$, $s[0] = 3$, and $(cbr[0][0], cbr[0][1], cbr[0][2]) = ((0,0), (1,0), (1,1))$. The loop started at line 13 iterates 3 times. For $q = 2$, we have $b[0] = 2$. At line 18, $cb = cbr[0][2] = (1,1)$. At line 19, 2 copies of $T_3[1]$ (see Fig. 4) are added to S.

Next, consider MergeUnlabeledTree(). Two unlabeled trees t and t' of i and $j \geq i$ roots respectively may be merged in 3 ways to produce a tree of $i+j$ roots:

M_1 Their leaf nodes are merged, which is how $T_2[0]$ is obtained from 2 copies of $T_1[0]$ (see Fig. 4).

M_2 The leaf of t' become the parent of the leaf of t, which is how $T_3[0]$ is obtained from $T_1[0]$ and $T_2[0]$. Note that roles of t and t' cannot be switched.

M_3 Both leaf nodes may become the parents of a new leaf node, which is how $T_2[0]$ and $T_3[1]$ are merged to produce $T_5[6]$.

When $k > 2$ subtrees are merged into a new tree, 2 subtrees are merged first and the remaining subtrees are merged into the intermediate tree one by one. Which of the 3 ways of merging is used at each step is critical. Incorrect choice produces some trees multiple times while omitting others. The resultant T_i will not be an enumeration. MergeUnlabeledTree() is detailed below:

Algorithm 2. *MergeUnlabeledTree(S)*
Input: a set S of $k \geq 2$ unlabeled (sub)trees.

1 sort trees in S in ascending order of number of roots as $(t_0, ..., t_{k-1})$;
2 if t_0 has one root and t_1 has one root, merge them to t by M_1;
3 else if t_0 has one root and t_1 has 2 or more roots, merge them to t by M_2;
4 else merge them to t by M_3;
5 for $i = 2$ to $k - 1$,
6   if t_i has one root, merge t and t_i to t' by M_1;
7   else merge t and t_i to t' by M_2;
8 $t = t'$;
9 return t;

EnumerateUnlabeledTree(n) correctly enumerates unlabeled trees of n roots. A formal proof of correctness is beyond the space limit. Our implementation generates list T_n whose cardinality is shown in Table 1 for $n \leq 10$:

Table 1. Cardinality of T_n for $n \leq 10$

n	1	2	3	4	5	6	7	8	9	10
$|T_n|$	1	1	2	5	12	33	90	261	766	2312

5   Enumerate Root-Labeled Trees Given Unlabeled Tree

To realize the second step of menu selection, we enumerate root-labeled trees for a given unlabeled tree of n roots. Since left-right order of causal events into the same NIN-AND gate does not matter, the number of root-labeled trees is less than $n!$. We propose the following algorithm based on the idea of assigning labels to each group of roots with mirror subtree (defined below) handling. It enumerates root-labeled trees correctly, although a formal proof is beyond space limit. The root-labeled trees are plotted as they are enumerated.

Algorithm 3. *EnumerateRootLabeledTree(t)*
Input: an unlabeled tree t of n roots.

1 if all root nodes have the same child;
2   label roots using n labels in arbitrary order;

3   plot the root-labeled tree;
4   grp = grouping of roots with the same child nodes;
5   search for mirror subtrees using grp;
6   if no mirror subtrees are found, EnumerateNoMirrorRLT(t, grp);
7   else EnumerateMirrorRLT(t, grp);

Lines 1 to 3 handle cases such as $T_3[1]$ (Fig. 4). Otherwise, roots of the same child nodes are grouped in line 4. For $T_5[6]$ (Fig. 4), we have 2 groups of sizes 2 and 3. Two labels out of 5 can be assigned to the left group of size 2 in $C(5,2) = 10$ ways and the right group can be labeled using remaining labels. Hence, $T_5[6]$ has 10 root-labeled trees. It contains no mirror subtrees (which we will explain later) and line 6 is executed, as detailed below:

Algorithm 4. *EnumerateNoMirrorRLT(t, grp)*
Input: an unlabeled tree t of r roots without mirror subtree; grouping grp of roots with common child nodes.
1   n = r;
2   g = number of groups in grp;
3   for i = 0 to g − 2,
4   k = number of roots in group i;
5   s[i] = C(n, k);
6   cb[i] stores k-combinations of n objects without repetition;
7   n = n − k;
8   count = 1;
*9   for i = 0 to g − 2, count = count * s[i];*
10 for i = 0 to count − 1,
11   initialize lab[0..r − 1] to r root labels;
12   convert i to a mixed base number b[0..g − 2] using base s[0..g − 2];
13   for j = 0 to g − 2,
14   get combination gcb = cb[j][b[j]];
15   for each number x in gcb, label a root in group j by lab[gcb[x]];
16   remove labels indexed by gcb from lab[];
17   label roots in group g − 1 using labels in lab[];
18   plot the root labeled tree;

In EnumerateNoMirrorRLT(), lines 3 to 7 process roots group by group. For each group (except the last one) of size k, the number of ways that k labels can be selected from n is recorded in $s[i]$. Here, n is initialized to the number of roots (line 1), and is reduced by k after each group of size k is processed (line 7). The labels in each selection are indexed by $cb[i]$. Lines 8 and 9 count the total number of root labeled trees isomorphic to t.

Lines 10 and onwards enumerate and plot each root-labeled tree. Tree index i is converted to a mixed base number $b[]$. Each $b[j]$ is then used to retrieve the label indexes in $cb[i]$ (line 14). The label list $lab[]$ is initialized in line 11, whose elements are used to label a root group in line 15, and the list is updated in line 16. The first $g − 1$ groups are labeled in the *for* loop of lines 13 to 16. The last group is labeled in line 17.

Next, we consider $T_4[2]$ in Fig. 4 (e). It has 2 root groups of size 2 and the left group can be assigned 2 labels in $C(4,2) = 6$ ways. However, half of them switch

Table 2. Number of mirror subtrees that may be present in a tree of n roots

n	1	2	3	4	5	6	7
$T_2[0]$	0	0	0	2	2	2 or 3	2 or 3
$T_3[0]$	0	0	0	0	0	2	2
$T_3[1]$	0	0	0	0	0	2	2

the labeling between left and right group in the other half. Hence, the number of root-labeled trees for $T_4[2]$ is 3, not 6. Applying EnumerateNoMirrorRLT() to $T_4[2]$ would be incorrect. We define *mirror subtrees* for such cases.

Definition 4. *A* subtree *s in an unlabeled tree t is a subgraph consisting of a non-root, non-leaf node of t (as the leaf of s) and all its ancestors in t.*

Two subtrees s and s' are mirror subtrees *if they are isomorphic, each has more than one root node, and the leaf of s and the leaf of s' have the same path length from the leaf of t in t.*

$T_4[2]$ in Fig. 4 (e) has 2 mirror subtrees, each of which is a copy of $T_2[0]$, and so does $T_5[9]$ in (g). $T_6[17]$ in (h) has 2 mirror subtrees, each of which is a copy of $T_3[0]$. $T_6[27]$ in (i) has 3 mirror subtrees, each of which is a copy of $T_2[0]$. None of the other trees in Fig. 4 has mirror subtrees. In general, a tree of $n \geq 4$ roots may have mirror subtrees from $T_{\lfloor n/2 \rfloor}$, where $[.]$ denotes the floor function. Table 2 shows the possible number and type of mirror subtrees for $n \leq 7$.

At line 5 of EnumerateRootLabeledTree(), mirror subtrees are searched. Most trees are rejected based on their *grp*. Otherwise, if the *grp* of a tree is compatible with a pattern in Table 2, its leaf is removed, splitting it into subtrees recursively, and possible mirror subtrees are detected. If mirror subtrees are found, at line 7, EnumerateMirrorRLT() is performed. We present its idea but not pseudocode.

Suppose an unlabeled tree t with mirror subtrees and n roots have $k \leq n$ roots in mirror subtrees. If $n - k > 0$, remaining roots are labeled first in the same way as EnumerateNoMirrorRLT(). Then, for each partially root-labeled tree, mirror subtrees are root-labeled. For example, for $T_5[9]$ in Fig. 4 (g), the left-most root can be labeled in 5 ways as usual, using up one label. The first mirror subtree is labeled in $C(4,2)/2 = 3$ ways, using up another 2 labels. The second mirror subtree is then labeled using the remaining labels. For $T_6[27]$ in (i), the first mirror subtree is labeled in $C(6,2)/3 = 5$ ways, using up 2 labels. The second mirror subtree is labeled in $C(4,2)/2 = 3$ ways, using up another 2 labels. The third mirror subtree is then labeled using the remaining labels. Table 3 shows the number of root-labeled trees given some unlabeled trees in Fig. 4 as enumerated by our implementation of EnumerateRootLabeledTree().

Table 3. Number of root-labeled trees for some given unlabeled trees

Unlabeled tree	$T_4[2]$	$T_5[6]$	$T_5[9]$	$T_6[17]$	$T_6[27]$
No. root-labeled trees	3	10	15	90	15

Table 4. Toal number of root-labeled trees with n roots

n	1	2	3	4	5	6	7
No. root-labeled trees	1	1	4	26	236	2752	39208

Table 4 shows the total number of root-labeled trees with n roots for $n \leq 7$. Since T_7 contains 90 unlabeled trees (Table 1), each has on average $39208/90 \approx$ 435 root-labeled trees. Its implication is discussed in the next section.

6   Remarks

As learning from data is limited by missing values, small samples, and cost in data collection, elicitation of CPT remains an alternative in constructing BNs when the expert is available. Due to conditional independence encoded in BNs, a CPT that involves more than 10 causes is normally not expected. Even so, the task of eliciting up to 2^{10} parameters is daunting. NIN-AND trees provide a causal model that reduces the number of parameters to be elicited to linear (10 for 10 binary causes), while capturing both reinforcing and undermining interactions among causes. A tree-shaped causal structure of a linear number of nodes (less than 20 for 10 causes), however, must be elicited in addition.

This contribution proposes the two-step menu selection for causal structure elicitation. The technique reduces the cognitive load on the expert, compared to structure elicitation from scratch or the single step menu selection. For the first step, we extend an idea of counting from phylogenetics into an algorithm to enumerate NIN-AND tree structures with unlabeled root nodes. For the second step, we develop an algorithm to enumerate completely labeled NIN-AND tree structures given a structure selected by the expert from the first step. Compared to off-line enumeration, our online enumeration is interactive. Even though the choice from the first step may be inaccurate, the two-step selection can be repeated easily (in seconds) until the expert's satisfaction.

As the average number of root-labeled trees is beyond 400 when the number of causes is beyond 7, we believe that the two-step menu selection is practical for elicitation of NIN-AND tree (and thus CPT) with up to 7 causes. For CPTs with 8 causes or beyond, we have developed an alternative technique to be presented elsewhere. Empirical evaluation of our proposed elicitation techniques with human experts is underway.

Acknowledgements

Financial support from NSERC, Canada through Discovery Grant to the first author is acknowledged. We thank reviewers for their comments.

References

1. Cayley, A.: A theorem on trees. Quarterly J. Mathematics, 376–378 (1889)
2. Felsenstein, J.: Inferring Phylogenies. Sinauer Associates, Sunderland (2004)

3. Galan, S.F., Diez, F.J.: Modeling dynamic causal interaction with Bayesian networks: temporal noisy gates. In: Proc. 2nd Inter. Workshop on Causal Networks, pp. 1–5 (2000)
4. Heckerman, D., Breese, J.S.: Causal independence for probabilistic assessment and inference using Bayesian networks. IEEE Trans. on System, Man and Cybernetics 26(6), 826–831 (1996)
5. Lemmer, J.F., Gossink, D.E.: Recursive noisy OR - a rule for estimating complex probabilistic interactions. IEEE Trans. on System, Man and Cybernetics, Part B 34(6), 2252–2261 (2004)
6. Moon, J.W.: Counting Labeled Trees. William Clowes and Sons, London (1970)
7. Pearl, J.: Probabilistic Reasoning in Intelligent Systems: Networks of Plausible Inference. Morgan Kaufmann, San Francisco (1988)
8. Riordan, J.: The enumeration of trees by height and diameter. IBM J., 473–478 (November 1960)
9. Xiang, Y., Jia, N.: Modeling causal reinforcement and undermining for efficient cpt elicitation. IEEE Trans. Knowledge and Data Engineering 19(12), 1708–1718 (2007)

Compiling the Lexicographic Inference Using Boolean Cardinality Constraints

Safa Yahi and Salem Benferhat

Université Lille-Nord de France,
Artois, F-62307 Lens, CRIL, F-62307 Lens
CNRS UMR 8188, F-62307 Lens
{yahi,benferhat}@cril.univ-artois.fr

Abstract. This paper sheds light on the lexicographic inference from stratified belief bases which is known to have desirable properties from theoretical, practical and psychological points of view. However, this inference is expensive from the computational complexity side. Indeed, it amounts to a Δ_2^p-complete problem. In order to tackle this hardness, we propose in this work a new compilation of the lexicographic inference using the so-called Boolean cardinality constraints. This compilation enables a polynomial time lexicographic inference and offers the possibility to update the priority relation between the strata without any re-compilation. Moreover, it can be efficiently extended to deal with the lexicographical closure inference which takes an important place in default reasoning. Furthermore, unlike the existing compilation approaches of the lexicographic inference, ours can be efficiently parametrized by any target compilation language. In particular, it enables to take advantage of the well-known prime implicates language which has been quite influential in artificial intelligence and computer science in general.

1 Introduction

Handling exceptions, information fusion and belief revision are fundamental tasks for a common sense reasoning agent. These activities are related in the sense that they all lead in general to reason with inconsistent pieces of information.

Many approaches have been proposed in order to reason under inconsistency such as the *coherence-based approaches* [17]. According to these approaches, the inference is considered as a two step process consisting first in generating some preferred consistent subbases and then using classical inference from some of them. When priorities attached to pieces of knowledge are available, the task of coping with inconsistency is simplified, because priorities give information about how to solve conflicts.

Examples of prioritized coherence-based approaches which select one consistent subbase are the possibilistic inference [11] and the linear order inference [15]. Examples of approaches which use several consistent subbases are the inclusion-based inference [5] and the lexicographic inference [2,13].

Y. Gao and N. Japkowicz (Eds.): Canadian AI 2009, LNAI 5549, pp. 171–182, 2009.
© Springer-Verlag Berlin Heidelberg 2009

In this work, we focus our attention on the lexicographic inference which is known to have expected theoretical properties. Indeed, in the case of default reasoning for instance, it has been shown to be satisfactory from a psychological side [4]. Besides, it has been used in several applications [21]. However, the lexicographic inference is expensive from a computational point of view. In [7], it has been shown to be a Δ_2^p-complete problem. Then, it requires a polynomial number of calls to an NP oracle to be achieved.

In order to tackle this problem, we adhere to knowledge compilation [18,6,9]. Roughly speaking, compiling a knowledge base consists in turning it off-line into a formula from which on-line query answering can be done more efficiently than from the original knowledge base.

More precisely, we propose in this work a flexible compilation of the lexicographic inference. This compilation permits a polynomial time lexicographic inference and offers the opportunity to update the priority relation between the strata of the belief base without requiring any re-compilation cost. This is helpful when the stratification changes with queries for instance. Moreover, contrarily to the existing approaches for compiling the lexicographic inference, ours can be parametrized by any target compilation language. Especially, it enables to take advantage of the well-known prime implicates language which takes an important place in knowledge compilation and artificial intelligence in general. Furthermore, our approach can be efficiently extended to handle the lexicographical closure inference which is an interesting approach in reasoning tolerant exceptions. Indeed, actual human reasoning has been shown to manifest properties which are similar to those of the lexicographical closure.

Our approach relies on the use of Boolean cardinality constraints which arise in many practical situations. Boolean cardinality constraints impose upper and lower bounds to the number of variables fixed to *True* in a set of propositional variables. Among the $\mathcal{CNF}$ encodings of Boolean cardinality constraints that we can find in the literature, we shed light on the one proposed in [1]. First, this encoding is satisfactory from a spatial point of view and unit propagation restores the generalized arc consistency property on the encoded constraints. Also, unlike the other encodings, it gives too needed flexibility to our compilation.

The remainder of the paper is organized as follows. We start by recalling the lexicographic inference from stratified belief bases and giving a brief background on knowledge compilation. After that, we review the notion of Boolean cardinality constraints and recall the UT-MGAC $\mathcal{CNF}$ encoding proposed in [1]. In the following section, we introduce our compilation approach and give the corresponding properties. Then, we extend it to deal with the lexicographical closure inference. Finally, we compare the proposed compilation approach with other related work before concluding.

2   A Refresher on the Lexicographic Inference

We consider a finite set of propositional variables V whose the elements are denoted by lower case Roman letters $a, b, \ldots$ The symbols $\top$ and $\bot$ denote tautology and contradiction respectively. Let PL_V be the propositional language built up from V, $\{\top, \bot\}$ and the connectives $\wedge, \vee, \neg, \Rightarrow$ in the usual way. Formulae, i.e, elements of PL_V are denoted by Greek letters $\phi, \varphi, \psi, \ldots \models$ denotes classical inference relation. A clause is a disjunction of literals where a literal is a propositional variable or its negation. A term is a conjunction of literals. A conjunctive normal form ($\mathcal{CNF}$) is a conjunction of clauses while a disjunctive normal form ($\mathcal{DNF}$) is a disjunction of terms.

A stratified belief base ($\mathcal{SBB}$ for short) is a set of formulae equipped with a total preorder reflecting the priority relation that exists between its formulae. An $\mathcal{SBB}$ Σ can be viewed as a finite sequence $S_1 \sqcup \ldots \sqcup S_m$ ($m \geq 1$) where each S_i is a stratum containing propositional formulae having the same priority level i and which are more reliable than the formulae of stratum S_j for any $j > i$.

The lexicographic preference relation between consistent subbases of an $\mathcal{SBB}$ is defined as follows:

Definition 1. *Given an $\mathcal{SBB}$ $\Sigma = S_1 \sqcup \ldots \sqcup S_m$, let $\mathcal{A}$ and $\mathcal{B}$ be two consistent subbases of Σ. Then, $\mathcal{A}$ is said to be lexicographically preferred to $\mathcal{B}$, denoted by $\mathcal{A} >_{\mathcal{L}ex} \mathcal{B}$, iff $\exists i, 1 \leq i \leq m$ such that $|S_i \cap \mathcal{A}| > |S_i \cap \mathcal{B}|$ and $\forall\, 1 \leq j < i$, $|S_i \cap \mathcal{A}| = |S_i \cap \mathcal{B}|$.*

Let $\mathcal{L}ex(\Sigma)$ denote the set of all the lexicographically preferred consistent subbases of Σ, i.e, those which are maximal with respect to $>_{\mathcal{L}ex}$. Then, the lexicographic inference from Σ is defined by:

Definition 2. *Let Σ be an $\mathcal{SBB}$ and ψ be a classical propositional formula. ψ is a lexicographic consequence of Σ, denoted by $\Sigma \vdash_{\mathcal{L}ex} \psi$, iff ψ is a classical consequence of any lexicographically preferred consistent subbase of Σ, namely*

$$\Sigma \vdash_{\mathcal{L}ex} \psi \Leftrightarrow \forall \mathcal{B} \in \mathcal{L}ex(\Sigma) : \mathcal{B} \models \psi.$$

Note that every lexicographically preferred consistent subbase is maximal with respect to set inclusion.

Example 1. *Let us consider the following $\mathcal{SBB}$ $\Sigma = S_1 \sqcup S_2 \sqcup S_3$ such that:*

- $S_1 = \{r \wedge q \wedge e\}$
- $S_2 = \{\neg r \vee \neg p, \neg q \vee p\}$
- $S_3 = \{\neg e \vee p\}$

Σ admits three maximal consistent subbases: $A = \{r \wedge q \wedge e, \neg r \vee \neg p\}, B = \{r \wedge q \wedge e, \neg q \vee p, \neg e \vee p\}$ and $C = \{\neg r \vee \neg p, \neg q \vee p, \neg e \vee p\}$.
One can easily see that:

- $|A \cap S_1| = 1, |A \cap S_2| = 1$ and $|A \cap S_3| = 0,$
- $|B \cap S_1| = 1, |B \cap S_2| = 1$ and $|B \cap S_3| = 1,$
- $|C \cap S_1| = 0, |C \cap S_2| = 2$ and $|C \cap S_3| = 1.$

So, $B >_{\mathcal{L}ex} A >_{\mathcal{L}ex} C$, i.e., $\mathcal{L}ex(\Sigma) = \{B\}$ and for instance, $\Sigma \vdash_{\mathcal{L}ex} p$.

In [7,12], the authors give an algorithm that answers the lexicographic inference using the following idea: to check whether a stratified belief base $\Sigma = S_1 \sqcup \ldots \sqcup S_m$ lexicographically entails a formula ψ, they first compute the profile of Σ. The profile of Σ, denoted in the following by K_Σ, is a vector of m elements where each element $K_\Sigma[i]$ corresponds to the number of formulae from the stratum S_i which are satisfied by each lexicographically preferred consistent subbase (since all the lexicographically preferred consistent subbases have always the same cardinality at each stratum). So, they introduce an NP-complete oracle called MAX-GSAT-ARRAY which is defined as follows:

Definition 3. MAX-GSAT-ARRAY *is given by the following decision problem:*

- *Instance: The pair (Σ, K) where Σ is an $\mathcal{SBB}$ and K is an array of size equal to the number of strata in Σ.*
- *Question: Is there an interpretation which satisfies at least $K[i]$ formulae for each stratum i in Σ?*

Then, once the profile K_Σ has been defined, $\Sigma \vdash_{\mathcal{L}ex} \psi$ if and only if there is no an interpretation that satisfies at least $K_\Sigma[i]$ formulae from each stratum S_i and does not satisfy ψ. Hence, they define another NP-complete oracle denoted by NGSAT-LEX as follows:

Definition 4. NGSAT-LEX *is given by the following decision problem:*

- *Instance: The tuple (Σ, K, ψ) where Σ is an $\mathcal{SBB}$, K is an array of size equal to the number of strata in Σ and ψ is a formula.*
- *Question: Is there an interpretation which satisfies at least $K[i]$ formulae for each stratum i in Σ and does not satisfy ψ?*

Algorithm 1.1 describes more formally this process. Then, it has been shown in [7] that the decision problem associated with the lexicographic inference is Δ_2^p-complete. Indeed, Algorithm 1.1 is deterministic polynomial and uses an NP oracle. Hence, the lexicographic inference belongs to the class Δ_2^p. Moreover, it has been proved to be complete for this class by exhibiting a polynomial reduction from a known Δ_2^p-complete problem to it.

3   Knowledge Compilation

According to knowledge compilation, a knowledge base is compiled off-line into a target language, which is then used on-line to answer a large number of queries more efficiently than from the original knowledge base. Thus, the key motivation behind knowledge compilation is to push as much of the computational overhead into the off-line phase, which is amortized over all on-line queries.

A target compilation language is a class of formulae which is at least tractable for clausal deduction. In [9], the authors provide a knowledge compilation map. Indeed, they identify a relatively large number of target compilation languages that have been presented in AI, formal verification and computer science in general, and they analyze them according to their succinctness and their polynomial

Algorithm 1.1. Lexicographic inference

Data. An $\mathcal{SBB}$ $\Sigma = S_1 \sqcup \ldots \sqcup S_m$
    A classical formula ψ
Result: Does $\Sigma \vdash_{\mathfrak{Lex}} \psi$?
begin
    $K_\Sigma \leftarrow\, <0,0,\ldots,0>$
    for $i \leftarrow 1$ *to* m **do**
        $k \leftarrow |S_i|$
        $Stop \leftarrow false$
        while $k \geq 0$ *and* $\neg Stop$ **do**
            $K_\Sigma[i] \leftarrow k$
            if MAX-GSAT-ARRAY(Σ, K_Σ) **then**
                $Stop \leftarrow true$

            else
                $k \leftarrow k - 1$

    if NGSAT-LEX(Σ, K_Σ, ψ) **then**
        return *false*

    else
        return *true*

end

time queries and transformations. The succinctness of a target language measures its spatial efficiency. More formally, a language $\mathcal{L}$ is at least as succinct as a language $\mathcal{L}'$ iff for every formula α from $\mathcal{L}$, there exists an equivalent formula β from $\mathcal{L}'$ such that the size of β is polynomial in the size of α.

Moreover, the authors show that these languages result from imposing some properties (like decomposability, determinism, smoothness, decision, order, etc) on the $\mathcal{NNF}$ (Negation Normal Form) language where an $\mathcal{NNF}$ formula is built up from literals, $\top$ and $\bot$ using only the conjunction and disjunction operators.

Then, the target compilation languages considered in [9] are $\mathcal{DNF}$ (Disjunctive Normal Form), $\mathcal{DNNF}$ (Decomposable NNF), $d\text{-}\mathcal{DNNF}$ (Deterministic DNNF), $sd\text{-}\mathcal{DNNF}$ (smooth d-DNNF), $\mathcal{FBDD}$ (Free Binary Decision Diagrams), $\mathcal{OBDD}$ (Ordered Binary Decision Diagrams), $\mathcal{MODS}$ (for models), $\mathcal{PI}$ (Prime Implicates) and $\mathcal{IP}$ (Prime Implicants).

For lack of space, we only briefly review $\mathcal{PI}$ and $\mathcal{DNNF}$ given their importance [1]. So, $\mathcal{PI}$ is the subset of $\mathcal{CNF}$ in which each clause entailed by a formula from this language is subsumed by a clause that appears in the formula, and no clause in the formula is subsumed by another. Now, a $\mathcal{DNNF}$ formula is an $\mathcal{NNF}$ formula satisfying the decomposability property which means that for any conjunction $\bigwedge_i \alpha_i$ appearing in the formula, no variables are shared by the conjuncts α_i.

We stress here that $\mathcal{DNNF}$ and $\mathcal{PI}$ play an important role in terms of succinctness. On the one hand, with the exception of $\mathcal{PI}$, $\mathcal{DNNF}$ is the most

---

[1] The reader can be referred to [9] for a refresher on all these target languages.

succinct among all the target compilation languages studied in [9]. On the other hand, $\mathcal{PI}$ it is not more succinct than $\mathcal{DNNF}$, but it is unknown whether $\mathcal{DNNF}$ is more succinct than it or not.

4   Boolean Cardinality Constraints

Boolean cardinality constraints impose upper and lower bounds to the number of variables fixed to *True* in a set of propositional variables. A Boolean cardinality constraint, denoted by $\#(\mu, \lambda, E)$, where μ and λ are positive integers and E is a set of propositional variables, is satisfied iff at least μ and at most λ of the variables in E can be assigned to *True*. The corresponding $\mathcal{CNF}$ encoding is defined as follows.

Definition 5. *Given a Boolean cardinality constraint $\#(\mu, \lambda, E)$, its $\mathcal{CNF}$ encoding consists in giving a $\mathcal{CNF}$ formula $\Psi_{\#(\mu,\lambda,E)}$ over a set of variables including E such that $\Psi_{\#(\mu,\lambda,E)}$ is satisfiable by an interpretation iff the values assigned to the variables in E by this interpretation satisfy the cardinality constraint.*

Among the approaches developed for this purpose, one can list [1,19,20]. In the following, we will focus on the UT-MGAC (for Unit Totalizer Maintaining Generalized Arc Consistency) $\mathcal{CNF}$ encoding proposed in [1] for several reasons. First, this latter is satisfactory from a spatial point of view. Indeed, it requires $O(|E|log([E|))$ additional variables and $O(|E|^2)$ additional clauses of length at most 3. This encoding is also efficient in the sense that unit propagation, which is implemented in almost complete SAT solvers and compilers, restores the generalized arc consistency on the variables of E, i.e, if a constraint is violated by a partial assignment, unit propagation generates an empty clause. Moreover, this encoding is interesting from a compilation point of view. Indeed, the bounds μ and λ can be modified without any recompilation cost.

The key feature of such encoding is a unary representation of integers: an integer v such that $0 \leq v \leq n$ is represented by a set $V = \{v_1, \ldots, v_n\}$ of n propositional variables. If the value of v is x then $v_1 = \ldots = v_x = True$ and $v_{x+1} = \ldots = v_n = False$. The main advantage of this representation is that the integer variable can be specified as belonging to an interval. Indeed, the inequality $x \leq v \leq y$ is specified by the partial instantiation of V that fixes to *True* any v_i such that $i \leq x$ and fixes to *False* any v_j such that $j \geq y + 1$.

An adder based on this representation can be used to deduce the interval to which an integer $c = a + b$ belongs given the intervals of a and b. So, a cardinality constraint $\#(\mu, \lambda, E)$ can be encoded as a totalizer structured as a pyramidal adder network $\mathcal{T}(E)$ extended by a comparator $\mathcal{C}(\mu, \lambda)$ which restricts the possible output values. The totalizer $\mathcal{T}(E)$ is a $\mathcal{CNF}$ formula defined on 3 sets of variables: $E = \{e_1, \ldots, e_n\}$: the set of input variables, $S = \{s_1, \ldots, s_n\}$: the set of output variables and L: a set of linking variables. So, an adder with inputs $a_1, \ldots, a_{m_1}, b_1, \ldots, b_{m_2}$ and output $r_1, \ldots, r_m$ is encoded as:

$\bigwedge_{i,j,k}(\neg a_i \vee \neg b_j \vee r_k) \wedge (a_{i+1} \vee b_{j+1} \vee \neg r_{k+1})$ for $0 \leq i \leq m_1$, $0 \leq j \leq m_2$, $0 \leq k \leq m$ and $k = i + j$ using the following notations: $a_0 = b_0 = r_0 = \top$ and $a_{m_1+1} = b_{m_2+1} = r_{m+1} = \bot$.

As to the comparator $\mathcal{C}(\mu, \lambda)$, it is given by the term $\mathcal{C}(\mu, \lambda) \equiv \bigwedge_{1 \leq i \leq \mu}(s_i) \wedge \bigwedge_{\lambda+1 \leq j \leq n}(\neg s_j)$.

5   Compiling the Lexicographic Inference

Clearly enough, according to Algorithm 1.1, a possible compilation of the lexicographic inference consists in computing the corresponding profile once for all. Then, the problem falls down from the Δ_2^p-complete class to the NP-complete one. However, the problem remains intractable. Moreover, the computation of the profile K_Σ requires a polynomial number of calls to an NP oracle. This last point is critical when the priorities between the strata change or in the case of the lexicographic closure since the profile changes.

The compilation we propose in this work enables a polynomial inference. In addition, it is flexible in the sense that the profile can be computed in polynomial time.

The first step consists in translating the oracles used in Algorithm 1.1 into a classical satisfiability problem using cardinality constraints. In the following, we need to handle Boolean cardinality constraints associated with classical propositional formulae. Thus, we start by giving the following definition.

Definition 6. *A $\mathcal{BCC}\text{-}\mathcal{FRM}$ formula is given by the pair $(\psi, \mathcal{C})$ where ψ is a propositional formula and $\mathcal{C}$ is a set of Boolean cardinality constraints. The $\mathcal{BCC}\text{-}\mathcal{FRM}$ formula $(\psi, \mathcal{C})$ is satisfiable iff there exists a model of ψ that satisfies each Boolean cardinality constraint from $\mathcal{C}$.*

Then, given an $\mathcal{SBB}$ Σ, we encode it under the form of a classical propositional formula Φ_Σ by introducing one new propositional variable per formula as follows.

Definition 7. *Let $\Sigma = S_1 \sqcup \ldots \sqcup S_m$ be an $\mathcal{SBB}$. The propositional base Φ_Σ associated with Σ is given by: $\Phi_\Sigma = \bigwedge_{i=1}^{m}(\bigwedge_{\phi_{ij} \in S_i}(\neg New_{ij} \vee \phi_{ij}))$*

In the following, we denote by N_i the set of New_{ij}'s associated with the formulae of the stratum S_i: $N_i = \{New_{ij}/\phi_{ij} \in S_i\}$.

This enables to translate the oracles MAX-GSAT-ARRAY and NGSAT-LEX used in Algorithm 1.1 into a $\mathcal{BCC}\text{-}\mathcal{FRM}$ satisfiability problem as follows.

1. the instance MAX-GSAT-ARRAY (Σ, K) is satisfiable iff the $\mathcal{BCC}\text{-}\mathcal{FRM}$ formula $(\Phi_\Sigma, \{\#(K[i], |N_i|, N_i) : i = 1..m\})$ is satisfiable.
2. the instance NGSAT-LEX (Σ, K, ψ) is satisfiable iff the $\mathcal{BCC}\text{-}\mathcal{FRM}$ formula $(\Phi_\Sigma \wedge \neg\psi, \{\#(K[i], |N_i|, N_i) : i = 1..m\})$ is satisfiable.

Since $\#(K[i], |N_i|, N_i)$ does not impose any upper bound on the number of variables over N_i that can be assigned to *True*, it will be denoted in the following by $\#(K[i], N_i)$ which simply means that the previous number is at least $K[i]$.

Now, based on a $\mathcal{CNF}$ encoding of cardinality constraints, we obtain a translation into a classical satisfiability problem as shown by this proposition.

Proposition 1. *Let $\Sigma = S_1 \sqcup \ldots \sqcup S_m$ be an $\mathcal{SBB}$, K be an integers array of size m and ψ be a formula. Let Φ_Σ be the propositional formula associated with Σ according to Definition 7. Then,*

1. *the instance* MAX-GSAT-ARRAY (Σ, K) *is satisfiable if and only if the formula $\Phi_\Sigma \wedge \bigwedge_{i=1}^{m} \Psi_{\#(K[i], N_i)}$ is satisfiable [2],*
2. *the instance* NGSAT-LEX (Σ, K, ψ) *is satisfiable if and only if the formula $\neg\psi \wedge \Phi_\Sigma \wedge \bigwedge_{i=1}^{m} \Psi_{\#(K[i], N_i)}$ is satisfiable.*

Let us now apply the UT-MGAC $\mathcal{CNF}$ encoding recalled previously. So, in order to compute the profile K_Σ, we make a polynomial number of satisfiability tests of the formula $\Phi_\Sigma \wedge \bigwedge_{i=1}^{m} (\mathcal{T}(N_i) \wedge \mathcal{C}(K_\Sigma[i]))$.

On the other hand, we have

$$\Phi_\Sigma \wedge \bigwedge_{i=1}^{m} (\mathcal{T}(N_i) \wedge \mathcal{C}(K_\Sigma[i]))$$

is satisfiable iff

$$\Phi_\Sigma \wedge \bigwedge_{i=1}^{m} \mathcal{T}(N_i) \nvDash \bigvee_{i=1}^{m} \neg\mathcal{C}(K_\Sigma[i]).$$

The good news is that the formula $\Phi_\Sigma \wedge \bigwedge_{i=1}^{m} \mathcal{T}(N_i)$ is independent from the profile K_Σ. Thus, it can be considered as a fixed part during the whole building of K_Σ.

Moreover, the formula $\bigvee_{i=1}^{m} \neg\mathcal{C}(K_\Sigma[i])$ is a clause (since a disjunction of negation of terms).

Consequently, in order to ensure a polynomial time computation of the K_Σ, it is sufficient to compile the formula $\Phi_\Sigma \wedge \bigwedge_{i=1}^{m} \mathcal{T}(N_i)$ into any target compilation language $\mathcal{L}$. Indeed, by definition, any target compilation language enables to achieve clausal entailment in polynomial time.

As for the second point, we have $\neg\psi \wedge \Phi_\Sigma \wedge \bigwedge_{i=1}^{m} (\mathcal{T}(N_i) \wedge \mathcal{C}(K[i]))$ is satisfiable iff $\Phi_\Sigma \wedge \bigwedge_{i=1}^{m} \mathcal{T}(N_i) \nvDash \bigvee_{i=1}^{m} \neg\mathcal{C}(K[i]) \vee \psi$. Thus, once again, if we compile $\Phi_\Sigma \wedge \bigwedge_{i=1}^{m} \mathcal{T}(N_i)$ into any target compilation language $\mathcal{L}$, then the previous inference test can be accomplished in polynomial time. We stress here that the formula $\bigvee_{i=1}^{m} \neg\mathcal{C}(K[i]) \vee \psi$ can be easily put in a $\mathcal{CNF}$ form in polynomial time provided that ψ is given under a $\mathcal{CNF}$ form. Hence, the lexicographic inference compilation we propose is defined as follows:

Definition 8. *Let Σ be an $\mathcal{SBB}$ and $\mathcal{L}$ be a target compilation language. The compilation of Σ with respect to $\mathcal{L}$, denoted by $Comp(\Sigma, \mathcal{L})$, is the compilation of $\Phi_\Sigma \wedge \bigwedge_{i=1}^{m} \mathcal{T}(N_i)$ into $\mathcal{L}$, namely*

$$Comp(\Sigma, \mathcal{L}) \equiv \mathcal{L}(\Phi_\Sigma \wedge \bigwedge_{i=1}^{m} \mathcal{T}(N_i))$$

---

[2] We recall that $\Psi_{\#(K[i], N_i)}$ denotes a $\mathcal{CNF}$ encoding of the constraint $\#(K[i], N_i)$.

The corresponding properties are given by the following proposition.

Proposition 2. *Let $Comp(\Sigma, \mathcal{L})$ be the compilation of an SBB Σ with respect to a target compilation language $\mathcal{L}$ and ψ be a $\mathcal{CNF}$ formula. Then,*

1. *computing the profile K_Σ can be achieved in polynomial time,*
2. *the lexicographic inference of ψ can be checked in polynomial time too.*

Consequently, the proposed compilation can be parametrized by any target compilation language. Moreover, it enables to update the priority between strata without any re-compilation cost since the profile is computed in polynomial time.

6   Compiling the Lexicographical Closure

In this section, we extend the lexicographic inference compilation that we have proposed in the previous section to the case of the lexicographical closure which applies to default bases. By a default rule we mean a generic rule of the form (generally, if α then β) having possibly some exceptions. These rules are denoted by $\alpha \rightarrow \beta$. Then, a default base is a set $\Upsilon = \{\alpha_i \rightarrow \beta_i : i = 1, n\}$ of default rules. The lexicographical closure represents an interesting way to define a non-monotonic inference relation from a default base Υ. It starts from an SBB Σ_Υ generated using System Z [16]. This stratification basically consists in assigning to each default rule a priority level: default rules with the lowest priority are the most general ones. They are such that assigning their antecedent to be true does not cause inconsistencies. Then, the lexicographical closure [14] is defined as follows:

Definition 9. *Let Υ be a default base and let Σ_Υ be the associated SBB using Pearl's System Z. Then, a default rule $\alpha \rightarrow \beta$ is said to be a lexicographic consequence of Υ, denoted by $\Upsilon \Vdash_{\mathcal{L}ex} \alpha \rightarrow \beta$, iff $\{\alpha\} \sqcup \Sigma_\Upsilon \vdash_{\mathcal{L}ex} \beta$.*

One can see that in the case of the lexicographical closure and contrarily to the classical lexicographic inference, the lexicographically preferred consistent subbases vary with respect to the query at hand. Then, we must compute the profile of Σ each time we want to check whether a new default rule $\alpha \rightarrow \beta$ is entailed. This requires a polynomial number of calls to an NP oracle as explained previously.

So, we propose a more efficient way to compute the profile of $\{\alpha\} \sqcup \Sigma_\Upsilon$. In fact, testing the satisfiability of $\alpha \wedge \Phi_{\Sigma_\Upsilon} \wedge \bigwedge_{i=1}^{m}(\mathcal{T}(N_i) \wedge \mathcal{C}(K_{\Sigma_\Upsilon}[i]))$ is equivalent to verify that

$$\Phi_{\Sigma_\Upsilon} \wedge \bigwedge_{i=1}^{m} \mathcal{T}(N_i) \nvDash \neg\alpha \vee \bigvee_{i=1}^{m} \neg\mathcal{C}(K_{\Sigma_\Upsilon}[i]).$$

Thus, if we use again the compilation $Comp(\Sigma_\Upsilon, \mathcal{L})$ given by Definition 8, we ensure a polynomial computation of K_{Σ_Υ} and hence a polynomial lexicographical closure inference under the assumption that $\neg\alpha$ is under a $\mathcal{CNF}$ form.

Note that if the observation α is given under the form of a term which is frequent, then putting its negation under a $\mathcal{CNF}$ form is polynomial. The same result holds when α is given under the form of a $\mathcal{DNF}$.

Example 2. *Let us consider the following default base* $\Upsilon = \{b \rightarrow f, p \rightarrow \neg f, b \rightarrow w, p \rightarrow s\}$ *which can be read as follows: "generally, birds fly, penguins do not fly, birds have wings and penguins swim".*

The associated $\mathcal{SBB}$ *using Pearl's System Z is* $\Sigma_\Upsilon = S_1 \sqcup S_2$ *such that* $S_1 = \{\neg p \vee \neg f, \neg p \vee s\}$ *and* $S_2 = \{\neg b \vee f, \neg b \vee w\}$ *which reflects that the most specific rules are preferred.*

Now, let us check whether $\Upsilon \Vdash_{\mathcal{L}ex} \alpha \rightarrow \beta$ *such that* $\alpha \equiv p \wedge b$ *and* $\beta \equiv \neg f \wedge w$. *First, we give the propositional encoding* Φ_{Σ_Υ} *associated with* Σ_Υ *according to Definition 7:* $\Phi_{\Sigma_\Upsilon} \equiv (\neg p \vee \neg f \vee \neg New_{11}) \wedge (\neg p \vee s \vee \neg New_{12}) \wedge (\neg b \vee f \vee \neg New_{21}) \wedge (\neg b \vee w \vee \neg New_{22})$. *So,* $N_1 = \{New_{11}, New_{12}\}$ *and* $N_2 = \{New_{21}, New_{22}\}$.

Then, the totalizer corresponding to N_1 *is given by:*
$\mathcal{T}(N_1) \equiv (\neg New_{11} \vee s_{11}) \wedge (New_{12} \vee \neg s_{12}) \wedge (\neg New_{12} \vee s_{11}) \wedge (New_{11} \vee \neg s_{12}) \wedge (\neg New_{11} \vee \neg New_{12} \vee s_{12}) \wedge (New_{11} \vee New_{12} \vee \neg s_{11})$.

As to the totalizer associated with N_2, *it is given by:*
$\mathcal{T}(N_2) \equiv (\neg New_{21} \vee s_{21}) \wedge (New_{22} \vee \neg s_{22}) \wedge (\neg New_{22} \vee s_{21}) \wedge (New_{21} \vee \neg s_{22}) \wedge (\neg New_{21} \vee \neg New_{22} \vee s_{22}) \wedge (New_{21} \vee New_{22} \vee \neg s_{21})$.

Now, let us compute the profile K_{Σ_Υ}. *Initially,* $K_{\Sigma_\Upsilon} = \ <0, 0>$.

1. *At the first iteration, we put* $K_{\Sigma_\Upsilon}[1] = |S_1| = 2$. *So,* $\mathcal{C}(K_{\Sigma_\Upsilon}[1]) \equiv s_{11} \wedge s_{12}$ *and* $\mathcal{C}(K_{\Sigma_\Upsilon}[2]) \equiv \top$. *Then, we check whether* $\Phi_{\Sigma_\Upsilon} \wedge \mathcal{T}(N_1) \wedge \mathcal{T}(N_2) \not\models (\neg p \vee \neg b) \vee (\neg s_{11} \vee \neg s_{12})$. *This is the case, so we move to the next stratum.*
2. *Now, we put* $K_{\Sigma_\Upsilon}[2] = |S_2| = 2$. *Thus,* $\mathcal{C}(K_{\Sigma_\Upsilon}[2]) \equiv s_{21} \wedge s_{22}$. *Then, we verify if* $\Phi_{\Sigma_\Upsilon} \wedge \mathcal{T}(N_1) \wedge \mathcal{T}(N_2) \not\models (\neg p \vee \neg b) \vee (\neg s_{11} \vee \neg s_{12}) \vee (\neg s_{21} \vee \neg s_{22})$. *This is not checked, hence we put* $K_{\Sigma_\Upsilon}[2] = 1$ *and we test whether* $\Phi_{\Sigma_\Upsilon} \wedge \mathcal{T}(N_1) \wedge \mathcal{T}(N_2) \not\models (\neg p \vee \neg b) \vee (\neg s_{11} \vee \neg s_{12}) \vee (\neg s_{21})$. *This is checked. So,* $K_{\Sigma_\Upsilon} = \ <2, 1>$.

Then we have $\Phi_{\Sigma_\Upsilon} \wedge \mathcal{T}(N_1) \wedge \mathcal{T}(N_2) \models (\neg p \vee \neg b) \vee (\neg s_{11} \vee \neg s_{12}) \vee (\neg s_{21}) \vee (\neg f \wedge w)$. *Consequently,* $\Upsilon \Vdash_{\mathcal{L}ex} \alpha \rightarrow \beta$.

Now, let us suppose compiling the formula $\Phi_{\Sigma_\Upsilon} \wedge \mathcal{T}(N_1) \wedge \mathcal{T}(N_2)$ *into any target compilation* $\mathcal{L}$. *This corresponds to the compilation of* Σ_Υ *with respect to* $\mathcal{L}$, *i.e.,* $Comp(\Sigma_\Upsilon, \mathcal{L})$. *Clearly enough, based on* $Comp(\Sigma_\Upsilon, \mathcal{L})$, *all the previous clausal entailment tests will be achieved in polynomial time since, by definition, a target compilation language supports clausal entailment in polynomial time.*

7   Related Work

A basic way to compile the lexicographic inference consists in computing all the lexicographically preferred consistent subbases and then compiling their disjunction into a target compilation language. However, unlike our approach, when the priority relation between the strata changes, we have to recompile the base from scratch. Moreover, this basic way is not convenient with respect to the lexico-graphical closure since the corresponding lexicographically preferred consistent subbases vary on-line with respect to the query at hand.

Another approach for compiling the lexicographic inference has been proposed in [3]. However, the corresponding objective is to make the lexicographic infer-ence coNP and not polynomial like our approach. Also, this approach suffers from the same drawback as the basic way. Indeed, it is not possible to update the priority without recompiling the base, and it is not suitable to compile the lexicographical closure.

Now, the closest compilation approaches to ours are those proposed in [7,8,10]. Like our approach, they introduce a new variable per formula and offer the pos-sibility to change priorities between the strata without any re-compilation. More precisely, the compilation developed in [7] is based on the $\mathcal{OBDD}$ language and the approach proposed in [8] is efficient with respect to the $\mathcal{DNF}$ language. As for the one given in [10], it takes advantage of both $\mathcal{DNF}$ and $\mathcal{DNNF}$ languages. However, our approach can be efficiently parametrized by any tar-get compilation language while the previous approaches are not. In particu-lar, contrarily to them, our compilation is efficient with respect to the prime implicates language $\mathcal{PI}$. Indeed, $\mathcal{PI}$ and $\mathcal{OBDD}$ are incomparable when it comes to succinctness. Hence, some propositional formulas have $\mathcal{OBDD}$ rep-resentations which are exponentially larger than some of their $\mathcal{PI}$ representa-tions, and the converse also holds. Similarly, $\mathcal{PI}$ and $\mathcal{DNF}$ are incomparable. Moreover, $\mathcal{PI}$ is not more succinct than $\mathcal{DNNF}$ but we do not know whether $\mathcal{DNNF}$ is more succinct than it. Furthermore, note that until today, we can not satisfy the decomposability property of $\mathcal{DNNF}$ without enforcing other properties like determinism or structured decomposability which reduces the succinctness of $\mathcal{DNNF}$. Consequently, our approach completes those proposed in [7,8,10].

In addition, contrarily to the previous related work, we have efficiently ex-tended our approach to deal with the lexicographical closure inference.

8   Conclusion

In this paper, we have proposed a new compilation for the lexicographic inference from stratified belief bases. This approach relies on an efficient $\mathcal{CNF}$ encoding of Boolean cardinality constraints. This encoding makes our compilation flexible. Indeed, it does not imply any re-compilation whenever the priority relation be-tween strata changes. Moreover, we have shown that it can be extended in order to enable an efficient lexicographical closure inference. Furthermore, it completes existing approaches in the sense that it offers the possibility to take advantage of any target compilation language. This work calls for several perspectives. For in-stance, we aim at extending this approach to handle the lexicographic inference from partially preordered belief bases introduced in [22].

References

1. Bailleux, O., Boufkhad, Y.: Efficient CNF encoding of boolean cardinality constraints. In: Rossi, F. (ed.) CP 2003. LNCS, vol. 2833, pp. 108–122. Springer, Heidelberg (2003)
2. Benferhat, S., Dubois, D., Cayrol, C., Lang, J., Prade, H.: Inconsistency management and prioritized syntaxbased entailment. In: IJCAI 1993, pp. 640–645 (1993)
3. Benferhat, S., Kaci, S., Le Berre, D., Williams, M.A.: Weakening conflicting information for iterated revision and knowledge integration. In: IJCAI 2001(2001)
4. Benferhat, S., Bonnefon, J.-F., Neves, R.D.S.: An experimental analysis of possibilistic default reasoning. In: KR 2004, pp. 130–140 (2004)
5. Brewka, G.: Preferred subtheories: an extended logical framework for default reasoning. In: Proceedings of IJCAI 1989, pp. 1043–1048 (1989)
6. Cadoli, M., Donini, F.M.: A survey on knowledge compilation. AI Communications 10, 137–150 (1997)
7. Cayrol, C., Lagasquie-Schiex, M.C., Schiex, T.: Nonmonotonic reasoning: From complexity to algorithms. Ann. Math. Artif. Intell. 22(3-4), 207–236 (1998)
8. Coste-Marquis, S., Marquis, P.: On stratified belief base compilation. Annals of Mathematics and Artificial Intelligence 42(4), 399–442 (2004)
9. Darwiche, A., Marquis, P.: A knowledge compilation map. Journal of Artificial Intelligence Research 17, 229–264 (2002)
10. Darwiche, A., Marquis, P.: Compiling propositional weighted bases. Artificial Intelligence 157(1–2), 81–113 (2004)
11. Dubois, D., Lang, J., Prade, H.: Possibilistic logic. Handbook of Logic in Articial Intelligence and Logic Programming 3, 439–513 (1994)
12. Lagasquie-Schiex, M.C.: Contribution à l'étude des relations d'inférence non-monotone combinant inférence classique et préférences. Phd Thesis (1995)
13. Lehmann, D.: Belief revision revisited. In: Proceedings of IJCAI 1995, pp. 1534–1539 (1995)
14. Lehmann, D.J.: Another perspective on default reasoning. Annals of Mathematics and Artificial Intelligence 15(1), 61–82 (1995)
15. Nebel, B.: Base revision operations and schemes: semantics, representation and complexity. In: Proceedings of ECAI 1994, pp. 341–345 (1994)
16. Pearl, J.: System Z: A natural ordering of defaults with tractable applications to default reasoning. In: Proceedings of TARK 1990, pp. 121–135 (1990)
17. Rescher, N., Manor, R.: On inference from inconsistent premises. Theory and Decision 1, 179–217 (1970)
18. Selman, B., Kautz, H.: Knowledge compilation and theory approximation. Journal of the ACM 43(2), 193–224 (1996)
19. Sinz, C.: Towards an optimal cnf encoding of boolean cardinality constraints. In: van Beek, P. (ed.) CP 2005. LNCS, vol. 3709, pp. 827–831. Springer, Heidelberg (2005)
20. Warners, J.P.: A linear-time transformation of linear inequalities into conjunctive normal form. Inf. Process. Lett. 68(2), 63–69 (1998)
21. Würbel, E., Jeansoulin, R., Papini, O.: Revision: an application in the framework of GIS. In: KR 2000, pp. 505–515 (2000)
22. Yahi, S., Benferhat, S., Lagrue, S., Sérayet, M., Papini, O.: A lexicographic inference for partially preordered belief bases. In: KR 2008, pp. 507–517 (2008)

Improving Document Search Using Social Bookmarking

Hamidreza Baghi and Yevgen Biletskiy

University of New Brunswick, Fredericton, New Brunswick, Canada
{hamid.baghi,biletskiy}@unb.ca

Abstract. During the last decade, the use of community-based techniques has emerged in various data mining and search systems. Nowadays, many web search engines use social networking analysis to improve the search results. The present work incorporates one of the popular collaborative tools, called Social Bookmarking, into search. In the present paper, a technique, which utilizes Social Bookmarking information into search, is discussed.

1 Introduction

By the growth of the Semantic Web and Web 2.0, there are many collaborative tools available for users to express their thoughts, interests, and opinions both explicitly and implicitly. People use social friend finding websites, wikis, weblogs, and social bookmarking websites to share their interests and information. This huge amount of personal information can be very helpful if it is utilized to improve the search results, personalization, and the adaptation of websites.

Social bookmarking is one of the collaborative tools, in which people share links to web pages that are interesting for them. People have been using bookmarking since the beginning of the Web to store the links that they are interested in or that they use frequently. Usually people classify the bookmarks in different folders or tag them. Some people also write a description for a bookmark. Although people can store their bookmarks locally on their browser, there are websites that people can use to put their bookmarks on in order to be able to access the desired websites from anywhere and share them with others.

Bookmark data can be used to improve the effectiveness of text search in information retrieval. This data can be utilized to find correlation between pages and words. It will bring the semantic relations between pages, which are introduced by the users, into search. In this paper, a method utilizing bookmark data in information retrieval systems is presented.

2 Related Work

There has been much research work on using Collaborative Filtering techniques for search improvement. Smyth et al. introduce a method of search improvement for specific sites and portals [1]. The goal of the search engine introduced is to provide

Y. Gao and N. Japkowicz (Eds.): Canadian AI 2009, LNAI 5549, pp. 183–186, 2009.

the community, which searches a specific portal, with better results. For example, if somebody is searching about "jaguar" on a motoring website, she should be provided with results about cars, but not cats. The provided solution is to create a hit matrix, H. Each element of the hit matrix, H_{ij}, is the number of times page p_j is selected for query q_i. In new searches, the relevance is calculated using this hit matrix. Similarly, Balfe et al. [2] present a system, I-SPY, which attempts to reuse past search behaviors to re-rank the search results. This happens according to the implied preferences of like-minded community of users. They presented a query similarity model for query expansion. Chau et al. have designed a multi-agent system that enables cross user collaboration in Web search and mining [3]. In this system, one user can annotate a search session and share it with others. This annotation is used to improve the search results of other users. This system needs the user to provide explicit feedback to the system. Glance describes a community search assistant, which enables communities of searchers to search in a collaborative fashion by using query recommendation [4]. A query graph is created between the related queries. A user can follow a path on the query graph to find the related queries. The related queries are determined by the actual returned results, but not by the keywords in the queries. Zhuang el al. present an algorithmic model based on network flow theory that creates a search network using search engine logs to describe the relationships between relevant queries, documents, and users [5]. Bender et al. discuss the approach of P2P web search [6]. A sample system called Minerva is developed using this method. The idea of Minerva is to have some servers which hold some information about the peers. Each peer has a local index about pages that are crawled by this peer. Each peer sends term summaries of its local index to the global server. When a query is issued, the query is searched in the local index. If the result is not satisfactory, the query should be sent to other peers that are likely to know the answer. The result from various peers are gathered, merged, and provided to the user. The paper also introduces the concept of query routing.

Social bookmarking is also of interest to be used in search, recommendation, and personalization. The work [7] describes the basics of social bookmarking and introduces the social bookmarking sites. It states that there are generally two ways, which are used by users to categorize the bookmarks: using a taxonomy and using tags. Markines et al. provide a comprehensive study of using bookmarks in search, personalization, and recommendation [8]. They have built the GiveALink system, which gathers the bookmarks donated by people and analyzes them. The work have also provided a method based on hierarchy of bookmarks in order to find similar and related bookmarks in the collection. They also present different application of using these bookmark collection and smilarity measure.

3   Methodology

In this section, we present the methodology which is used to utilize social bookmarking to improve document search. Many techniques of Web Search, such as Collaborative Filtering and Link Analysis techniques, use implicit or explicit information created by the community of web users. One major part of this information created by users is the relation between information entities on the web. Users provide the relation information in various ways. A source of relation information, which is important and we think can be used to improve search, is *social bookmarking*. This includes social bookmarking websites such as delicious[1] and websites like digg[2], which are regularly updated with tagged links submitted by its users.

In social bookmarking, people bookmark a page in order to keep it for later use. They usually tag the page with one or more keywords related to the content of the page, which the link refers to, in order to organize the bookmarks. Each bookmark usually consists of a link to the page and a text describing the content of the page which is normally set to the title of the page. Some social bookmarking schemes, use hirerchical folders instead of tags. This is not very different from using tags because it can be easily translated to the tag scheme by considering each folder as a tag for all of its direct or indirect children.

Formally, there are: a set of tags (T), a set of bookmarks (B), and a set of users (U). Each bookmark is tagged with some tags by some users. Hence, a triple relation between elements of these sets exists. We show this relation as $\mathcal{G}$:

$$\mathcal{G} \subseteq U \times T \times B \qquad (1)$$

Each member of this relation (u, t, b) represents a single user u, which tagged bookmark b with tag t. These relations can be described as a tag-bookmark matrix. Tag-bookmark matrix can be built using different schemes. For example, the simplest way is to use a binary scheme. In this case each each member of the matrix indicates whether a bookmark is tagged by someone using a particular tag. Other schemes can take into account the number of user who did assign the same tag to the same bookmark. The amount that a user can be trusted can be measured and incorporated in the matrix.

The tag-bookmark matrix can be used to find relations that can not be elicited only by processing the text contents of the bookmarked pages. This matrix can find semantically related pages. This information helps improving search by including the semantic relations that are created by a large society of users.

4   Conclusion and Future Work

The work presented in this paper has described a collaborative information retrieval techniques using Social Bookmarking, which is used to improve results of information search in repositories and/or the web, based on identification and utilizing semantic relations between repository/web pages introduced by users and search keywords.

Bookmarks data can also be used for personalized search in order to improve the relevance of retrieved results to users' preferences, interests, and goals. The system should consider one's bookmarks and searching and browsing history. These bookmark files are shared on a social bookmarking server which is able

---

[1] http://delicious.com
[2] http://digg.com/

to analyze them. When a user searches for a document, the system takes into account her bookmarked web pages, the pages' contents, her browsing history, and the other bookmarked web sites on the web to bring more personalized results to the users. Based on the classification and tagging information and bookmarked web pages' contents, the system finds similar web pages from the bookmarked pages of the other users. The system either retrieves those pages (if they are relevant to the user's query) or uses the pages' content to expand the query and filter and reorder the results.

References

1. Smyth, B., Balfe, E., Boydell, O., Bradley, K., Briggs, P., Coyle, M., Freyne, J.: A Live-User Evaluation of Collaborative Web Search. International Joint Conference on Artificial Intelligence 19, 1419 (2005)
2. Balfe, E., Smyth, B.: An Analysis of Query Similarity in Collaborative Web Search. In: Losada, D.E., Fernández-Luna, J.M. (eds.) ECIR 2005. LNCS, vol. 3408, pp. 330–344. Springer, Heidelberg (2005)
3. Chau, M., Zeng, M., Chen, H., Huang, M., Hendriawan, D.: Design and evaluation of a multi-agent collaborative web mining system. Decis. Support Syst. 35(1), 167–183 (2003)
4. Glance, N.S.: Community search assistant. In: IUI 2001: Proceedings of the 6th international conference on Intelligent user interfaces, pp. 91–96. ACM, New York (2001), doi:10.1145/359784.360293
5. Zhuang, Z., Cucerzan, S., Giles, C.L.: Network Flow for Collaborative Ranking. In: Fürnkranz, J., Scheffer, T., Spiliopoulou, M. (eds.) PKDD 2006. LNCS, vol. 4213, p. 434. Springer, Heidelberg (2006)
6. Bender, M., Michel, S., Triantafillou, P., Weikum, G., Zimmer, C.: Minerva: collaborative p2p search. In: VLDB 2005: Proceedings of the 31st international conference on Very large data bases, pp. 1263–1266. VLDB Endowment (2005)
7. Hammond, T., Hannay, T., Lund, B., Scott, J.: Social bookmarking tools (i). D-Lib Magazine 11, 1082–9873 (2005)
8. Markines, B., Stoilova, L., Menczer, F.: Bookmark Hierarchies and Collaborative Recommendation. In: Proceedings of the National Conference on Artificial Intelligence, vol. 21(2), p. 1375. AAAI Press, MIT Press, Menlo Park, Cambridge (2006)

Rank-Based Transformation
in Measuring Semantic Relatedness

Bartosz Broda[1], Maciej Piasecki[1], and Stan Szpakowicz[2,3]

[1] Institute of Informatics, Wrocław University of Technology, Poland
{bartosz.broda,maciej.piasecki}@pwr.wroc.pl
[2] School of Information Technology and Engineering, University of Ottawa, Canada
szpak@site.uottawa.ca
[3] Institute of Computer Science, Polish Academy of Sciences, Poland

Abstract. Rank weight functions had been shown to increase the accuracy of measures of semantic relatedness for Polish. We present a generalised ranking principle and demonstrate its effect on a range of established measures of semantic relatedness, and on a different language. The results confirm that the generalised transformation method based on ranking brings an improvement over several well-known measures.

1 Introduction

The ability to measure relatedness between words is useful in many Natural Language Processing applications – from word sense disambiguation [1] to thesaurus construction [2]. Relatedness can be calculated using manually built lexical resources [3] such as WordNet [4], or extracted directly from a corpus. The former appears more reliable, considering that human experts have created the source of knowledge, but there are drawbacks. The coverage of any lexical resource is inevitably limited, and its construction costs time and money. This is not suitable: it is the main motivation of our research to facilitate the manual construction of a wordnet for Polish [5].

The corpus-based approach boils down to the construction of a *Measure of Semantic Relatedness* (MSR) – a real-valued function of pairs of *lexical units*[1].

Many ways of measuring semantic relatedness have been proposed. A typical successful scheme would base a description of an LU's meaning on the frequencies of its co-occurrences with other LUs in lexico-syntactic contexts [2,6,7,8]. A lexico-syntactic relation links a given LU with a specific lexical element via a specific syntactic connection. For example, we can describe the LU *bird* by subject_of(*bird,sing*). The influence of lexico-syntactic information on the accuracy of MSRs has been analysed in [6,9]. Each LU is described by a vector of features that correspond to all context types considered. The initial value of a feature is the frequency of occurrences of the given LU in the corresponding context. Raw frequencies must be *transformed* before the final computation of the

---

[1] We take a lexical unit (LU), a little informally, to be a lexeme – a one-word or multi-word expression that names a dictionary entry.

Y. Gao and N. Japkowicz (Eds.): Canadian AI 2009, LNAI 5549, pp. 187–190, 2009.

MSR value. A transformation – usually a combination of filtering and weighting – is necessary because some features deliver little or no information, and the raw frequencies can be accidental and biased by the corpus used. Transformations proposed in the literature mostly combine simple heuristics for frequencies and weighting. [2] presents a measure based on the ratio of the information shared and total description. [10] calculates semantic similarity of LUs using Pointwise Mutual Information (PMI) modified by a discounting factor.

[8] introduced a *Rank Weight Function* (RWF) as a means of generalising from the exact corpus frequencies in the calculation of feature values to the relative scores of importance assigned to features. In this work we want to show a generalisation of the RWF-based transformation to a model in which RWF follows the application of a weighting scheme. We also propose to remove the assumption of the linear order of features. Also, we want to test whether rank-based transformation also performs well on different data – for a different language using different language tools and resources.

2   Generalised RWF-Based Transformation

We assume that what contributes most information to a LU's description is not the feature's exact frequency. Instead of the (possibly weighted) frequencies, in RWF the feature value represents the difference of its relative *relevance* to the given LU in comparison to other features that describe this LU [8]. The relevance is measured using *z-score* (a *t-score* variant). *Generalised RWF* (GRWF) differs from RWF in two ways. We do not assume that features are linearly ordered, and we look at the application of weighting schemes other than z-score. This allow as to emphasise that GRWF can be seen as a general wrapper for other MSRs. We now describe the transformation based on the GRWF.

1. A co-occurrence matrix is created. We denote this matrix as $\mathbf{M}[w_i, c_j]$ – the co-occurrence frequency in a corpus of word w_i with feature c_j.
2. The frequencies in matrix $\mathbf{M}$ are transformed using a weighting function.
3. The subset F_{sel} of the most relevant features of the row vector is selected.
4. Values of every cell of $\mathbf{M}$ are quantised by applying a function f_Q.
5. Features from the set $F_{sel}[w_i]$ are sorted in the ascending partial order on the weighted values.
6. For each selected feature c_j a new value is calculated:
   $\mathbf{M}[w_i, c_j] = f_{top}(\mathbf{M}[w_i, \bullet]) - f_{por}(c_j)$, where
   - f_{top} defines the highest rank – two strategies possible. In an *absolute strategy* $f_{top} = k$, where k is high enough to be greater or equal than the highest number of relevant features over rows. In a *relative strategy*, $f_{top}(F_{sel}[w_i]) = size(F_{sel}[w_i]) + 1$, so the value of the best feature can vary across LUs and depend on the number of relevant features.
   - $f_{por}(c_j)$ calculates the position of c_j – two strategies possible: *natural*, when subsequent positions are numbered consecutively, and *with ties*, in which position j can be occupied by m features and the next position after such a tie is $j + m$.

Table 1. The accuracy of the MSRs for English in relation to nouns of different frequency (tf). There are 2103 questions for nouns with frequency $tf > 1000$ in BNC, 3343 for $tf > 500$, 7652 for $tf > 100$ and 14376 for $tf > 10$.

	Unmodified MSR				RWF				GRWF			
MSR / tf	1000	500	100	10	1000	500	100	10	1000	500	100	10
Lin	79.16	78.47	75.29	68.04	79.16	78.83	75.95	68.67	**79.73**	**79.25**	76.37	**69.41**
PMI	73.17	72.73	71.69	65.86	77.55	77.33	76.16	68.98	79.16	78.83	**76.65**	69.27
z-score	—	—	—	—	77.83	76.56	74.25	67.00	77.16	76.26	74.64	67.54

3   Experiments

For evaluation purposes, we used the *WordNet-Based Synonymy Test* (WBST) [11]. The test mimics the synonymy part of the *Test of English as a Foreign Language*, which was first used to evaluate an MSR in [12]. WBST consists of a set of questions that are automatically generated from a wordnet. For a pair of a near-synonyms, three other words, drawn randomly from a wordnet, act as *detractors*. For example, given *investigator* choose one of *critic*, **detective**, *east*, *offering*. Detailed analyses of the test's properties appear in [11,13].

In experiments we used well-known and popular resources and tools: WBST instances were generated from WordNet 3.0. Co-occurrence data was gathered from the British National Corpus [14] processed with MiniPar [15]. GRWF is parametrised in a few ways. We wanted to find whether there is any universal combination of parameters, and to avoid bias by over-fitting. That us why we split the test into two parts. Parameter adjustment was done on one part. The best combination of parameters was relative ranking with a high number of features ($k > 5000$). We found that simple rounding was best for a quantisation function f_Q (for PMI before rounding dividing values by 10 gave the best results). The results for the test part of WBST are shown in Table 1. In all cases the difference between an original MSR and GRWF is statistically significant (using McNemar test). Detailed result can be found on the Web[2].

4   Conclusions and Further Work

The RWF transformation had been successfully applied to Polish data. By performing similar experiments in a well-studied setting we showed that this success was not limited to one language and one set of language tools and resources. On WBST – for English and for Polish – GRWF achieved a significantly better accuracy than MSRs amply discussed in the literature: Lin's measure and PMI. The advantage of GRWF is becoming more apparent with the decrease of the frequency of words included in WBST.

The influence of the parameter values on MSR accuracy must be investigated more deeply. Most noticeably, investigation of other quantisation schemes should

---

[2] ⟨http://plwordnet.pwr.wroc.pl/msr⟩ contains the results for Polish and for the best measure analysed in [7]. The conclusions are similar.

be performed. Other open research questions include a systematic method of finding the best parameter values for a given application, a study of the formal properties of the GRWF transformation, applications to more languages, and a comparison of the GRWF-based MSR with additional MSRs. It is also worth noting that we have proposed a remarkably simple transformation, and turned away from the fine-tuning of weights. This *sui generis* stepping back from statistical processing has brought about a significant improvement, and has posed interesting questions about the appropriate way of representing lexical meaning in the spirit of distributional semantics.

References

1. Agirre, E., Edmonds, P. (eds.): Word Sense Disambiguation: Algorithms and Applications. Springer, Heidelberg (2006)
2. Lin, D.: Automatic retrieval and clustering of similar words. In: Proc. COLING 1998, ACL, pp. 768–774 (1998)
3. Budanitsky, A., Hirst, G.: Evaluating wordnet-based measures of semantic distance. Computational Linguistics 32(1), 13–47 (2006)
4. Fellbaum, C. (ed.): WordNet – An Electronic Lexical Database. MIT Press, Cambridge (1998)
5. Derwojedowa, M., Piasecki, M., Szpakowicz, S., Zawisławska, M.: plWordNet – The Polish Wordnet. Project homepage, http://www.plwordnet.pwr.wroc.pl
6. Ruge, G.: Experiments on linguistically-based term associations. Information Processing and Management 28(3), 317–332 (1992)
7. Weeds, J., Weir, D.: Co-occurrence retrieval: A flexible framework for lexical distributional similarity. Computational Linguistics 31(4), 439–475 (2005)
8. Piasecki, M., Szpakowicz, S., Broda, B.: Automatic selection of heterogeneous syntactic features in semantic similarity of Polish nouns. In: Matoušek, V., Mautner, P. (eds.) TSD 2007. LNCS (LNAI), vol. 4629, pp. 99–106. Springer, Heidelberg (2007)
9. Piasecki, M., Broda, B.: Semantic similarity measure of Polish nouns based on linguistic features. In: Abramowicz, W. (ed.) BIS 2007. LNCS, vol. 4439, pp. 381–390. Springer, Heidelberg (2007)
10. Lin, D., Pantel, P.: Concept discovery from text. In: Proc. COLING 2002, Taipei, Taiwan, pp. 577–583 (2002)
11. Freitag, D., Blume, M., Byrnes, J., Chow, E., Kapadia, S., Rohwe, R., Wang, Z.: New experiments in distributional representations of synonymy. In: Proc. Ninth Conf. on Computational Natural Language Learning, pp. 25–32 (June 2005)
12. Landauer, T., Dumais, S.: A solution to Plato's problem: The latent semantic analysis theory of acquisition. Psychological Review 104(2), 211–240 (1997)
13. Piasecki, M., Szpakowicz, S., Broda, B.: Extended similarity test for the evaluation of semantic similarity functions. In: Vetulani, Z. (ed.) Proc. 3rd Language and Technology Conference, Poznań, Poznań, Wydawnictwo Poznańskie Sp. z o.o (2007)
14. BNC: The British National Corpus, version 2 (BNC World). Distributed by Oxford University Computing Services on behalf of the BNC Consortium (2001)
15. Lin, D.: Principle-based parsing without overgeneration. In: Proc. 31st Meeting of the ACL, pp. 112–120 (1993)

Optimizing a Pseudo Financial Factor Model with Support Vector Machines and Genetic Programming

Matthew Butler and Vlado Kešelj

Faculty of Computer Science, Dalhousie University,
6050 University Ave., Halifax, NS, Canada B3H 1W5
{mbutler,vlado}@cs.dal,ca

Abstract. We compare the effectiveness of Support Vector Machines (SVM) and Tree-based Genetic Programming (GP) to make accurate predictions on the movement of the Dow Jones Industrial Average (DJIA). The approach is facilitated though a novel representation of the data as a pseudo financial factor model, based on a linear factor model for representing correlations between the returns in different assets. To demonstrate the effectiveness of the data representation the results are compared to models developed using only the monthly returns of the inputs. Principal Component Analysis (PCA) is initially used to translate the data into PC space to remove excess noise that is inherent in financial data. The results show that the algorithms were able to achieve superior investment returns and higher classification accuracy with the aid of the pseudo financial factor model. As well, both models outperformed the market benchmark, but ultimately the SVM methodology was superior in terms of accuracy and investment returns.

Keywords: support vector machines, genetic programming, financial forecasting, principle component analysis.

1 Introduction

We concentrate on and compare the results from a SVM and tree-based GP, performed in LIBSVM [1] and lilgp [2], respectively. To demonstrate the effectiveness of using the pseudo financial factor model the algorithm outputs are compared to models developed from using only the monthly changes of the inputs. The information that is used to generate the predictions is macro-economic data such as information on inflation and corporate bond ratings. The relationship between market movements and macro-economic data is not linear or monotonic. To assist in modeling these relationships a financial factor model is created that represents correlations between the market and each indicator in the input set. The combination of financial factor modeling with machine learning was explored by Azzini and Tettamanzai [3], where they implemented an evolving

Y. Gao and N. Japkowicz (Eds.): Canadian AI 2009, LNAI 5549, pp. 191–194, 2009.

neural network to create a factor model to explain the returns in the DJIA. The canonical form of a linear financial factor model is shown below:

$$r_i = b_{i1} \cdot f_1 + b_{i2} \cdot f_2 + \ldots + b_{im} \cdot f_m + \epsilon_i$$

r_i is the return on asset i, m is the number for factors, b_{ij} is the change in return of asset i per unit change in factor j, f_j is the change in return of factor j, and ϵ_i is the portion of the return in asset i not related to m factors. Traditionally, a financial factor model would be used to explain the returns of an asset by an equation and when the model output and the actual return begin to diverge, appropriate investments are made under the assumption that the two will converge again in the near future. As the name of the paper suggests, we are not using the equation in the traditional sense but changing the left hand side of the equation to be a class rather than a price level. A class of 1 indicates the DJIA will rise over the next month and -1 suggests it will fall over the same time period. The new pseudo equation is thus:

$$r_i = b_{i1} \cdot f_1 + b_{i2} \cdot f_2 + \ldots + b_{im} \cdot f_m$$

where $r_i \in \{1, -1\}$.

2   Data Description and Preprocessing

The data used to train the models was based on macro-economic data that was utilized by Enke and Thawornwong [4], where they created a market prediction model that outperformed the S&P 500 market index with the aid of a multilayer perceptron. The monthly changes in the indicators were combined with their respective β to create the inputs. The β calculation is shown below:

$$\beta(DJIA, X_i) = \frac{cov(DJIA, X_i)}{var(X_i)}$$

The β value is an indication of how much the market would move based on a unit movement of 1 in a given indicator; it was calculated on a rolling 10-year period.

Both algorithms were trained on data from 1977 to 2001 and then tested for 84 months or 7 years up until June 2008. The justification for the extended training period was to expose each model to market reactions during each stage of the business cycle. Initially the data is projected into principle component (PC) space where 99% of the variance is equated for, than the data is projected back into attribute space with a proper rank and excess noise removed.

3   Trading Strategy and Results

3.1   Trading Strategy

The experiment is setup as a semi-active trading strategy where at the beginning of each month a prediction is made as to whether or not the DJIA will contract or

Table 1. Testing results for each algorithm and data set

	SVM		GP	
	Factors	**No Factors**	**Factors**	**No Factors**
Overall Accuracy	69.05%	59.52%	66.67%	57.72%
Precision (contractions)	68.60%	60.00%	63.41%	55.55%
# of contraction predictions	35	25	41	27
Precision (expansions)	69.40%	59.30%	69.76%	57.89%
Yearly investment yield (%)	21.70%	4.13%	16.00%	4.04%
Cumulative return ($)	$4505	$1335	$3042	$1327
Excess return to the market[1]	20.60%	3.03%	14.90%	2.94%
Sharpe ratio[2]	3.994	0.822	2.813	0.795

expand over the coming month. If the prediction is for the market to go up, than the model will take a long position; conversely, if the market is predicted to fall, than a short position will be taken. Several financial instruments are available to short the DJIA, but essentially they all profit from market contractions.

3.2   Testing Results

The best model for both algorithms from each data set, determined by the training results, was supplied the out-of-sample data that spanned 84 months from 2001 up until June of 2008. In Table 1 we display the testing results for each algorithm. The investment returns are based off an initial investment of $1000 and for simplicity reasons transaction costs are ignored. Reported in the results is the Sharpe Ratio, which is a gauge of how much additional return the trading system generates for the extra risk it is exposed to—the higher the Sharpe ratio the better the risk-adjusted performance.

The testing results in Table 1 clearly show the advantages of using the financial factor model to create the inputs for the SVM and GP algorithms, where the overall accuracy and investment return were superior.

4   Discussion and Conclusions

In this study we compared the effectiveness of a novel data representation to optimize SVM and GP trading models to make accurate predictions on the movement of the DJIA. In each of the performance measures the algorithms achieved superior performance when the inputs reflected the pseudo financial factor model. Precision for contraction predictions was of particular interest in this study due to the trading strategy. Since we are investing directly in the DJIA and also using it as the benchmark the only way to overperform is to avoid market contractions. This can be done in one of two ways, exiting the market and investing in a risk-free rate or alternatively short-selling the market

---

[1] DJIA yearly investment return over testing period was 1.10%.
[2] The risk-free rate in the calculation was replaced by the market rate.

to profit directly from its decline. The later is a much more aggressive approach and was the one utilized in our study; therefore incorrect market contractions will lead to the investment losing money while the market is increasing. As a result a precision for contraction predictions of less that 50% will most likely lead to inferior investment returns.

The effectiveness of using the factor model could be explained by the fact that the algorithms are given more information about the problem with this type of data representation. Not only is the model training on the returns of the indicators but they are also supplied a ratio that describes the relationship between said indicator and the market. This enables the algorithm to have a more complete picture and therefore is able to create a more robust market model. Each of the models presented in this paper were able to outperform the DJIA, however the non-financial factor models did so by a much smaller margin. Ultimately the SVM proved to be the most effective in terms of risk and return, it's Sharpe ratio was the highest reflecting the most efficient use of the extra risk the model took on to achieve the excess returns. The obtained results for investment returns are not entirely accurate as transaction costs were ignored. However, because the trading strategy was semi-active and only made trades on a month to month basis, and only if required, the transaction costs would be less inhibitory to overall profits than that of other more active trading approaches.

References

1. Chang, C.C., Lin, C.J.: LIBSVM: A library for support vector machines (2001), http://www.csie.ntu.edu.tw/~cjlin/libsvm
2. Punch, B., Zongker, D.: Michigan State University GARAGe, Web site of Genetic Algorithms Research and Applications Group, GARAGe (1998), http://garage.cse.msu.edu/
3. Azzini, A., Tettamanzi, A.: A neural evolutionary approach to financial modeling. In: GECCO 2006: Proceedings of the 2006 Conference on Genetic and Evolutionary Computation, pp. 1605–1612 (2006)
4. Enke, D., Thawornwong, S.: The use of data mining and neural networks for forecasting stock market returns. Expert Systems with Applications 29, 927–940 (2005)

Novice-Friendly Natural Language Generation Template Authoring Environment

Maria Fernanda Caropreso[1], Diana Inkpen[1], Shahzad Khan[2], and Fazel Keshtkar[1]

[1] School of Information Technology and Engineering, University of Ottawa
{caropres,diana,akeshtka}@site.uottawa.ca
[2] DISTIL Interactive
s.khan2@distilinteractive.com

Abstract. Natural Language Generation (NLG) systems can make data accessible in an easily digestible textual form; but using such systems requires sophisticated linguistic and sometimes even programming knowledge. We have designed and implemented an environment for creating and modifying NLG templates that requires no programming knowledge, and can operate with a minimum of linguistic knowledge. It allows specifying templates with any number of variables and dependencies between them. It internally uses SimpleNLG to provide the linguistic background knowledge. We test the performance of our system in the context of an interactive simulation game.

1 Introduction

Natural Language Generation (NLG) is the process of constructing outputs from non-linguistic inputs [1]. NLG is useful in systems in which textual interaction with the users is required, as for example Gaming, Robotics, and Automatic Help Desks.

However, the use of the available NLG systems is far from simple. The most complete systems often require extensive linguistic knowledge, as in the case of the KMLP system [2]. A simpler system, SimpleNLG [6], requires Java programming knowledge. This knowledge cannot be assumed for the content and subject matter experts who develop eLearning systems and serious games. However, these individuals do need to interact with the NLG system in order to make use of the message generation capability to support their product development efforts. It is then necessary to provide them with an environment that will allow them to have access in a simpler way to the features they need of a specific NLG system.

We present an environment that provides simple access to the use of SimpleNLG in order to generate sentences with variable parts or templates. We developed this NLG Template Authoring Environment guided by the need of templates required for generating content in the ISO 14001[1] game developed by DISTIL Interactive[2]. The goal of this project was to provide the game content designers with an accessible tool

---

[1] The ISO 14001 game's formal title is 'Business in Balance: Implementing an Environmental Management System'.
[2] http://www.distilinteractive.com/

Y. Gao and N. Japkowicz (Eds.): Canadian AI 2009, LNAI 5549, pp. 195–198, 2009.

they could use to create and manipulate the NLG templates, and thus generate sentences that would support the narrative progression of the game.

In the rest of this paper we first introduce general concepts of NLG and some of the tools available. We then introduce Serious Games (or training games) and their need for NLG. With this we motivate the developing of our NLG Template Authoring Environment. We then describe its design, implementation, evaluation and capabilities to cover different aspects of the templates. We finish the paper presenting our conclusions and what is pending as future work.

2   Natural Language Generation and SimpleNLG

There are two widely adopted approaches to NLG, the 'deep-linguistic' and the 'template-based' [4]. The deep-linguistic approach attempts to build the sentences up from a wholly logical representation. An example of this type of system is KMPL [2]. The template-based NLG systems provide scaffolding in the form of templates that contain a predefined structure and some of the final text. A commonly quoted example is the Forecast Generator (FOG) system designed to generate weather reports [5].

SimpleNLG is an NLG system that allows the user to specify a sentence by giving its content words and their grammatical roles (such as subject or verb). It is implemented as a java library and it requires java programming knowledge to be used.

SimpleNLG allows the user to define flexible templates by using programming variables in the sentence specification. The variable parts of the templates could be filled with different values. When templates are used without an NLG system, they are called canned-text, and they have the disadvantage of not being very flexible, as only the predefined variables can change. When templates are defined using SimpleNLG, however, they keep all the functionality of the NLG system (for example, being able to modify the verb features or the output format, and making use of the grammatical knowledge), while also allowing for the variable values to change.

3   Serious Games and the Need for NLG

The term serious games refer to a sub-category of interactive simulation games in which the main objective is to train the player in a particular subject matter. The player is typically presented with challenging situations and is encouraged to practice different strategies at dealing with them, in a safe, virtual environment. Through tips and feedback provided during and at the end of the game, the player develops an understanding of the problem and what are the successful ways of confronting it [3].

The game ISO 14001 from DISTIL Interactive is an example of a serious game. The objective of this game is to train the player in the process of implementing an environmental management system (EMS) ISO 14001. The player controls the main character of the game, who manages the implementation of a standards-based process in a simulated fictional organization. All the other characters of the game are controlled by the computer. Feedback is provided as e-mail messages from other characters to the main character.

The amount of textual information required in serious games can be a burden on the game designers. It is then desirable to include templates that will statically provide the basic information, combined with variable parts that adapt the narrative to the circumstances. The templates were hard-coded in the game ISO 14001. In our current work, we propose the use of a more flexible way of generating templates for the dialog of the games. We present a NLG Template Authoring Environment that takes advantage of the grammatical knowledge of SimpleNLG in a simpler way. It does not require the user to have either advance linguistic or programming knowledge.

4 NLG Template Authoring Environment

The NLG Template Authoring Environment allows the user to input a model sentence template, to indicate its variables with their type and possible values, and to specify dependencies between variables. It then shows the user all the possible sentences that could be generated from the given template by calculating all the possible combinations of variable values that respect the specified dependencies. The user can then refine the template by changing either the given example or the specified variables and dependencies, in order to adjust the generated sentences to the needs of the game.

Figure 1 shows a graphical design for the NLG Template Authoring Environment and a simple example of a sentence with specified variables and dependencies. In this figure, the variables are indicated by circles and a dependency between variables is indicated by an arc connecting two circles.

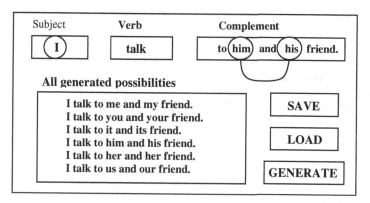

Fig. 1. Graphical Design for the NLG Template Authoring Environment

A prototype of the NLG Template Authoring Environment has been implemented in Java. It allows variables of different type, such as pronouns and some predefined noun subsets (e.i.: employee type). The SimpleNLG library was used to automatically generate correct sentences and provide the user with the possibility of exploring different attributes to the verb such as tense, form and mood.

In order to verify the correct functioning of the NLG Template Authoring Environment, we selected a set of sentence templates from the game ISO 14001. The templates were selected manually, while keeping in mind the need to cover different aspects, as for example the number and type of the variables and dependencies.

Templates that presented dependencies between variables were successfully generated with our system by declaring the variables and establishing the dependencies. In
templates with variations to the verb the user has to be aware and specify them, as for
example its tense and whether or not it uses modals such as "could" or "would". Another feature that the user needs to be aware of is the possibility of specifying dependencies between the main verb and a variable in the subject of the sentence.

5 Conclusions and Future Work

We have identified the need for a NLG Template Authoring Environment that allows
game content designers without linguistic and programming background to experiment with and finally design language templates. We have designed a system that
allows the user to specify an example sentence together with variables, its dependencies, and verb options that complete the template. This system shows the user all the
possible sentences that could be generated with the specified template. It can be used
to refine the template until it satisfies the user's needs. We have implemented a simple prototype that makes use of the SimpleNLG java library which provides us with
correct sentences and the possibility of including many verb variations. We have
evaluated our NLG Template Authoring Environment in a set of sentence templates
from the game ISO 14001 covering different characteristics.

In this version, we enforced some limitations to make the prototype manageable. In
particular, the current system is text only and does not allow for refinements. In the
future we will provide a user-friendly intuitive graphical interface that will allow the
user to iteratively make changes to the sentence, variables and dependencies definitions. In addition, in a future version a module will be added in order to allow users to
create their own subset of nouns or adjectives to be added as variables' type.

Acknowledgements

This work is supported by the Ontario Centres of Excellence (OCE) and Precarn
Incorporated.

References

[1] Bateman, J.A.: Natural Language Generation: an introduction and open-ended review of
the state of the art (2002)
[2] Bateman, J.A.: Enabling technology for multilingual natural language generation: the
KPML development environment. Journal of Natural Language Engineering 3(1), 15–55
(1997)
[3] French, D., Hale, C., Johnson, C., Farr, G.: Internet Based Learning: An introduction and
framework for higher education and business. Kogan Page, London (1999)
[4] Gagné, R.M., Briggs, L.J.: Principles of instructional design, 4th edn. Harcourt Brace-
Jovanovich, Fort Worth (1997)
[5] Goldberg, E., Driedger, N., Kittredge, R.I.: Using Natural-Language Processing to Produce Weather Forecasts. IEEE Expert: Intelligent Systems and Their Applications 9(2),
45–53 (1994)
[6] Reiter, E.: SimpleNlg package (2007),
http://www.csd.abdn.ac.uk/ereiter/simplnlg

A SVM-Based Ensemble Approach to Multi-Document Summarization

Yllias Chali[1], Sadid A. Hasan[2], and Shafiq R. Joty[3]

[1] University of Lethbridge
Lethbridge, AB, Canada
chali@cs.uleth.ca
[2] University of Lethbridge
Lethbridge, AB, Canada
hasan@cs.uleth.ca
[3] University of British Columbia
Vancouver, BC, Canada
rjoty@cs.ubc.ca

Abstract. In this paper, we present a Support Vector Machine (SVM) based ensemble approach to combat the extractive multi-document summarization problem. Although SVM can have a good generalization ability, it may experience a performance degradation through wrong classifications. We use a committee of several SVMs, i.e. Cross-Validation Committees (CVC), to form an ensemble of classifiers where the strategy is to improve the performance by correcting errors of one classifier using the accurate output of others. The practicality and effectiveness of this technique is demonstrated using the experimental results.

Keywords: Multi-Document Summarization, Support Vector Machines, Ensemble, Cross-Validation Committees.

1 Introduction

Although SVMs achieve high generalization with training data of a very high dimension, it may degrade the classification performance by making some false predictions. To overcome this drawback, a SVM ensemble is clearly suitable. The main idea of an ensemble is to construct a set of SVM classifiers and then classify unseen data points by taking a majority voting scheme. In this paper, we concentrate on performing query relevant and extractive multi-document summarization task as it is defined by DUC-2007 (Document Understanding Conference). We use the cross-validation committees [1] approach of constructing an ensemble to inject differences into several SVM classifiers. We then compare the ensemble system's performance with a single SVM system and a baseline system. The evaluation result shows the efficiency of ensemble approaches in this problem domain.

Y. Gao and N. Japkowicz (Eds.): Canadian AI 2009, LNAI 5549, pp. 199–202, 2009.
© Springer-Verlag Berlin Heidelberg 2009

1. Divide whole training data set D into v–fractions $d_1, \ldots, d_v$
2. Leave one fraction d_k and train classifier c_k with the rest of the data $(D - d_k)$
3. Build a committee from the classifiers using a simple averaging procedure.

Algorithm 1: Cross-Validation Committees Method

2   Cross-Validation Committees (CVC)

In this research, we use the *cross-validation committees* approach to build a SVM ensemble. CVC is a training set sampling method where the strategy is to construct the training sets by leaving out disjoint subsets of the training data [2]. The typical algorithm of the CVC approach [1] is presented in Algorithm 1.

3   Experimental Setup

3.1   Problem Definition

The task at DUC-2007 is defined as: *"Given a complex question (topic descrip-tion) and a collection of relevant documents, the task is to synthesize a fluent, well-organized 250-word summary of the documents that answers the question(s) in the topic"*. We consider this task and employ a SVM-based ensemble ap-proach to generate topic-oriented 250-word extract summaries for 25 topics of DUC-2007 document collection using DUC-2006 data set for training.

3.2   Data Labeling

We use an automatic labeling method to label our large data sets (DUC-2006 data) using ROUGE [3]. For each sentence in a topic, we calculate its ROUGE score corresponding to the given abstract summaries from DUC-2006. Then based on these scores, we choose the top N sentences to have the label $+1$ (summary) and the rest to have the label -1 (non-summary).

3.3   Feature Extraction

Each of the document-sentences is represented as a vector of feature-values. We extract several query-related features and some other important features from each sentence. The features we use are: n-gram overlap, Longest Common Sub-sequence (LCS), Weighted LCS (WLCS), skip-bigram, exact word overlap, syn-onym overlap, hypernym/hyponym overlap, gloss overlap, Basic Element (BE) overlap, syntactic tree similarity measure, position of sentences, length of sen-tences, Named Entity (NE), cue word match and title match [4,5,6].

3.4   SVM Ensemble

We use the cross-validation committees (CVC) approach to build a SVM en-semble. We divide the training data set (DUC-2006 data) into 4 equal-sized fractions and according to the CVC algorithm, each time we leave separate 25%

data out and use the rest 75% data for training. Thus, we generate 4 different SVM models. Next, we present the test data (25 topics of DUC-2007 data) before each of the generated SVM models which produces individual predictions to those unseen data. Then, we create the SVM ensemble by combining their predictions by simple weighted averaging. We increment a particular classifier's decision value, the normalized distance from the hyperplane to a sample by 1 (giving more weight) if it predicts a sentence as positive and decrement by 1 (imposing penalty), if the case is opposite. The resulting prediction values are used later for ranking the sentences. During training steps, we use the third-order polynomial kernel keeping the value of the trade-off parameter C as default. For our SVM experiments, we use SVM^{light1} package [7]. We perform the training experiments in WestGrid[2].

3.5   Sentence Ranking

In the Multi-Document Summarization task at DUC-2007, the word limit was 250 words. To meet this criteria, we rank sentences in a document set, then select the top N sentences. We use the combined decision values of the 4 different SVM classifiers to rank the sentences. Then, we choose the top N sentences until the summary length is reached.

3.6   Evaluation Results

In DUC-2007, each topic and its document cluster were given to 4 different NIST assessors, including the developer of the topic. The assessor created a 250-word summary of the document cluster that satisfies the information need expressed in the topic statement. These multiple "reference summaries" are used in the evaluation of our summary content. We evaluate the system generated summaries using the automatic evaluation toolkit ROUGE [3]. We generate summaries for the first 25 topics of the DUC-2007 data and tested our SVM ensemble's performance with a single SVM system and a baseline system.

Table 1. ROUGE measures for SVM Ensemble

Measures	R-1	R-L	R-W	R-SU
Precision	0.4081	0.3359	0.1791	0.1621
Recall	0.3705	0.3051	0.0877	0.1334
F-score	0.3883	0.3197	0.1177	0.1463

In Table 1, we present the ROUGE scores of our SVM ensemble system in terms of Precision, Recall and F-scores. We show the four important ROUGE metrics in the results: ROUGE-1 (unigram), ROUGE-L (LCS), ROUGE-W (weighted LCS with weight=1.2) and ROUGE-SU (skip bi-gram). Table 2 shows

---

[1] http://svmlight.joachims.org/
[2] http://westgrid.ca/

Table 2. ROUGE F-Scores for Diff. Systems

Systems	R-1	R-L	R-W	R-SU
Baseline	0.3347	0.3107	0.1138	0.1127
Single	0.3708	0.3035	0.1113	0.1329
Ensemble	0.3883	0.3197	0.1177	0.1463

the F-scores for a single SVM system, one baseline system and the SVM ensemble system. The single SVM system is trained on the full data set of DUC-2006 and the baseline system's approach is to select the lead sentences (up to 250 words) from each topic's document set. Table 2 clearly suggests that the SVM ensemble system outperforms both the single SVM system and the baseline system with a decent margin in all the ROUGE measures.

4   Conclusion

In this paper, we applied an effective supervised framework to confront query-focused text summarization problem based on Support Vector Machine (SVM) ensemble. The experimental results on the 25 document sets of DUC-2007 show that the SVM ensemble technique outperforms the conventional single SVM system and the baseline system which proves the effectiveness of the ensemble approach in the domain of supervised text summarization.

References

1. Parmanto, B., Munro, P.W., Doyle, H.R.: Improving committee diagnosis with re-sampling techniques. In: Advances in Neural Information Processing Systems, vol. 8, pp. 882–888 (1996)
2. Dietterich, T.G.: Ensemble methods in machine learning, pp. 1–15. Springer, Heidelberg (2000)
3. Lin, C.-Y.: ROUGE: A Package for Automatic Evaluation of Summaries. In: Proceedings of Workshop on Text Summarization Branches Out, Post-Conference Workshop of Association for Computational Linguistics, Barcelona, Spain, pp. 74–81 (2004)
4. Chali, Y., Joty, S.R.: Selecting sentences for answering complex questions. In: Proceedings of EMNLP (2008)
5. Edmundson, H.P.: New methods in automatic extracting. Journal of the ACM 16(2), 264–285 (1969)
6. Sekine, S., Nobata, C.A.: Sentence extraction with information extraction technique. In: Proceedings of the Document Understanding Conference (2001)
7. Joachims, T.: Making large-Scale SVM Learning Practical. In: Advances in Kernel Methods - Support Vector Learning (1999)

Co-Training on Handwritten Digit Recognition

Jun Du and Charles X. Ling

Department of Computer Science
The University of Western Ontario, London, Ontario, N6A 5B7, Canada
jdu42@csd.uwo.ca, cling@csd.uwo.ca

Abstract. In this paper, we apply a semi-supervised learning paradigm — co-training to handwritten digit recognition, so as to construct high-performance recognition model with very few labeled images. Experimental results show that, based on arbitrary two types of given features, co-training can always achieve high accuracy. Thus, it provides a generic and robust approach to construct high performance model with very few labeled handwritten digit images.

1 Introduction

Handwritten digit recognition is an active topic in pattern recognition research. To construct a high performance recognition model, a large amount of labeled handwritten digit images are always required. Labeling these images is boring and expensive, while obtaining unlabeled images is much easier and cheaper. Thus, it is crucial to develop a high performance model with very few labeled (and a large amount of unlabeled) handwritten digit images.

In this paper, we apply a semi-supervised learning paradigm — co-training to handwritten digit recognition. We extract different sets of features as different views from images, and apply co-training on these features to construct recognition model. The experiments on real-world handwritten digit images show that, based on arbitrary two types of given features, co-training can always construct high performance recognition model.

2 Co-Training on Few Labeled Handwritten Digit Images

To construct handwritten digit recognition model, some image features need to be extracted beforehand, such as Karhunen-Loeve features, Zernike moment features, Fourier features, and so on [1]. Most of these features are sufficient to construct a high performance recognition model (given enough labeled training images). Thus we can directly consider each type of features as a view of the data, and easily obtain two views by extracting two types of features from handwritten digit images. In addition, obtaining these features is also very cheap in handwritten digit scenario — we just need run different feature extraction programs on the same group of training images.

To illustrate the co-training framework on handwritten digit images, we consider handwritten digit recognition as a binary classification problem for two

Y. Gao and N. Japkowicz (Eds.): Canadian AI 2009, LNAI 5549, pp. 203–206, 2009.

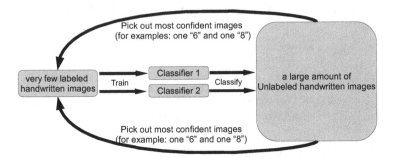

Fig. 1. Co-training framework to construct recognition model (classifier)

handwritten digits ("6" and "8"). Figure 1 shows the co-training framework. Given two independent and sufficient views (i.e., two sets of features), co-training uses very few labeled and a large amount of unlabeled handwritten digit images to incrementally construct high-performance recognition model. Specifically, it consists of three general steps.

In step 1, based on few labeled digit images, two classifiers are initially constructed according to two views (two types of features). These two classifiers are always very weak, because only small amount of the labeled images are initially included in the training set. In step 2, each classifier classifies all the unlabeled digit images, chooses the few images whose labels it predicts most confidently, and adds them (with the predicted labels) to the training set. For instance, to recognize handwritten "6" and "8", "Classifier 1" (as in Figure 1) selects one unlabeled image which it predicts to be "6" most confidently and one it predicts to be "8" most confidently, and adds them back to the training set, so does "Classifier 2" (as in Figure 1). In step 3, the two classifiers are re-constructed based on the updated training set, and the whole process repeats until some stopping criterion is met.

In this co-training framework, two classifiers "teach" each other with the unlabeled digit images which they predict most confidently, thus augmenting the recognition accuracy incrementally.

3   Experiments

We apply co-training to a real-world handwritten digit dataset (MFEAT) from the UCI Machine Learning Repository [2]. For each digit image, six types of features are extracted. Information on these features is tabulated in Table 1.

To simplify, we conduct the experiments on 200 handwritten "6" images and 200 handwritten "8" images for recognition, as they are relatively similar to each other.[1] we randomly pick out only 2 images (one "6" and one "8") as the training

---

[1] Utilizing co-training to recognize all the 10 handwritten digits is a multi-class form of co-training, which is essentially very similar to recognize only two digits (binary-class form of co-training). We simplify the recognition problem here, and will further study the performance of co-training for 10-digit recognition in our future work.

Table 1. The properties of six types of features for handwritten digit images. "No. of Features" lists the number of features in each type. "Accuracy (2 training images)" lists the accuracy of the recognition model constructed on only two labeled digit images (one "6" and one "8"); "Accuracy (300 training images)" lists the accuracy of the recognition model constructed on 300 labeled digit images (150 "6" and 150 "8").

		Accuracy	
Feature Type	No. of Features	2 training images	300 training images
Fourier Coefficients	76	73%	99%
Karhunen-Loeve Coefficients	64	74%	94%
Profile Correlations	216	77%	100%
Zernike Moments	47	39%	91%
Morphological Features	6	86%	99%
Pixel Averages	240	72%	98%

data, 100 images (50 "6" and 50 "8") as the test set, and consider all the rest (149 "6" and 149 "8") as the unlabeled data. In each iteration, each classifier (from each view) picks out one "6" and one "8", and adds them to the training set. The whole process repeats for 75 iterations to include all the unlabeled images into the training set. The naive Bayes (from the WEKA package [3]) is used to construct the classifiers and pick the most confident images. Only two of the given six types of features are required to apply co-training, thus we conduct the experiments on arbitrary two types of features — which yields a total of 15 combinations.

Table 1 lists the properties of these six types of features. The accuracy is evaluated in two cases. One considers that digit images in training set and unlabeled set are all labeled images (i.e., totally 300 training images) for constructing the model; the other considers that only two digit images in training set are labeled images (i.e., only two training images). We can see from Table 1 that, for all these six types of features, the recognition accuracy is always very high when given 300 training images. However, when given only two training images, the accuracy is much lower. In the following experiments, we expect that co-training could achieve a high recognition accuracy with only two labeled images and 298 unlabeled images.

The co-training experiments show that co-training works quite well with all the 15 feature combinations, and Figure 2 shows the performance of recognition model on two representative combinations (performance on the rest 13 feature combinations are similar to these two). Specifically, with some combinations (such as "Fourier Coefficients & Karhunen-Loeve Coefficients", "Fourier Coefficients & Zernike Moments", "Fourier Coefficients & Morphological Features", and so on), the accuracy increases promptly and significantly within few iterations, and then gradually approaches (or even exceeds) the upper bound, as shown in the left subfigure of Figure 2. With the other combinations (such as "Fourier Coefficients & Profile Correlations", "Fourier Coefficients & Pixel Averages", "Zernike Moments & Pixel Averages", and so on), the accuracy decreases in few iterations, but then immediately increases and approaches (or

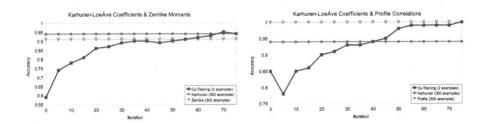

Fig. 2. Recognition accuracy of the co-training on two representative feature combinations. "Co-Training (2 examples)" represents the accuracy of the recognition model with co-training given only two labeled images and 298 unlabeled images. The other two lines represent the accuracy on the corresponding types of features given 300 labeled images.

even exceeds) the upper bound, as shown in the right subfigure of Figure 2. Consequently, the accuracy of the recognition model is always very high with co-training when all the unlabeled images are included in the training set (i.e., after 75 iterations).

4   Conclusion

We apply co-training to handwritten digit recognition in this paper. Specifically, we first extract different types of features from handwritten digit images as different views, and then use co-training to construct recognition model based on these views. Experimental results show that, based on arbitrary two types (among the given six types) of features, co-training can always achieve high recognition accuracy. Thus it provides a generic and robust approach to construct high performance recognition model with very few labeled (and a large amount of unlabeled) handwritten digit images.

References

1. Jain, A.K., Duin, R.P.W., Mao, J.: Statistical pattern recognition: a review. IEEE Transactions on Pattern Analysis and Machine Intelligence 22(1), 4–37 (2000)
2. Blake, C., Keogh, E., Merz, C.J.: UCI repository of machine learning databases, Department of Information and Computer Science, University of California, Irvine, CA (1998), http://www.ics.uci.edu/~mlearn/MLRepository.html
3. Witten, I.H., Frank, E.: Data Mining: Practical Machine Learning Tools and Techniques, 2nd edn. Morgan Kaufmann Series in Data Management Systems. Morgan Kaufmann, San Francisco (2005)

Evaluation Methods for Ordinal Classification

Lisa Gaudette and Nathalie Japkowicz

School of Information Technology and Engineering, University of Ottawa
lgaud082@uottawa.ca, nat@site.uottawa.ca

Abstract. Ordinal classification is a form of multi-class classification where there is an inherent ordering between the classes, but not a meaningful numeric difference between them. Little attention has been paid as to how to evaluate these problems, with many authors simply reporting accuracy, which does not account for the severity of the error. Several evaluation metrics are compared across a dataset for a problem of classifying user reviews, where the data is highly skewed towards the highest values. Mean squared error is found to be the best metric when we prefer more (smaller) errors overall to reduce the number of large errors, while mean absolute error is also a good metric if we instead prefer fewer errors overall with more tolerance for large errors.

1 Introduction

Ordinal classification, sometimes referred to as ordinal regression, represents a type of multi-class classification where there is an inherent ordering relationship between the classes, but where there is not a meaningful numeric difference between them [1]. This type of problem occurs frequently in human devised scales, which cover many domains from product reviews to medical diagnosis.

In this type of scenario, some errors are worse than others. A classifier which makes many small errors could be preferable to a classifier that makes fewer errors overall but which makes more large errors. This paper is motivated by work in ordinal sentiment analysis of online user reviews. In this domain, small errors are not so important as larger errors; humans are not perfect at detecting a 1 star difference on a 4 or 5 star scale [2], while classifying a 1 star review as a 5 star review is a very serious problem. In addition, this domain is highly imbalanced - a great deal of reviewers will rate a product as 5 stars. This paper will examine various evaluation measures in the context of this scenario.

2 Related Work

In recent years, there has been much discussion on the flaws of accuracy as a metric for comparing performance on machine learning tasks (see [3], among others). In addition to the flaws inherent in using accuracy for binary problems, in the ordinal case, accuracy tells us nothing about the severity of the error and in many applications this is important. Most papers on ordinal classification that

Y. Gao and N. Japkowicz (Eds.): Canadian AI 2009, LNAI 5549, pp. 207–210, 2009.

we have found simply use accuracy as an error measure, without considering whether or not it is an appropriate measure, such as [2, 1] among others. A few papers do use more interesting techniques; the "normalized distance performance measure" is used in [4] for work with image retrieval, and an AUC type measure for the ordinal case is introduced in [5], representing a volume under a surface.

While we have found little work discussing evaluation metrics for ordinal problems, there has been more work done comparing metrics for binary problems. A recent work which includes some multi-class problems is [6], which looks at correlations over a large variety of datasets and algorithms and then tests the metrics for sensitivity to different kinds of noise on an artificial binary problem.

3   Measures

Accuracy (ACC), Mean Absolute Error (MAE), and Mean Squared Error (MSE) are used as they are common, simple, evaluation metrics. Accuracy represents the number of correctly classified examples as a proportion of all examples, while MAE and MSE represent the mean of the absolute or squared difference between the output of the classifier and the correct label over all examples.

Linear Correlation (Correl) measures a linear relationship between two sets of numbers. A strong correlation between the classifier output and the actual labels should represent a good classifier.

Normalized Distance Performance Measure (NDPM) is a measure introduced by [7] that is designed for information retrieval in cases where a user has established relative preferences for documents, without necessarily referring to a predefined scale. NDPM is based on how often the user ranking and the system ranking are contradictory, and how often the user and system rankings are "compatible" - when the user ranking establishes a preference between two documents but the system ranking does not. In this sense it is a truly ordinal measure, and does not assume anything about the magnitude of the error between two classes.

Accuracy within n (ACC1, ACC2, etc) represents a family of measures which are similar to accuracy, however, they allow for a wider range of outputs to be considered "correct". In the case where the correct output is 4 stars, outputs of 3 stars and 4 stars would be considered accurate within 1. When there are k classes, accuracy within $k - 2$ includes all outputs as correct except for the worst possible kind of error, that is, mistaking class k with class 1 or vice versa. Traditional accuracy can be referred to as ACC0 in this context.

Used together, these measures provide a more qualitative picture of the performance of the classifier and the magnitude of it's errors, while greatly summarizing the information in a confusion matrix. However it should be noted that used alone these measures suffer from many of the same problems as accuracy, and most often we want a single number to summarize the performance of a system. One thing these measures can do is allow us to define a situation in which a classifier is indisputably superior to another - when accuracy within n is higher for each n, including 0, (or higher for at least one and equal for others).

While this concept is not entirely novel, it is not used frequently. For example, [2] uses a similar idea they call "Rating Difference" when describing human performance on their task.

Accuracy + Correlation (ACC+Correl) is simply the mean of accuracy and correlation. The motivation behind this combination is that including the information provided by correlation should improve on accuracy by providing some indication of the severity of the errors. We are not aware of any other work using this combination as a metric.

4   Experiments

In order to test the performance of the measures, we build classifiers using the DVD reviews in the dataset from [8]. We use reviews from the large "unlabeled" file, which does still include labels of 1, 2, 4, and 5 stars. In order to create features, we first use an algorithm which scores the words found in a set of 2500 randomly selected reviews based on how often they appear in positive vs. negative documents, and keep the top and bottom 25% of words as features. We then create a bag of words feature set based on the appearance of those words in a different set of 2500 documents. We then train classifiers on the features using WEKA [9] with default settings to perform 10 fold cross validation. We used 4 different classifiers: SMO, J48, and the OrdinalClassClassifier method introduced in [1] with each of SMO and J48 as base classifiers.

We also performed tests on artificial data, however, these results have been omitted for space reasons.

5   Results

Correlations between the different measures are shown in Table 1. MSE correlates most strongly with ACC1, ACC2, and ACC3, which shows that is is best at

Table 1. Correlations (absolute values) between measures - 4 classifiers, imbalanced dataset

	MAE	MSE	Correl	NDPM	ACC + Correl	ACC	ACC1	ACC2	ACC3
MAE	1	0.944	0.256	0.065	0.384	0.523	0.904	0.874	0.775
MSE	0.944	1	0.234	0.014	0.272	0.226	0.923	0.944	0.898
Correl	0.256	0.234	1	0.821	0.955	0.307	0.006	0.122	0.364
NDPM	0.065	0.014	0.821	1	0.819	0.366	0.285	0.117	0.222
ACC+Correl	0.384	0.272	0.955	0.819	1	0.576	0.069	0.138	0.343
ACC	0.523	0.226	0.307	0.366	0.576	1	0.205	0.104	0.096
ACC1	0.904	0.923	0.006	0.285	0.069	0.205	1	0.924	0.696
ACC2	0.874	0.944	0.122	0.117	0.138	0.104	0.924	1	0.744
ACC3	0.775	0.898	0.364	0.222	0.343	0.096	0.696	0.744	1

capturing differences in the accuracies within n while combining the information into a single measure. MAE also correlates reasonably well with the accuracies within n while correlating better with simple accuracy. However, all of the other measures are very far behind, while in the artificial tests they seemed much closer. In particular, correlation appears to be a terrible measure in practice while it was promising in the artificial tests.

6   Conclusions

Given the imbalanced dataset studied, MSE and MAE are the best performance metrics for ordinal classification of those studied. MSE is better in situations where the severity of the errors is more important, while MAE is better in situations where the tolerance for small errors is lower. This is despite the fact that neither of these measures are truly ordinal by design.

For future work, we would like to expand this analysis across a wider collection of datasets and methods in the domain of ordinal sentiment classification. We are also interested in exploring the evaluation of this problem in a more formal framework.

References

1. Frank, E., Hall, M.: A simple approach to ordinal classification. Technical Report 01/05, Department of Computer Science, University of Waikato (2001)
2. Pang, B., Lee, L.: Seeing stars: Exploiting class relationships for sentiment categorization with respect to rating scales. In: Proceedings of the 43rd Annual Meeting on Association for Computational Linguistics (ACL 2005) (2005)
3. Provost, F., Fawcett, T., Kohavi, R.: The case against accuracy estimation for comparing induction algorithms. In: Proceedings of the Fifteenth International Conference on Machine Learning (ICML 1998) (1998)
4. Wu, H., Lu, H., Ma, S.: A practical SVM-based algorithm for ordinal regression in image retrieval. In: Proceedings of the eleventh ACM international conference on Multimedia (MM 2003) (2003)
5. Waegeman, W., Baets, B.D., Boullart, L.: ROC analysis in ordinal regression learning. Pattern Recognition Letters 29, 1–9 (2008)
6. Ferri, C., Hernández-Orallo, J., Modroiu, R.: An experimental comparison of performance measures for classification. Pattern Recognition Letters 30, 27–38 (2009)
7. Yao, Y.Y.: Measuring retrieval effectiveness based on user preference of documents. Journal of the American Society for Information Science 46, 133–145 (1995)
8. Blitzer, J., Dredze, M., Pereira, F.: Biographies, Bollywood, boom-boxes and blenders: Domain adaptation for sentiment classification. In: Proceedings of the 45th Annual Meeting of the Association of Computational Linguistics (ACL 2007) (2007)
9. Witten, I.H., Frank, E.: Data Mining: Practical machine learning tools and techniques, 2nd edn. Morgan Kaufmann, San Francisco (2005)

STFLS: A Heuristic Method for Static and Transportation Facility Location Allocation in Large Spatial Datasets

Wei Gu[1], Xin Wang[1], and Liqiang Geng[2]

[1] Department of Geomatics Engineering, University of Calgary, Calgary, AB (Canada)
{wgu,xcwang}@ucalgary.ca
[2] NRC Institute for Information Technology, Fredericton, NB (Canada)
Liqiang.Geng@nrc-cnrc.gc.ca

Abstract. This paper solves a static and transportation facility location alloca-tion problem defined as follows: given a set of locations Loc and a set of de-mand objects D located in Loc, the goal is to allocate a set of static facilities S and a set of transportation facilities T to the locations in Loc, which minimizes both the average travelling distance from D to S and the maximum transporta-tion travelling distance between D and S through T. The problem is challenging because two types of facilities are involved and cooperate with each other. In this paper, we propose a static and transportation facility location allocation al-gorithm, called STFLS, to solve the problem. The method uses two steps of searching for static facility and transportation facility locations Experiments demonstrate the efficiency and practicality of the algorithm.

Keywords: Facility location problem, Static facility, Transportation facility.

1 Introduction

Facility location problem is an important research topic in spatial analysis [1]. The objective is to determine a set of locations for the supply so as to minimize the total supply and assignment cost. For instance, city planners may have a question about how to allocate facilities such as hospitals and fire stations for new residence area. The decision will be made based on the local populations and the capability of the limited resources. Various methods for the single type of facility location problem have been proposed for the above applications [1, 2, 3].

In reality, we often face two types of facilities location problem when the number of the single type of facilities within a service area is inefficient. For example, for emergency medical services, we can locate the hospital locations in such a way that it achieves full coverage of a service with the minimum total travelling distance. This usually ends up the hospital locations close to the dense community. However, for the residents in the sparse and remote area, since the number of hospitals is limited, in order to offer fast response time, the ambulance should be located to shorten the time to access medical services. In this application, two types of facilities need to be located in a region, static facilities (e.g. hospitals) and transportation facilities

Y. Gao and N. Japkowicz (Eds.): Canadian AI 2009, LNAI 5549, pp. 211–214, 2009.

(e.g. ambulances). The service is supplied to the customers by the cooperation of these two types of facilities. However, none of the current methods can apply to the two types of facilities location problem directly.

In the paper, the average travelling distance and transportation facility travelling distance are defined as follows:

Definition 1. Given a set of demand objects D and a set of static facilities S, the *average travelling distance (ATD)* is defined as:

$$ATD = \frac{\sum_{d_j \in D} dist(d_j, s_i) * d_j.w}{\sum_{d_j \in D} d_j.w}$$, where $s_i \in S$, $d_j \in D$ and s_i is d_j's assigned static

facility. $d_j.w$ is a positive number representing the demand of the demand object d_j.

Definition 2. Given a set of demand objects D, a set of static facilities S and a set of transportation facilities T, the *transportation travelling distance (TTD)* of the demand object d_j is defined as:

$$TTD(d_j) = dist(d_j, s_i \| t_k) + dist(d_j, s_i) \quad , \quad \text{where } s_i \in S \ , \ d_j \in D \ , \ t_k \in T \ \text{and}$$

$dist(d_j, s_i \| t_k)$ is the distance from a location of a demand object d_j to its assigned static facility location s_i or the closest transportation facility location t_k , whichever is shorter.

The main contribution of the paper is that we introduce a new type of facility location problem regarding to static and transportation facilities and propose a novel heuristic algorithm to solve the problem. Instead of only minimizing the average travelling distance between the static facilities and demand objects, the new problem also considers the constraint between the transportation facilities and static facilities and minimizes the maximum travelling distance of transportation facilities.

2 STFLS: Static and Transportation Facility Location Searching Algorithm

Static and transportation optimal facility locations problem is a NP-hard problem. In this section, we propose a heuristic method called Static Transportation Facilities Location Searching Algorithm (STFLS). The algorithm contains two steps (shown in Fig. 1): static facility location searching and transportation facility location searching. In the following, we will give a detail discussion on these two steps.

2.1 Static Facility Location Searching

Searching optimal locations for static facilities is a NP-hard problem. Various methods have been proposed [1, 3]. However, most methods are not efficient enough to deal with large spatial datasets. In this step, we propose a heuristic method to find local optimal locations using clustering. Clustering is the process of grouping a set of objects into classes so that objects within a cluster have high similarity to one another,

```
STFLS(D, S, T)
Input: a set of demand objects D, a set of static facili-
ties S with unknown locations, and a set of transporta-
tion facilities T with unknown locations
Output: locations of S and T
  /* static facilities locations searching step */
  1 SearchStaticFacilityLocations (D,S)
 /* transportation facilities locations searching step*/
  2 SearchTransportationFacilitiesLocations(D,S,T,
    threshold)
```

Fig. 1. Pseudo code of STFLS

but are dissimilar to objects in other clusters [4]. The clustering process is used to reduce the searching area.

In each iteration, there are two searching steps, intra-cluster searching and inter-cluster searching. Each static facility and the demand objects assigned to it would be seen as a cluster. In intra-cluster searching step, every static facility's optimal location in its cluster is found out. In the inter-cluster searching step, we compare the local optimal static facilities' location in every cluster and select one which can reduce the average distance most.

2.2 Transportation Facilities Location Searching

Locations of transportation facilities depend on both locations of demand objects and static facilities. To reduce the computation time, we use a greedy method in this step. The strategy is that it changes a transportation facility to the location whichever reduces the maximal transportation reachability distance most within each loop and stops if the exchange cannot bring the reduction of transportation reachability distance or the iteration time reaches the redefined times.

3 Experiments

3.1 Comparison between STFLS and Optimal Solution

Synthetic datasets for demand objects were created in a $300 \otimes 300$ area. The values in the following experiments are the average of the results which are from running the algorithm six times. To compare the performance of STFLS with the optimal solution, we generate a dataset with 100 demand objects and locate 3 static facilities and 2 transportation facilities. The weight of each demand object is 30 and the capability of

Table 1. The comparison between STFLS and the optimal solution

	STFLS	Optimal solution
Average traveling distance	48.7	46.5
Maximum traveling distance	199.0	192.3
Execution time (s)	1	200

each static facility is 1000. Table 1 shows that the optimal solution has a better performance than STFLS but it is time consuming.

3.2 Comparison between STFLS and Capacitated p-Median

In this section, we compare STFLS with a capacitated p-median algorithm in [5]. The comparisons are taken under a with 10000 demand objects and locate 10 static facilities and 5 transportation facilities. The weight of each demand object is 30 and the capability of each static facility is 40000. STFLS can reduce the average travelling distance from 29.7 to 27.8 and reduce the maximum transportation travelling distance from 238.3 to 187.6.

Table 2. The comparison between STFLS and the capacitated p-median algorithm

	STFLS	Capacitated p-median
Average traveling distance	27.8	29.7
Maximum traveling distance	187.6	238.3

4   Conclusion and Future Work

In this paper, we introduce a new type of facility location problem with both static and transportation facilities and propose a novel heuristic algorithm STFLS to solve it. STFLS assigns two types of facilities to the local optimal locations based on the spatial distribution of demand objects and the dependency of the facilities. To our knowledge, STFLS is the first algorithm to handle the location allocation problem about two types of facilities. Experimental results show that STFLS can allocate two types of facilities into an area in a reasonable accuracy and efficiency.

References

1. Owen, S.H., Daskin, M.S.: Strategic facility location: A review. European Journal of Operational Research 111(3), 423–447 (1998)
2. Longley, P., Batty, M.: Advanced Spatial Analysis: The CASA Book of GIS. ESRI (2003)
3. Arya, V., Garg, N., Khandekar, R., Pandit, V., Meyerson, A., Mungala, K.: Local search heuristics for k-median and facility location problems. In: Proceedings of the 33rd Annual ACM Symposium on the Theory of Computing, pp. 21–29 (2001)
4. Han, J., Kamber, M., Tung, A.K.H.: Spatial Clustering Methods in Data Mining: A Survey. In: Miller, H., Han, J. (eds.) Geographic Data Mining and Knowledge Discovery. Taylor and Francis, Abington (2001)
5. Ghoseiri, K., Ghannadpour, S.F.: Solving Capacitated P-Median Problem using Genetic Algorithm. In: Proceedings of International Conference on Industrial Engineering and Engineering Management (IEEM), pp. 885–889 (2007)

An Ontology-Based Spatial Clustering Selection System

Wei Gu[1], Xin Wang[1], and Danielle Ziébelin[2]

[1] Department of Geomatics Engineering, University of Calgary (Canada)
{wgu,xcwang}@ucalgary.ca
[2] STEAMER-LIG, University Joseph Fourier (France)
Danielle.Ziebelin@imag.fr

Abstract. Spatial clustering, which groups similar spatial objects into classes, is an important research topic in spatial data mining. Many spatial clustering methods have been developed recently. However, many users do not know how to choose the most suitable spatial clustering method to implement their own projects due to lack of expertise in the area. In order to reduce the difficulties of choosing, linking and executing appropriate programs, we build a spatial clustering ontology to formalize a set of concepts and relationships in the spatial clustering domain. Based on the spatial clustering ontology, we implement an ontology-based spatial clustering selection system (OSCS) to guide users selecting an appropriate spatial clustering algorithm. The system consists of the following parts: a spatial clustering ontology, an ontology reasoner using a task-model, a web server and a user interface. Preliminary experiments have been conducted to demonstrate the efficiency and practicality of the system.

Keywords: Ontology, Spatial clustering, Knowledge representation formalism.

1 Introduction

With the rapid growth of volume of spatial datasets, spatial clustering becomes an important topic in knowledge discovery research. It aims to group similar spatial objects into classes and is useful in exploratory pattern-analysis, grouping, decision-making, and machine-learning [1]. However, most existing clustering algorithms do not consider semantic information during the clustering process. Thus, spatial clustering users need to be familiar with spatial clustering methods in order to choose a most suitable spatial clustering method for their applications. Thus, providing knowledge support in clustering will be helpful for common users.

Classification of spatial clustering methods is not easy since the categorization of spatial clustering methods is not unique [1]. In addition, there is no clear line between different categories and constraints appear among the concepts. An *ontology* is a formal explicit specification of a shared conceptualization [2]. It provides domain knowledge relevant to the conceptualization and axioms [3].

In this paper, we propose an ontology-based spatial clustering selection system. The purpose of the system is to guide users selecting an appropriate spatial clustering algorithm. Through designing a spatial clustering ontology, we can provide knowledge support to common users in a semantic level.

Y. Gao and N. Japkowicz (Eds.): Canadian AI 2009, LNAI 5549, pp. 215–218, 2009.

2   Related Works

As for spatial clustering ontology, Hwang introduced a high-level conceptual framework which includes a user-interface, metadata, a domain ontology, a task ontology, and an algorithm builder [4]. The advantage of the framework is shown by two examples: the first one contrasts two task ontologies at different levels of scale; the second one contrasts two domain ontologies, one with an on-water constraint and another without. However, no formal spatial ontology has been used in specifying the domain ontology. Wang and Hamilton proposed an ontology-based spatial clustering framework [3]. With the framework, users give their goals in natural language. The framework uses the ontology query language which is translated from users' goals to identify the proper clustering methods and the appropriate datasets based on a spatial clustering ontology. However, they did not discuss how to generate spatial clustering ontology in the framework.

3   OSCS: A Ontology-Based Spatial Clustering Selection System

In this section, we propose a novel system called Ontology-based Spatial Clustering Selection System (OSCS). The system helps users selecting an appropriate spatial clustering algorithm. The architecture of the system is shown in Fig. 1. The system consists of four components: A spatial clustering ontology, an ontology reasoner, a web server and a user interface.

The system performs spatial clustering under the following steps. First, the spatial clustering ontology is generated in a web ontology language and is posted on the Internet. Secondly, the users' goal as clustering demands is translated into an objective task that can perform reasoning on the ontology by Ontology Reasoner component. Then, the appropriate spatial clustering algorithms are selected from the reasoning results. Thirdly, the selected clustering algorithm performs clustering on the datasets and clustering results are returned to users.

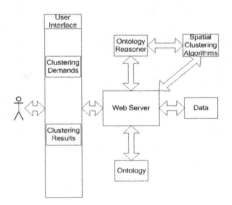

Fig. 1. The architecture of OSCS

3.1 Spatial Clustering Ontology

Spatial clustering ontology in the system is a formal representation of a set of concepts and relationships within the spatial clustering domain. The ontology is used to explicitly represent the meaning of terms in vocabularies and make the information be easily accessed by computers. In the system, we use OWL [5] to generate a spatial clustering ontology. The spatial clustering methods in the ontology are organized by a hierarchical categorization of spatial clustering algorithms and spatial clustering characteristics.

The hierarchical classification of spatial clustering algorithms is developed in terms of clustering techniques [1]. Every spatial clustering algorithm is stored under a right node in the tree structure.

Some spatial clustering algorithms can be categorized into more than one category. As a supplement of a hierarchical classification, we define a set of characteristics to classify spatial clustering algorithms. The characteristics include: *Assignment Way*, *Attributes Type*, *Constraint*, *Dataset Size*, *Dimensionality*, *Distance Measure*, *Measure Way*, *Noise Points Influence*, *Search Way*.

3.2 Ontology Reasoner

The ontology reasoner in the system is used to reason the useful knowledge in the ontology. The input of the reasoner is the semantics information given by users, and the output is a set of appropriate spatial clustering algorithms. The reasoner in the system is based on Pellet [6] which is an open-source Java based OWL-DL reasoner. The reasoner performs the reasoning using the following steps. First, we build a task model [7] to associate the reasoning to the spatial clustering ontology and user's clustering goals. Secondly, each present spatial clustering algorithm of the ontology may be considered as an inference (elementary task) directly associate with a function or as a complex task which combines with a decomposition method represented in a problem solving strategy. Finally we propose to recursively decompose each complex task into elementary sub-tasks.

3.3 User Interface and Web Server

In OSCS, a user interface is implemented to receive the semantics requirements from users, transfer the information to web server and return clustering results to users. Currently, we implement two types of requirement input interfaces. For experts, we supply them a tree view interface which the spatial clustering algorithms are organized in a hierarchical organization. Based on their background knowledge, professional users can explore in the tree to find out the suitable algorithms under the tree nodes. From the characteristics view interface, the users' requirements are described by type of attributes, scalability requirement, dimensionality of data, assignment method, outliers handling, defined parameters and constraints. Users can select the characteristics from the interface, then the system would find out a set of suitable algorithms as the requirement.

4 An Example Application

In this section, we use an application for facility location problem to test the system. The data used in this application is a population data in South Carolina. The data set consists of 867 census tracts (Census2000). In the application, we use a spatial clustering algorithm to find suitable locations for building five hospitals and assign their service area.

The system recommends to use a spatial clustering algorithm, Capability K-MEANS [8]. Table 1 shows the comparison between Capability K-MEANS and K-MEANS (the most popular algorithm in facility location problem). The Capability K-MEANS gets a better performance by reducing the Average traveling distance and Maximum traveling distance.

Table 1. The comparison between Capability K-MEANS and K-MEANS

	Capability K-MEANS	K-MEANS
Average traveling distance	59.7 km	61.3km
Maximum traveling distance	322km	328 km

5 Conclusion and Future Work

The major contribution of our work is to develop an ontology-based spatial clustering reasoning system. Based on the spatial clustering ontology, the system can guide users to choose the most suitable spatial clustering algorithm without mastering the knowledge in spatial clustering domain. Through some preliminary experiments, it is proved that the system can select proper clustering methods for different applications. In the future, we will evaluate the system on a wide range of problems.

References

1. Han, J., Kamber, M., Tung, A.K.H.: Spatial Clustering Methods in Data Ming: A Survey. In: Geographic Data Mining and Knowledge Discovery, pp. 1–29. Taylor & Francis, Abington (2001)
2. Gruber, T.R.: A Translation Approach to Portable Ontologies Specifications. Knowledge Acquisition 5(2) (1993)
3. Wang, X., Hamilton, H.J.: Towards An Ontology-Based Spatial Clustering Framework. In: Proceedings of 18th Conference of the Canadian Society for Computational Studies of Intelligence, pp. 205–216 (2005)
4. Owl Web Ontology Language Overview, http://www.w3.org/TR/owl-features
5. Hwang, J.: Ontology-Based Spatial Clustering Method: Case Study of Traffic Accidents. Student Paper Sessions, UCGIS Summer Assembly (2003)
6. Pellet website, http://pellet.owldl.com
7. Crubézy, M., Musen, M.: Ontologies in Support of Problem Solving. In: Staab, S., Studer, R. (eds.) Handbook on Ontologies, pp. 321–341. Springer, Heidelberg (2004)
8. Ng, M.K.: A note on constrained k-means algorithms. Pattern Recognition, 515–519 (2000)

Exploratory Analysis of Co-Change Graphs for Code Refactoring

Hassan Khosravi and Recep Colak

School of Computing Science, Simon Fraser University Vancouver, Canada

Abstract. Version Control Systems (VCS) have always played an essential role for developing reliable software. Recently, many new ways of utilizing the information hidden in VCS have been discovered. Clustering layouts of software systems using VCS is one of them. It reveals groups of related artifacts of the software system, which can be visualized for easier exploration. In this paper we use an Expectation Maximization (EM) based probabilistic clustering algorithm and visualize the clustered modules using a compound node layout algorithm. Our experiments with repositories of two medium size software tools give promising results indicating improvements over many previous approaches.

Keywords: Clustering, Software artifacts, Expectation Maximization.

1 Introduction

VCS are used to manage the multiple revisions of the same source of information and are vital for developing reliable software. They are mainly used in medium to large size projects so that many developers can simultaneously alter different parts of the code without interfering with each other's work. VCS repositories keep many versions of the code and has the ability to resolve conflicts on parts of the code that several people have changed simultaneously. On top of their regular operational duties, VCS contain very useful hidden information, which can be used to reverse engineer and refactor the codes. The underlying idea behind such methods is the fact that dependent software artifacts, which are usually elements of a submodule, co-occur in VCS transactions. Although this is the intended behavior, quality of code usually decays over time. In such cases, software artifacts of unrelated modules start to co-occur or co-change together. This is a strong indication of quality decay, which must be monitored and treated. Several approaches, most notably from the pattern mining community, have been introduced to attack this problem. Clustering is among the most famous approaches used.

The main contribution of this paper is using a probabilistic clustering algorithm on VCS repositories. Existing VCS repository mining algorithms mostly use hard clustering techniques, whereas we propose using a probabilistic soft clustering algorithm followed by a powerful compound graph based layout method to help the end user to easily comprehend the results.

Y. Gao and N. Japkowicz (Eds.): Canadian AI 2009, LNAI 5549, pp. 219–223, 2009.

2 Method

In this paper we use the concept of co-change graphs as introduced in [2]. A co-change graph is an artificial graph constructed from VCS repository that abstracts the information in the repository. Let $G = (V, E)$ be a graph in which V is the set of software artifacts. An undirected edge $(v_1, v_2) \in E$ exists, if and only if artifacts v_1 and v_2 co-occur in at least a predefined number of commit transactions. This graph is the input of the probabilistic graph clustering algorithm.

The exploited clustering algorithm which is based on Expectation Maximization (EM) framework, has only recently been applied to graph clustering. The intuition behind Expectation maximization is very similar to K-means, the most famous clustering technique [4]. Both EM and K-means algorithms have two steps; an expectation step followed by a maximization step. The first step is with respect to the unknown underlying model parameters using the current estimate of the parameters. The maximization step then provides a new estimate of the parameters. Each step assumes that the other step has been solved. Knowing the assignment of each data points, we can estimate the parameters of the cluster and the parameters of the distributions, assign each point to a cluster. Unlike kmeans, expectation maximization can use soft clustering in which variable are assigned to each cluster with a probability equal to the relative likelihood of that variable belonging to the class.

We first define the algorithm as a black box. It takes an adjacency graph A_{ij} of a graph G and the number of clusters c as input. The algorithm outputs the posterior probabilities q_{ir} denoting the probability of node v_i belonging to cluster r. Finally, assignment of nodes to clusters is done using these posterior probabilities.

Let Π_r denote the fraction of vertices in cluster r. We initialize the probability of each π_r with $(n/c + noise)$, as recommended by authors. Let θ_{ri} denote the probability that an edge from a particular node in group r connects to node i. This matrix shows how nodes in a certain cluster are connected to all the nodes of the graph. The aim of the algorithm is to represent a group of nodes for each cluster that all have similar patterns of connection to others. The parameters π_r, θ_{ri} and q_{ir} satisfy the normalization conditions:

$$\sum_{r=1}^{c} \pi_r = \sum_{i=1}^{n} \theta_{ri} = \sum_{r=1}^{c} q_{ir} = 1 \tag{1}$$

In the expectation (E) step of the algorithm the model parameter q_{ir} is updated assuming all the other model parameters is fixed.

$$q_{ir} = \frac{\pi_r \prod_j \theta_{rj}^{A_{ij}}}{\sum_s \pi_s \prod_j \theta_{sj}^{A_{ij}}} \tag{2}$$

In the Maximization (M) step of the algorithm, model parameters π_r and and θ_{rj} are updated assuming fixed posterior probabilities. The EM framework guarantees that the algorithm converges. The convergence happens when the log likelihood ceases to increase. Equations 2 and 3 show how the two steps are dependent on each other. To find the variable assignments we need to know the values of the model's parameters, and to update the parameters of the model the variable assignments are required.

$$\pi_r = \frac{1}{n} \sum_i q_{ir}, \qquad \theta_{rj} = \frac{\sum_i A_{ij} q_{ir}}{\sum_i k_i q_{ir}} \tag{3}$$

EM based clustering algorithm requires the number of cluster c to be given as input. This is a challenging problem. A practical solution is the Bayesian Information Criterion (BIC) which suits well with model based approaches Given any two estimated models, the model with the lower value of BIC is preferred. In our problem definition, n is the number of software artifacts. The free parameters are θ and π. Formula 4 is used to compute the BIC score. By changing the value of c, different models and different scores are achieved. we select the c that gives the minimum BIC score.

$$BIC = -2 \sum_{i=1}^{n} [\ln(\pi_{g_i}) + \sum_{j=1}^{n} A_{ij} \ln_{\theta_{g_ij}}] + (c + c*n)n \tag{4}$$

Finally, we use the compound layout algorithm as defined in [3] to visualize the result. We generate a compound node for every cluster and add all nodes to it as member nodes. Then we run the compound graph layout algorithm, which is an improved force directed layout algorithm tailored for nested graphs, using the Chisio[1] tool. We remove the compound nodes from the visualization Thus, compound nodes serves only for better layout and they are never shown to the end user.

3   Evaluation

We evaluate our clustering method by applying it to the VCS repositories of two software systems and comparing the results to authoritative decompositions. The clustering results are presented using compound node layouts that show software artifacts and their dependencies at class level. The two softwares have different sizes and different number of clusters. Because the evaluation requires the knowledge of authoritative decompositions, we chose systems that we are familiar with. We use Crocopat2.1 and Rabbit 2.1 as our examples and Figure 1 right presents some information about them.

Crocopat 2.1 is an interpreter for the language RML(relational manipulation language). It takes as input an RML program and relations, and outputs resulting relations. We have experimented different number of clusters 100 time each. Figure 1 shows the average results. The results indicates 4 clusters must be used confirming the authoritative decomposition. According to the authoritative decomposition, Crocopat has four major subsystems. The output of the probabilistic graph clustering algorithm clustering for Crocopat is shown in Figure 2 left.

---

[1] Chisio: Compound or Hierarchical Graph Visualization Tool.

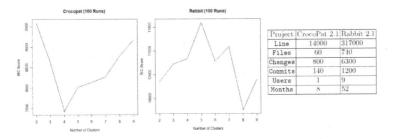

Project	CrocoPat 2.1	Rabbit 2.1
Line	14000	317000
Files	60	740
Changes	800	6300
Commits	140	1200
Users	1	9
Months	8	52

Fig. 1. BIC is used on Crocopat and Rabit in order to find the appropriate number of clusters to be fed to the EM algorithm. The table on the right presents some information about the Crocopat and Rabbit.

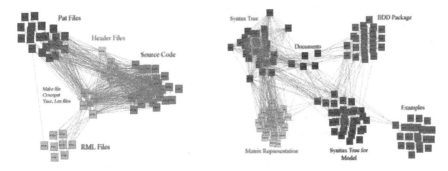

Fig. 2. Artifacts of Crocopat (one the right) and Rabbit(on the left) clustered using probabilistic fuzzy algorithm and visualized using compound nodes

Rabbit 2.1 is a model checking tool for modular timed automata. It is a command line program which takes a model and a specification file as input and writes out verification results. Figure 1 represents the average score for each number of clusters. Authoritative decomposition indicates that the number of clusters should be 6 whereas BIC score designate that 6 is the second best score and the best score is given by 8 clusters. We clustered the software using 6 clusters and the result achieved is illustrated in Figure 2. The clusters have been labeled on the Figure.

The result for clustering of both softwares were approved by the developers of both tools and have been found to be as good as, if not better than the existing approaches. Detailed analysis is omitted due to space considerations.

4   Conclusion

In this work, we proposed a probabilistic clustering approach followed by a compound graph visualizing in order to reverse engineer a software project. The algorithm is independent of the platforms, languages and tools the project was built on. It relies only on VCS files and it has internal tools for estimating best number of components of the system . We have shown that our approach gives useful results in finding components and the relations between them, which can be used for quality monitoring and code refactoring.

Acknowledgments. This article is the result of a student project in the software-engineering course CMPT 745 in fall 2008 at Simon Fraser University, and we thank Dirk Beyer for interesting discussions.

References

1. Newman, M., Leicht, E.: Mixture models and exploratory analysis in networks. In: PNAS 2007 (2007)
2. Beyer, D., Noack, A.: Clustering Software Artifacts Based on Frequent Common Changes. In: IWPC 2005 (2005)
3. Dogrusoz, U., Giral, E., Cetintas, A., Civril, A., Demir, E.: A Compound Graph Layout Algorithm for Biological Pathways. In: Pach, J. (ed.) GD 2004. LNCS, vol. 3383, pp. 442–447. Springer, Heidelberg (2005)
4. Dubes, R., Jain, A.: Algorithms for Clustering Data. Prentice-Hall, Englewood Cliffs (1998)

Classifying Biomedical Abstracts Using Committees of Classifiers and Collective Ranking Techniques

Alexandre Kouznetsov[1], Stan Matwin[1], Diana Inkpen[1], Amir H. Razavi[1], Oana Frunza[1], Morvarid Sehatkar[1], Leanne Seaward[1], and Peter O'Blenis[2]

[1] School of Information Technology and Engineering (SITE), University of Ottawa
akouz086@uottawa.ca, {stan,diana,araza082,ofrunza,
mseha092}@site.uottawa.ca, lspra072@uottawa.ca
[2] TrialStat Corporation, 1101 Prince of Wales Drive, Ottawa, ON, CA, K2C 3W7
poblenis@trialstat.com

Abstract. The purpose of this work is to reduce the workload of human experts in building systematic reviews from published articles, used in evidence-based medicine. We propose to use a committee of classifiers to rank biomedical abstracts based on the predicted relevance to the topic under review. In our approach, we identify two subsets of abstracts: one that represents the top, and another that represents the bottom of the ranked list. These subsets, identified using machine learning (ML) techniques, are considered zones where abstracts are labeled with high confidence as relevant or irrelevant to the topic of the review. Early experiments with this approach using different classifiers and different representation techniques show significant workload reduction.

Keywords: Machine Learning, Automatic Text Classification, Systematic Reviews, Ranking Algorithms.

1 Introduction

Evidence-based medicine (EBM) is an approach to medical research and practice that attempts to provide better care with better outcomes by basing clinical decisions on solid scientific evidence [1]. Systematic Reviews (SR) are one of the main tools of EBM. Building SRs is a process of reviewing literature on a specific topic with the goal of distilling a targeted subset of data. Usually, the reviewed data includes titles and abstracts of biomedical articles that could be relevant to the topic. SR can be seen as a text classification problem with two classes: a positive class containing articles relevant to the topic of review and a negative class for articles that are not relevant.

In this paper we propose an algorithm to reduce the workload of building SRs while maintaining the required performance of the existing manual workflow. The number of articles classified by the ML algorithm with high confidence can be considered a measure of workload reduction.

Y. Gao and N. Japkowicz (Eds.): Canadian AI 2009, LNAI 5549, pp. 224–228, 2009.
© Springer-Verlag Berlin Heidelberg 2009

2 Ranking Method

Ranking Algorithm. The proposed approach is based on using committees of classification algorithms to rank instances based on their relevance to the topic of review. We have implemented a two-step ranking algorithm. While the first step, called local ranking, is used to rank instances based on a single classifier output, the second step, named collective ranking, integrates the local ranking results of individual classifiers and sorts instances based on all local ranking results.

The local ranking process is a simple mapping:

$$R_j(w_{ij}^+, w_{ij}^-) \to s_{ij} \tag{1}$$

where R_j is the local ranking function for classifier j; w_{ij}^+ and w_{ij}^- are decision weights for the positive and the negative class assigned by classifier j to instance i; s_{ij} is the local ranking score for instance i based on classifier's j output. Depending on what the classifier j is using as weights, s_{ij} are calculated as the ratio or normalized difference of the weights.

All instances to be classified (test set instances) are sorted based on the local ranking scores s_{ij}. A sorted list of instances is built for each classifier j. As a result, each instance i is assigned a local rank l_{ij} that is the position (the rank) of the current instance in the ordered list of instances with respect to the current classifier j:

$$s_{ij} \to l_{ij}, \qquad l_{ij} \in \{1, 2, \ldots, N\} \tag{2}$$

where N is the total number of instances to be classified.

In the second step, the collective ranking score g_i is calculated for each instance i over all the applied classifiers, as follows:

$$g_i = \sum_j (N - l_{ij} + 1) \tag{3}$$

All instances to be classified are in the end based on the collective ranking scores. The collective ordered list of instances is a result of this sorting. Finally, we get the collective rank r_i for each instance as the number associated with that instance in the collective ordered list (the position in the list):

$$g_i \to r_i, \qquad r_i \in \{1, 2, \ldots, N\} \tag{4}$$

An instance with a higher collective rank is more relevant to the topic under review than another instance with a lower collective rank.

Classification rule for the committee of classifiers. The classification decision of the committee is based on the collective ordered list of instances. The key point is to establish two thresholds:

T - top threshold (number of instances to be classified as positive);
B - bottom threshold (number of instances to be classified as negative).

We propose a special Machine Learning (ML) technique to determine T and B for new test data by applying our classifiers on the labeled data (the training set). Since human experts have assigned the labels for all training set instances, top and bottom thresholds on the training set could be created with respect to the required level of prediction confidence (which in our case is the average recall and precision level of individual human experts). As top and bottom thresholds for the training set are assigned, we simply project them on the test set, while adjusting them to the new distribution of the data, the proportions of the size of the prediction zones and gray zone are maintained. After the thresholds are determined, the committee classification rule is as follows:

$$(r_i \leq T) => i \in Z^+, \quad c_i = relevant$$
$$(r_i > (N - B)) => i \in Z^-, \quad c_i = irrelevant \tag{5}$$
$$(T < r_i \leq (N - B)) => i \in Z^N$$

where c_i is final class prediction on instance i; r_i represents the collective rank of the instance i, N is a number of instances in the test set; Z^+ is the positive prediction zone the subset of the test set including all instances predicted to be positive with respect to required level of prediction confidence; Z^- is the negative prediction zone, the subset of the test set that consists of all instances predicted to be negative with respect to the required level of prediction confidence. The prediction zone is built as the union of Z^+ and Z^-. Test set instances that do not belong to the prediction zone belong to what we call the gray zone Z^N.

3 Experiments

The work presented here was done on a SR data set provided by TrialStat Corporation [2]. The source data includes 23334 medical articles pre-selected for the review. While 19637 articles have title and abstract, 3697 articles have only the title. The data set has an imbalance rate (the ratio of positive class to the entire data set) of 8.94%.

A stratified repeated random sampling scheme was applied to validate the experimental results. The data was randomly split into a training set and a test set five times. On each split, the training set included 7000 articles (~30%), while the test set included 16334 articles (~70%) The results from each split were then averaged.

We applied two data representation schemes to build document-term matrices: Bag-of-words (BOW) and second order co-occurrence representation [3]. CHI2 feature selection was applied to exclude terms with low discriminative power.
In order to build the committee, we used the following algorithms[1]: (1) Complement Naïve Bayes [4]; (2) Discriminative Multinomial Naïve Bayes[5]; (3) Alternating Decision Tree [6]; (4) AdaBoost (Logistic Regression)[7]; (5)AdaBoost (j48)[7].

---

[1] We tried a wide set of algorithms with good track record in text classification tasks, according to the literature. We selected the 5 which had the best performance on our data.

4   Results

By using the above described method to derive the test set thresholds from the training set, the top threshold is set to 700 and the bottom threshold is set to 8000. Therefore, the prediction zone consists of 8700 articles (700 top-zone articles and 8000 bottom-zone articles) that represent 37.3% of the whole corpus. At the same time, the gray zone includes 7634 articles (32.7% of the corpus). Table 1 presents the recall and precision results calculated for the positive class. (Only prediction zone articles are taken into account.) Table 1 also includes the average recall and precision results for human expert predictions[2], observed SR data from the TrialStat Inc.

The proposed approach includes two levels of ensembles: the committee of classifiers and an ensemble of data representation techniques. We observed that using the ensemble approach has brought significant impact on performance improving (possible because it removes the variance and the mistakes of individual algorithms).

Table 1. Performance Evaluation

Performance measure	Machine Learning results on the Prediction Zone	Average Human Reviewer's results
Recall on the positive class	91.6%	90-95%
Precision on the positive class	84.3%	80-85%

5   Conclusions

The experiments show that a committee of ML classifiers can rank biomedical research abstracts with a confidence level similar to human experts. The abstracts selected with our ranking method are classified by the ML technique with a recall value of 91.6% and a precision value of 84.3% for the class of interest. The human workload reduction that we achieved in our experiments is 37.3% over the whole data.

Acknowledgments. This work is funded in part by the Ontario Centres of Excellence and the Natural Sciences and Engineering Research Council of Canada.

References

1. Sackett, D.L., Rosenberg, W.M., Gray, J.A., Haynes, R.B., Richardson, W.: Evidence based medicine: what it is and what it isn't. BMJ 312(7023), 71–72 (1996)
2. TrialStat corporation web resources, http://www.trialstat.com/
3. Pedersen, T., Kulkarni, A., Angheluta, R., Kozareva, Z., Solorio, T.: An Unsupervised Language Independent Method of Name Discrimination Using Second Order Co-occurrence Features. In: Gelbukh, A. (ed.) CICLing 2006. LNCS, vol. 3878, pp. 208–222. Springer, Heidelberg (2006)

---

[2] Experts are considered working individually. A few reviewers usually review each article. We partially replace one expert with a ML algorithm.

4. Rennie, J., Shih, L., Teevan, J., Karger, D.: Tackling the poor assumptions of naive bayes text classifiers. In: ICML 2003, Washington DC (2003)
5. Su, J., Zhang, H., Ling, C.X., Matwin, S.: Discriminative Parameter Learning for Bayesian Networks. In: ICML 2008 (2008)
6. Freund, Y., Mason, L.: The alternating decision tree learning algorithm. In: Proceeding of the 16th International Conference on ML, Slovenia, pp. 124–133 (1999)
7. Freund, Y., Schapire, R.: Experiments with a new boosting algorithm. In: Thirteenth International Conference on ML, San Francisco, pp. 148–156 (1996)

Large Neighborhood Search Using Constraint Satisfaction Techniques in Vehicle Routing Problem

Hyun-Jin Lee, Sang-Jin Cha, Young-Hoon Yu, and Geun-Sik Jo

Department of Computer Science & Information Engineering, Inha University
5nchon@naver.com, {aldehyde7,yhyu}@eslab.inha.ac.kr,
gsjo@inha.ac.kr

Abstract. Vehicle Routing Problem(VRP) is a well known NP-hard problem, where an optimal solution to the problem cannot be achieved in reasonable time, as the problem size increases. Due to this problem, many researchers have proposed a heuristic using a local search. In this paper, we propose the Constraint Satisfaction Problem(CSP) model in Large Neighborhood Search. This model enables us to reduce the size of local search space. In addition, it enables the easy handling of many constraints in the real-world.

Keywords: Vehicle Routing Problem, Constraint Satisfaction Problem.

1 Introduction

Minimizing the cost is always the main target of commercial businesses. To reduce the cost, we need to find a solution to this problem. The Vehicle Routing Problem is introduced as approach for solving this problem. The objective of VRP is delivery to a set of customers with known demands along minimum-cost vehicle routes originating and terminating at a depot[2]. VRP is an NP-hard problem and the optimum or near optimum solution can be best obtained by heuristics. Heuristic approaches have been addressed in a very large number of studies[3]. Among of them, the Large Neighborhood Search(LNS) has more good solutions than other approaches[4]. We construct a Constraint Satisfaction Problem(CSP) model of exploration in LNS. The benefit of using the CSP technique is that it can easily use the constraints. Thus, we expect to use constraints more easily than previous studies on VRP.

2 Large Neighborhood Search

LNS have been introduced for VRP with time windows, in [1]. The central idea of LNS is very simple. The LNS is based on continual relaxation and re-optimization. For VRP, the positions of some customer visits are relaxed (the visits are removed from the routing plan), and then the routing plan is re-optimised over the relaxed positions (by re-inserting these visits). A single iteration of removal and re-insertion can be considered as an examination of a neighborhood. If a re-insertion is found such that the best routing plan has a low cost, this new solution is retained as the current

Y. Gao and N. Japkowicz (Eds.): Canadian AI 2009, LNAI 5549, pp. 229–232, 2009.

one. We propose a CSP model for re-insertion process. This model is advantageous when applying and adding constraints, as the model is based on CSP. The model is examined in detail in Section 3.

3   CSP Model for Reinsertion Problems

To develop the reinsertion model, we need to define symbols of graph expression. The symbols for the graph expression are as follows.

G_i is an initial cluster for each truck. G_i' is a partial graph including *depot*. It means that a cluster has a starting deport and there is no return depot. G_j'' is a partial graph including *virtual depot*. V is a customer's set, v_i is deleted vertex, and *Virtual vertex* is subgraph without *depot* and *virtual depot*. And V_i is set to the union of the virtual vertex and v_i, and E (arc) is a path of the truck. Figure 1 draws a schematic diagram of the graph expression.

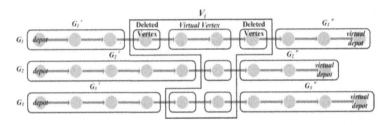

Fig. 1. A schematic diagram of the graph expression

Prior to modeling, we define a global time table. The time table is organized as a two-dimensional boolean table, to check if it is possible to move from node to node. In this table, we use a loose constraint to ensure that the two nodes are never connected when the constraint is violated. By deciding to use a loose constraint, a satisfactory means to ensure that the two nodes are not connected when the constraint is violated can be achieved. But, if the constraint is not violated, there is the possibility that the two nodes are unconnected.

In the reinsertion model, we construct a variable, domain and constraint that are essential for constructing a reinsertion model. The variables include a combination of G_i' and G_j''. This means that the variable is a combination of clusters from disconnected arc components. This setting is available for a newly transformed cluster combination from the initial cluster. Then, a domain is constructed for the subsets of V_i where it is the union of deleted vertices and virtual vertices. This domain represents all possible combinations of deleted vertices. The variable and domain that are used for this methodology can involve many elements such as the number of clusters and the deleted nodes. Next, we execute the variable reduction phase and domain reduction phase. After the reduction phase, the variable will be a final variable and the domain will be the final domain.

The next process involves setting constraints. First constraint is a small vehicle's capacity compared to the sum of vertices' capacity,   including variables (G_i' and G_j'') and also the domain. The second constraint is the possibility of valid combinations of variable and the domain. In other words, it enables the insertion of deleted vertices into a new cluster, which is created by combining subclusters. But, despite constraint satisfaction, the full route allows the possibility of an invalid configuration, due to the loose constraint in the time table. To overcome this problem, we add the function *Valid_Route_Check* to check the component's constraint. This function only checks the final route, to determine whether it is a valid or invalid route.

Using the aforementioned variable, domain and constraint, we proceed to our searching method. We use the search technique Forward Checking in [5]. Here, we modified some of the code, because this search technique has only one return value, so the modified code can have all possible solutions. This search technique is also used in the variable reduction phase and domain reduction phase.

The variable reduction phase also uses a model that has the constraint satisfaction technique. However, the domain in this phase differs from reinsertion modeling. We used a new variable and domain. First, we set the variable G_i' including *depot*. Second, we set the domain G_j'' including *virtual depot*. Then, the constraints involve two factors. The first is that the vehicle's capacity is less than the sum of the capacity of G' and G''. The second is that the *Valid_Route_Check* function must returns true.

Here, we perform the first reduction phase using NC[5] to describe the variable, domain, and constraints. This reduction phase is complete after searching using modified Forward Checking that uses the variable, domain(these are the results from NC) and new constraint. The new constraint is "The selected domain for each variable should not be duplicated". After executing the modified forward checking, we obtain a return value. The return value is used as the main reinsertion model's variable.

Next, we reduce the element's number of the domain by starting the domain reduction phase. The domain reduction phase involves the first step for setting the. variable and domain as elements of V_i, and the constraints are set according to two factors. The first factor is whether the time table's value is true or false, and the second factor is whether the vehicle's capacity is less than the sum of the selected variable's capacity and domain's capacity. The other step's settings differ from the first step's settings. The step's variable is set to combine with the previous variable and domain. And the step's domain is set as the elements of V_i, and the step's constraints are set by three factors. The first factor is that the vehicle's capacity must be less than the sum of the selected variable's capacity and the domain's capacity. The second factor is that the domain's element must avoid duplication of the variable's elements. The third factor is that the time table's value must be true for the last element in the variable and ahead of the element in the domain.

Finally, the combination of each step's variable and domain will be the final domain for the main reinsertion model.

4 Computational Results

The data used for comparison is from Solomon's instances. Each problem has 100 customers, a central depot, time windows and capacity constraints. Simulations are performed using random removal request in LNS for cases where removed size is four. And, we compared the local search time for *Forward_Checking* without variable and domain reduction, with the local search time for our CSP model. Each table number is an average value, obtained from 50 iterations. We set the unit of time as the second. In this paper, we don't presented quality of solutions because both of the models have same quality that is optimal solution in local search space.

Table 1. The column on the left provides the search time without variable and domain reduction, and the one on the right provides the search time using our CSP model

	C1		C2		R1		R2		RC1		RC2	
1	0.91	0.05	0.81	0.06	1.18	0.08	32.98	1.50	57.49	2.42	78.75	1.72
2	0.61	0.11	0.51	0.13	73.29	10.92	25.06	2.48	74.85	8.57	2.28	0.02
3	0.98	0.64	0.96	0.16	71.89	0.08	7.78	2.24	59.14	8.49	47.32	0.38
4	0.69	0.53	0.69	0.51	90.23	14.48	4.07	2.01	22.56	3.78	1.86	1.34
5	0.53	0.03	0.94	0.06	88.58	5.10	21.39	10.23	57.88	5.93	78.18	5.63
6	0.53	0.03	1.07	0.08	74.78	5.87	3.78	1.95	89.87	5.10	24.75	3.12
7	0.42	0.04	1.67	0.35	90.71	21.09	1.38	0.41	56.39	10.28	99.83	22.07
8	1.26	0.14	1.45	0.22	74.73	17.38	2.42	1.83	42.31	9.32	2.08	1.71
9	0.81	0.31			73.99	7.15	2.00	0.95				
10					72.42	8.45	3.68	0.11				

5 Conclusion

This paper presented reinsertion modeling using a constraint satisfaction technique in LNS. We can reduce unnecessary searching in the reinsertion phase using the constraint satisfaction technique. It is obvious that reduction has an effect on the execution time of LNS. Real-world models have many side constraints. So, implementing a constraint satisfaction technique in our model achieves good performance for real-world VRP, by increasing the number of constraints. Lastly, we presented the execution time without variable and domain reduction, and the execution time for our CSP model. So, we conclude that our CSP model achieves a result much faster than a search without variable and domain reduction.

References

1. Shaw, P.: Using Constraint Programming and Local Search Methods to Solve Vehicle Routing Problems. In: Maher, M., Puget, J.-F. (eds.) Principles and Practice of Constraint (1998)
2. Laporte, G., Osman, I.H.: Routing Problem-A Bibliography. Annals of Operation Research 61, 227–262 (1995)
3. Bräysy, O., Gendreau, M.: Vehicle routing problem with time windows, Part I: Route construction and local search algorithms. Transportation Science 39, 104–139 (2005)
4. Ropke, S., Pisinger, D.: An Adaptive Large Neighborhood Search Heuristic for the Pickup and Delivery Problem with Time Windows. Transportation Science 40, 455–472 (2006)
5. Tsang, E.: Foundations of Constraint Satisfaction, Department of Computer Science University of Essex, UK (1996)

Valuable Change Detection in Keyword Map Animation

Takuya Nishikido[1], Wataru Sunayama[1], and Yoko Nishihara[2]

[1] Hiroshima City University, Hiroshima 731-3194, Japan
[2] University of Tokyo, Tokyo 113-8656, Japan

Abstract. This paper proposes a map animation interface that supports interpretations of difference between two keyword relationships with varying viewpoints. Finding keywords whose relationships drastically change is crucial because the value of keywords mainly consists inof their relationships in networks. Therefore, this interface outputs marks for keywords that drastically change in their relations when it shows animations between two relationships.

1 Introduction

This paper proposes a map animation interface that marks objectively significant keywords in relationship transition. Significant keywords denote objectively important words in a map when it changes as the viewpoint is changed. Viewpoint denotes a criterion for calculating relationships among keywords. For example, "horses" and "cars" are related in terms of vehicles, but not related in terms of animals.

2 Animation Interface

The proposed animation interface is explained, as shown in Figure 1. As input, the interface requires keywords that a user wants to analyze and viewpoint keywords as criteria for calculating relationships. The interface outputs a keyword map after calculating the relationship, the clustering[1], and the value. A map is created for each viewpoint, and animation showing the differences between the two maps is realized in the interface.

2.1 Keyword Value Calculation

Keyword value $value(w)$ that represents connections among networks is evaluated by Eq.(1), where W denotes a keyword set link connected with a keyword w. Link value $link\ value(w_i, w_j)$ for the links between different clusters are defined as 3, the links connected in the same cluster whose mutual information value is larger than the chi square value [1] as 2, and the rest as 1. By using this formula, widely connected links are highly evaluated like as transportation expenses.

---

[1] Mutual information and chi square values[2] are compared by their T-scores.

Y. Gao and N. Japkowicz (Eds.): Canadian AI 2009, LNAI 5549, pp. 233–236, 2009.

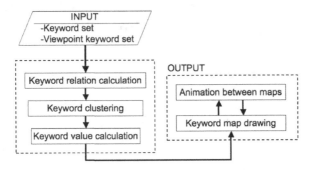

Fig. 1. Structure of animation interface

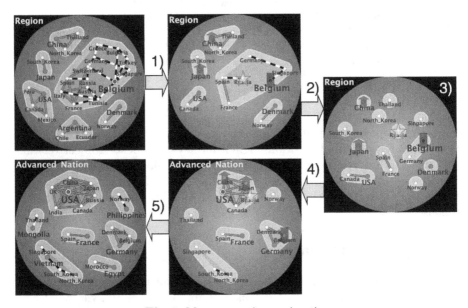

Fig. 2. Map comparison animation

$$value(w) = \sum_{w_i \in W} link\ value(w, w_i) \qquad (1)$$

2.2   Functions and Animation

Users display transition animation from one map to another. This animation consists of the five steps illustrated in Figure 2 to clarify the differences between maps.

1) Disappearance: keywords disappear that appear in the initial map but not in the goal map.
2) Separation: keywords not included in the same island in the goal map are separated from the current island.

Table 1. Data set for experiments

Theme	Viewpoint 1	Viewpoint 2	40 Keywords
Animal	carnivore	herbivore	bear, lion, tiger, shark, dog,...
Actor	costar	bad terms	Ryoko Hirosue, Joe Odagiri,...
Car	maker	replacement	crown, skyline, civic, alto,...
Occupation	popularity	high-income	pilot, lowyer, baseball player, ...
Adjective	man	woman	beautiful, cute, little, kind,...

3) Value exchange: exchange link values from initial map to goal map.
4) Connection: connect keywords not in the same island but in the same island in the goal map.
5) Appearance: keywords appear that are not included in the initial map.

The most important illustration in this study is marks for drastically value change points. Keywords with significant value change in the animation are identified and given marks, arrows and stars, as in Figure 2. A maximum of three keywords are marked whose increase or decrease of keyword value calculated by Eq.(2) is the largest. Also, marks are given to a maximum of three keywords whose absolute values of increase and decrease are given as Eq.(3) [2].

$$value\ updown(w) = \sum_{w_i \in W} (link\ value_{end}(w, w_i) - link\ value_{init}(w, w_i)) \quad (2)$$

$$value\ diff(w) = \sum_{w_i \in W} |link\ value_{end}(w, w_i) - link\ value_{init}(w, w_i)|. \quad (3)$$

3 Experiments

This experiment evaluated the number and the interpreted differences among two displayed maps. Five themes were prepared and 40 keywords and two viewpoint keywords designed for each, as in Table 1. The test subjects were 20 information science majors who were each assigned four themes and two interfaces. Each subject executed the following two tasks sequentially.

1) Compare two maps for each viewpoint without the animation function, find keywords whose position or connection with the other keywords is largely different, and justify your reasons in three minutes.
2) The same as Task 1) with animation, in seven minutes.

Figure 3 shows the number of answers including more than one viewpoint keyword. This result shows that only the proposed interface increased the number of answers for the marked keywords. Since the number of answers for no

---

[2] Set W is defined as $W\ init \cup W\ end$, $W\ init$ denotes a keyword set linked with keyword w in the initial map, and $W\ end$ denotes a keyword set linked with keyword w in the goal map.

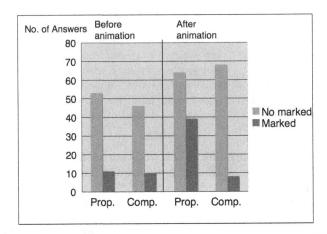

Fig. 3. Number of answers including a viewpoint keyword(s)

marked keywords was not so different, the proposed interface helped increase the quantity of meaningful answers. Since the number of marked keywords was fewer than in Table 1, most answers were for no marked keywords, as in Figure 3. In such a situation, the proposed interface could provide clues for new interpretations that might be overlooked when using a comparative interface.

4 Conclusion

In this paper, an animation interface was proposed that displays marked remarkable points. As experimental results, users could interpret unnoticed keywords when marks were not given.

References

1. Newman, M.E.J.: Detecting community structure in networks. The European Physical Journal B 38, 321–330 (2004)
2. Matsuo, Y., Sakaki, T., Uchiyama, K., Ishizuka, M.: Graph-based Word Clustering using Web Search Engine. In: Proc. 2006 Conference on Empirical Methods in Natural Language Processing (EMNLP 2006), pp. 542–550 (2006)

The WordNet Weaver: Multi-criteria Voting for Semi-automatic Extension of a Wordnet

Maciej Piasecki[1], Bartosz Broda[1], Michał Marcińczuk[1], and Stan Szpakowicz[2,3]

[1] Institute of Applied Informatics, Wrocław University of Technology, Poland
{maciej.piasecki,michal.marcinczuk,bartosz.broda}@pwr.wroc.pl
[2] School of Information Technology and Engineering, University of Ottawa, Canada
szpak@site.uottawa.ca
[3] Institute of Computer Science, Polish Academy of Sciences, Poland

Abstract. The WordNet Weaver application supports extension of a new wordnet. One of its functions is to suggest lexical units semantically close to a given unit. Suggestions arise from activation-area attachment – multi-criteria voting based on several algorithms that score semantic relatedness. We present the contributing algorithms and the method of combining them. Starting from a manually constructed core wordnet and a list of over 1000 units to add, we observed a linguist at work.

1 Motivation, Related Work and Main Ideas

The top levels of a wordnet hypernymy hierarchy must be built manually. This is what was done in Polish wordnet[1] (plWN). Our goal was an automated method of producing useful suggestions for adding lexical units (LUs) to plWN.

Several projects have explored extending an existing wordnet. [1] and [2] assign a distributionally motivated meaning representation to synsets, and treated the hypernymy structure so labelled as a decision tree for slotting in a new LU. [3] represents the meaning of a LU by *semantic neighbours* – k most similar units. To attach a new LU is to find a site in the hierarchy with a concentration of its semantic neighbours. [4] cast wordnet hypernymy extension probabilistically. Attaching a LU to a structure transforms it variously; the chosen addition maximises the probability of the change in relation to the evidence.

We analysed several methods of extracting lexical-semantic relations (LSR): a measure of semantic relatedness based on Rank Weight Functions (MSR_{RWF}) [5], post-filtering LU pairs produced by MSR_{RWF} with C_H, a classifier trained on data from plWN [6], three manually constructed lexico-syntactic patterns in the style of [7], and the *Estratto* method [8] which automatically extracts the patterns and next LSR instances. The accuracy of all methods in distinguishing pairs of synonyms, close hypernyms and meronyms is around 30%, too low for linguists' use, but the LU pairs shared by all methods are much more accurate.

We present an automatic method of *Activation-Area Attachment* (AAA): LUs attach to a small areas in the hypernymy graph rather than to one synset. For

---

[1] www.plwordnet.pwr.wroc.pl

Y. Gao and N. Japkowicz (Eds.): Canadian AI 2009, LNAI 5549, pp. 237–240, 2009.
© Springer-Verlag Berlin Heidelberg 2009

a new LU u, we seek synsets which give the strongest evidence of LUs in a close hyponym/hypernym/synonym relation with u, ideally synsets including u's near synonyms. We assume, though, that the intrinsic errors in data preclude certainty about the attachment point for u. Even if synset t appears to fit, we consider the evidence for t's close surroundings in the graph – all synsets located no further than some upper-bound distance d_h, as measured in hypernymy links to traverse from t. Evidence for the surroundings is less reliable than for the central synset t; its weight decreases in proportion to the distance.

We consider several sources of heterogeneous evidence: sets of LU pairs produced by patterns and C_H; the list $MSRlist(y, k)$ of the k units most related to y according to MSR_{RWF}; and the same list restricted to *bidirectional relations*: $MSR_{BiDir}(y, k) = \{y' : y' \in MSRlist(y, k) \wedge y \in MSRlist(y', k)\}$.

2   The Algorithm of Activation-Area Attachment

The AAA algorithm is based on *semantic fit* between two lemmas (representing LUs linked by a LSR), and between a lemma and a synset (defining a LU). We group synsets that fit the input lemma into *activation areas*. Lemma-to-lemma fit is the function $fit : \mathbf{L} \times \mathbf{L} \rightarrow \{0, 1\}$. $\mathbf{L}$ is a set of lemmas. $fit(x, y) = 1$ if and only if $x \in C_H(MSRlist(y, k))$ or $x \in MSR_{BiDir}(y, k)$, or one of $\langle y, x \rangle$ and $\langle x, y \rangle$ belongs to at least two sets among $MSR_{RWF}(x, k)$ and sets extracted by patterns. The function *score*: $\mathbf{L} \times \mathbf{L} \rightarrow R$ has the value 1 if $fit(x, y) = 1$; $MSR_{RWF}(x, y)$ if $x \in MSRlist(y, k)$; 0.5 or 0.3 if one of $\langle y, x \rangle$ and $\langle x, y \rangle$ has been extracted by a pattern with higher or lower accuracy, respectively.

Different sources are differently reliable; we estimated this, for example, by manual evaluation of the accuracy of the extracted pairs. To trust each source to a different degree, we introduce weighted voting present in fit and $score$.

Phase I of the algorithm finds synsets that fit the new lemma based on $fit(x, y)$ and synset contexts. **Phase II** groups the synsets it finds into connected subgraphs – *activation areas*. For each area, the linguist is shown the local maximum of the scoring function. Only MSR is defined for any lemma pair, so we also introduced a *weak fit* based just on MSR to help fill gaps in the description of strong fit. The weak fit helps avoid fragmented and small activation areas, but it depends on the predefined threshold $hMSR$ (below).

Let x and y be lemmas, S – a synset, $hypo(S, n)$ and $hyper(S, n)$ – sets of hyponym or hypernym synsets of S up to n levels, r ($=2$) – the context radius, $hMSR$ ($=0.4$) – the threshold for highly reliable MSR values; $minMSR = 0.1$.

Phase I. Lemma-to-synset calculation
1. $votes(x, S) = \sum_{y \in S} fit(x, y)$
2. $fit(x, S) = \delta_{(h=1)} \left(votes(x, S) + \sum_{S' \in hypo(S,r) \cup hyper(S,r)} \frac{votes(x, S')}{2 * dist(S, S')} \right)$
   where $\delta : N \times N \rightarrow \{0, 1\}$, such that $\delta(n, s) = 1$
   if and only if $(n \geq 1.5 * h$ and $s \leq 2)$ or $(n \geq 2 * h$ and $s > 2)$
3. $weak\_fit(x, S) =$
   $\delta_{(h=hMSR)} \left(\sum_{y \in S} score(x, y) + \sum_{S' \in hypo(S,r) \cup hyper(S,r)} \frac{\sum_{y \in S'} score(x, y)}{2 * dist(S, S')} \right)$

Phase II. Identify lemma senses: areas and centres

1. $synAtt(x) = \{\mathbf{S} : \mathbf{S} = \{S : fit(x, S) \vee weak\_fit(x, S)\}$, and $\mathbf{S}$ is a *connected hypernymy graph*$\}$
2. $maxScore(x, \mathbf{S}) = score\,(x, max_{S \in \mathbf{S}}\,score(x, S))$
3. Remove from $synAtt(y)$ all $\mathbf{S}$ such that $maxScore(x, \mathbf{S}) < minMSR$
4. Return the top 5 subgraphs from $synAtt(x)$ according to their $maxScore$ values; in each, mark the synset S with the highest $score(x, S)$

We require more *yes* votes for larger, fewer for smaller synsets. The parameter h of the δ template relates the function to what is considered a 'full vote'; for weak fit, $h = hMSR$. **Phase II** identifies connected subgraphs (in the hypernymy graph) which fit the new lemma x. For each area we find the local maximum of the score function for x. All subgraphs based on the strong fit are kept, plus those based on the weak fit, whose score is above $minMSR$ – up to $maxSens$.

3   The WordNet Weaver and Its Evaluation

The algorithm works in the *WordNet Weaver* [WNW], an extension of a wordnet editor in the *plWN* project. A linguist sees a list of LUs not yet in the wordnet, and suggestions generated for the selected LU. The strength of the fit and attachment areas are shown on the screen. The linguist can accept, change or reject a proposal. WNW was designed to facilitate the actual process of wordnet extension. Its primary evaluation was based on a linguist's work with 1360 new LUs divided into several subdomains, plus a sample of 161 randomly drawn LUs. We assumed two types of evaluation: monitor and analyse the linguist's recorded decisions, and automatically assess piecemeal re-building of the wordnet.

Recall (*all:* 75.24%, *random:* 55.77%) is the ratio of the accepted attachments of a lemma to all senses which the linguist added. Precision is calculated in relation to [1] *acceptable distances:* P_1 (14.86%, 13.10%) – exact attachment to the local maximum synset (by synonymy or hyponymy); [2] *any distance* accessible by links in a given subgraph: P_H (34.58%, 27.62%); [3] *acceptable types of links:* P_{H+} (36.35%, 30.48%) – meronymic links are counted as positive because they too are in the linguist's focus; [4] *measure of success:* $P_{\geq 1}$ (80.36%, 59.12%) is the percentage of new LUs for which at least one suggestion was successful as $H+$. Precision and recall were much higher for coherent subdomains such as plants or animals than for the random sample. At least one proper attachment area was found for most LUs, see $P_{\geq 1}$. Even for the worst random sample, the linguist considered proposals for 59.12% of new LUs promising.

Automatic evaluation checks how AAA reconstructs parts of plWN for 1527 LUs in the lower parts of the hypernymy structure. One step removes and reinserts 10 LUs. The C_H classifier is trained without the removed LUs. We have three strategies for evaluating AAA's proposals: *All* , *One* (the highest-scoring attachment site), $Best_{P \geq 1}$ (one closest attachment site). Results – in Table 1. As expected, suggestions based on strong fitness are significantly more accurate. Encouragingly, almost half of the suggestions based on strong fitness are near the correct place. The result for $Best_{P \geq 1}$ strategy shows more meaningful support for the linguist: the number of words with at least one useful suggestion.

Table 1. The accuracy [%] of plWN reconstruction. L – the distance.

L	All_{strong}	All_{weak}	$All_{strong+weak}$	One_{strong}	One_{weak}	$One_{strong+weak}$	$Best_{P \geq 1}$
0	26.65	7.90	16.46	45.80	16.24	34.96	42.81
1	35.76	14.50	24.21	58.73	28.96	47.81	61.19
2	42.87	21.39	31.20	67.69	40.51	57.72	75.02
3	48.31	27.36	36.93	**73.58**	51.08	65.33	**81.96**

4   Conclusions and Further Research

Our primary goal was to construct a tool that facilitates and streamlines the linguist's work. Still, in comparison to the evaluation in [3] our results seem better, e.g., 34.96% for the highest-scored proposal, while [3] reports a 15% best accuracy. Our result for the top 5 proposals is even higher, 42.81%. The best results in [1,2], also at the level of about 15%, were achieved in tests on a much smaller scale. [2] also performed tests only in two selected domains. The algorithm of [4], unlike ours, can be applied only to probabilistic evidence. We made two assumptions: attachment based on activation area and the simultaneous use of multiple knowledge sources. The assumptions appear to have succeeded in boosting the accuracy above the level of the MSR-only decisions (roughly represented in our approach by weak fitness). WNW appears to improve the linguist's efficiency a lot, but longer observations are necessary for a reliable justification.

References

1. Alfonseca, E., Manandhar, S.: Extending a lexical ontology by a combination of distributional semantics signatures. In: Gómez-Pérez, A., Benjamins, V.R. (eds.) EKAW 2002. LNCS, vol. 2473, pp. 1–7. Springer, Heidelberg (2002)
2. Witschel, H.F.: Using decision trees and text mining techniques for extending taxonomies. In: Proc. of Learning and Extending Lexical Ontologies by using Machine Learning Methods, Workshop at ICML 2005 (2005)
3. Widdows, D.: Unsupervised methods for developing taxonomies by combining syntactic and statistical information. In: Proc. Human Langauge Technology / North American Chapter of the ACL (2003)
4. Snow, R., Jurafsky, D., Ng., A.Y.: Semantic taxonomy induction from heterogenous evidence. In: COLING 2006 (2006)
5. Piasecki, M., Szpakowicz, S., Broda, B.: Automatic selection of heterogeneous syntactic features in semantic similarity of Polish nouns. In: Matoušek, V., Mautner, P. (eds.) TSD 2007. LNCS (LNAI), vol. 4629, Springer, Heidelberg (2007)
6. Piasecki, M., Szpakowicz, S., Marcinczuk, M., Broda, B.: Classification-based filtering of semantic relatedness in hypernymy extraction. In: Nordström, B., Ranta, A. (eds.) GoTAL 2008. LNCS, vol. 5221, pp. 393–404. Springer, Heidelberg (2008)
7. Hearst, M.A.: Automated Discovery of WordNet Relations. In: Fellbaum, C. (ed.) WordNet – An Electronic Lexical Database, pp. 131–153. MIT Press, Cambridge (1998)
8. Kurc, R., Piasecki, M.: Automatic acquisition of wordnet relations by the morphosyntactic patterns extracted from the corpora in Polish. In: 3rd International Symposium Advances in Artificial Intelligence and Applications (2008)

Active Learning with Automatic Soft Labeling for Induction of Decision Trees

Jiang Su[1], Sayyad Shirabad Jelber[1], Stan Matwin[1,2], and Jin Huang[1]

[1] SITE, University of Ottawa. Ottawa, Ontario K1N 6N5, Canada
[2] Institute of Computer Science, Polish Academy of Sciences, Warsaw, Poland

Abstract. Decision trees have been widely used in many data mining applications due to their interpretable representation. However, learning an accurate decision tree model often requires a large amount of labeled training data. Labeling data is costly and time consuming. In this paper, we study learning decision trees with lesser labeling cost from two perspectives: data quality and data quantity. At each step of active learning process we learn a random forest and then use it to label a large quantity of unlabeled data. To overcome the large tree size caused by the machine labeling, we generate weighted (soft) labeled data using the prediction confidence of the labeling classifier. Empirical studies show that our method can significantly improve active learning in terms of labeling cost for decision tree learning, and the improvement does not sacrifice the size of decision trees.

1 Introduction

In applied supervised learning, frequently labeling data is more costly than obtaining the data. For instance, one can obtain a large amount of text from web at a low cost, however labeling this data by humans can be practically prohibitive.Therefore training a sufficiently accurate model with a limited amount of labeled data is very desirable. Decision trees are one of the most popular data mining tools. The popularity of decision trees is mainly due to their interpretable representation, which can be easily be transformed into human understandable rules. However, as observed in [1], compared to other learning algorithms, decision trees require a large amount of labeled training data. Thus, reducing the labeling cost for decision tree learning is an important challenge.

In this paper, we present a new active decision tree learning algorithm, called *ALASoft* (**A**ctive **L**earning with **A**utomatic **Soft** labeling for induction of decision **T**rees). The goal is to reduce labeling cost for learning from both data quality and data quantity perspectives. ALASoft uses a random forest [2] classifier to select high quality unlabeled data to be labeled by experts in an active learning process [3]. In each iteration of active learning, a large amount of unlabeled data is labeled by a random forest, and then a decision tree is learned from training data labeled by human and random forest. To overcome the large tree size resulted from the machine labeled data [4,5], we propose to use a soft

Y. Gao and N. Japkowicz (Eds.): Canadian AI 2009, LNAI 5549, pp. 241–244, 2009.

Algorithm 1. ALASoft

Inputs:

- L_h a set of human labeled instances
- U a set of unlabeled instances
- S the data selection classifier
- M the modeling classifier

Until the termination condition is met, Do

1. Compute the entropy for each instance using Equation 1 and the predicted probability from S
2. Select the unlabeled instance with the largest entropy, remove it from U, get an expert to label it, and add to L_h
3. Retrain the classifier S using L_h
4. Assign a weighted label (Equation 2) to each instance in U using the predicted probability generated by S.
5. add all machine labeled instances in U into L_m
6. Train M using weighted data $L_m \cup L_h$.

labeling method that labels data by the prediction confidence of the labeling classifier, i.e. the decision forest.

Combining active learning and automatic labeling of data has been studied by other researchers such as [6,7]. Our method differs from previous research in that we only scan the unlabeled data once for each iteration of active learning algorithm. Additionally, to the best of our knowledge such methods have not been used to learn decision trees.

2   The ALASoft Algorithm

We show pseudo-code for ALASoft in Algorithm 1. The results reported here are for S being a random forest and M a decision tree. In active learning, the most informative unlabeled instance is the instance with the maximum prediction uncertainty. If all the unlabeled instances have the same uncertainty, we randomly select one. We use *entropy* to measure the prediction uncertainty of an unlabeled instance. The following is the definition of entropy:

$$En(E) = -\sum_{i=1}^{|C|} P(c_i|E)logP(c_i|E) \qquad (1)$$

where $|C|$ is the number of classes. $P(c|E)$ is the posterior probability which indicates what we know about an instance E in terms of the class label.

$$w(E) = En(E^u) - En(E^l) \qquad (2)$$

In Equation 2, both E^u and E^l are computed by applying their conditional probability $P(c|E)$ to Equation 1. For $En(E^u)$, entropy of an unlabeled instance,

Table 1. Comparisons of Data Utilization

Dataset	Active+R+S	Random	Active+D	Active+R		Active+R+H
Cmc	99± 145	131± 165	166± 190	116± 154	81±	99
Optdigits	24± 29	333± 279 o	301± 245 o	89± 134	16±	7
Car	232± 129	919± 347 o	472± 268 o	411± 2240	125±	37 •
Letter	40± 23	428± 346 o	562± 431 o	90± 124	31±	12
Pendigits	14± 5	533± 414 o	565± 434 o	47± 74	11±	3
Splice	176± 219	742± 738 o	719± 696 o	312± 2500	97±	78
Kr-vs-kp	294± 85	1716± 718 o	1615± 672 o	411± 393	264±	70
Waveform	70± 150	434± 458 o	450± 563 o	400± 5470	49±	107
Hypothyroid	72± 28	912± 597 o	581± 348 o	89± 66	77±	27
Sick	200± 258	1336± 931 o	1652± 1292 o	261± 294	222±	273
Page-blocks	53± 28	1064± 967 o	1253± 1481 o	82± 450	73±	171
Mushroom	57± 14	1521± 820 o	978± 667 o	83± 240	59±	14

Active+R+S has lower o or higher •
labeling cost compared to other active learners using paired t-test at significance level 0.05

we have $P(c|E^u) = \frac{1}{|C|}$, while for $En(E^l)$, entropy of the instance after assigning the class label, $P(c|E^l)$ is just the predicted probability from S. Equation 2 shows the expected entropy reduction as a result of labeling an unlabeled instance.

3 Experiment

We used the WEKA system for our experiments [8]. To better simulate the disproportionate amount of unlabeled versus labeled data we selected 12 large UCI data sets. The multi-class datasets are converted into two classes by keeping only the instances of the two largest classes. While we use decision forests for labeling unlabeled data, in principal other algorithms can be used for this task. We use the following abbreviations when discussing our experiments.

Random: the data selection classifier is a random classifier.

Active+D(ecision Trees). The data selection classifier is J48 [9].

Active+R(andomForest). The data selection classifier is a random forest [2].

Active+R+H(ard labeling). The data selection classifier is a random forest which is also used to assign labels to the unlabeled data.

Active+R+S(oft labeling). The data selection classifier is a random forest which is also used to assign weighted labels to the unlabeled data.

In each experiment, we divide a data set into disjoint training data Tr and test data Ts. We assume that a sufficiently accurate model $M_{optimal}$ is the one trained on Tr, and also define the *target accuracy* as the accuracy that $M_{optimal}$ achieves on $Test$. To evaluate our labeling approach, we treat Tr as a pool of unlabeled instances U. We start with an empty labeled dataset L. In each iteration, the data selection classifier selects an unlabeled instance from U without replacement, labels it (by human) and adds it to L. A decision tree model M_{active} can be learned from the human (and additional machine) labeled data, and then tested on Ts. The active learning process will stop when M_{active} achieves the *target accuracy* on Ts. Then, we record the number of labeled data in L. Note that M_{active} is now as accurate as $M_{optimal}$. The performance of each algorithm was averaged over 10 runs of 10-fold cross-validations.

Table 1 shows that learners that use machine labeled data significantly outperform active learners without unlabeled data. Table 1 shows that both Active+R+S and Active+R+H outperform Active+R in 4-5 datasets. Additionally, in a separate set of experiments, not reported in this paper, we observed that the ALASoft algorithm does not increase decision tree size. In 11 out of 12 datasets, the sizes of decision trees learned from data generated by Active+R+S are significantly smaller than the ones learned from Active+R+H.

4   Conclusion

In data mining applications, reducing labeling cost for decision tree learning is an important challenge. This paper describes a method that uses ensemble classifiers to select high quality data, and to enlarge the quantity of labeled data by soft labeling the unlabeled data. Our experiments show that while random forests can indeed select high quality data for decision tree learning, the usage of unlabeled data can further reduce the total labeling cost without sacrificing the tree size. Our method can be seen as a general framework, since the data selection classifier and modeling classifier can be easily replaced with other classifiers.

References

1. Kohavi, R.: Scaling up the accuracy of naive-Bayes classifiers: A decision-tree hybrid. In: Proceedings of the Second International Conference on Knowledge Discovery and Data Mining, pp. 202–207. AAAI Press, Menlo Park (1996)
2. Breiman, L.: Random forests. Mach. Learn. 45(1), 5–32 (2001)
3. Seung, H.S., Opper, M., Sompolinsky, H.: Query by committee. In: COLT, pp. 287–294 (1992)
4. Domingos, P.: Knowledge acquisition from examples via multiple models. In: Proc. 14th International Conference on Machine Learning, pp. 98–106. Morgan Kaufmann, San Francisco (1997)
5. Zhou, Z.H., Jiang, Y.: Nec4.5: Neural ensemble based c4.5. IEEE Trans. Knowl. Data Eng. 16(6), 770–773 (2004)
6. McCallum, A., Nigam, K.: Employing em and pool-based active learning for text classification. In: ICML, pp. 350–358 (1998)
7. Muslea, I., Minton, S., Knoblock, C.A.: Active + semi-supervised learning = robust multi-view learning. In: ICML, pp. 435–442 (2002)
8. Witten, I.H., Frank, E.: Data Mining –Practical Machine Learning Tools and Techniques with Java Implementation. Morgan Kaufmann, San Francisco (2000)
9. Quinlan, J.: C4.5: Programs for Machine Learning. Morgan Kaufmann, San Mateo (1993)

A Procedural Planning System for Goal Oriented Agents in Games

Yingying She and Peter Grogono

Concordia University, Montreal QC H3G1M8, Canada
{yy_she,grogono}@cse.concordia.ca

Abstract. This paper explores a procedural planning system for the control of game agents in real time. The planning methodology we present is based on offline goal-oriented behavioral design, and is implemented as a procedural planning system in real-time. Our design intends to achieve efficiency of planning in order to have smooth run-time animation generation. Our experimental results show that the approach is capable of improving agent control for real-time goal processing in games.

1   Introduction and Background

Planning is a basic technique for the agent to reason about how to achieve its objectives [1]. In this paper, planning is implemented in an agent architecture with an embedded Procedural Planning System(PPS). The PPS is a part of the AI module of Gameme. Gameme is a game design API that we have developed. In Gameme, we consider game design as a goal-oriented design. So goal oriented design can be the starting point of the entire game design. On the other hand, behaviors are the underlying module of the game system. The behavior-based AI(BBAI) is the decomposition of intelligence into simple, robust, reliable modules [2]. So, we provide a symbolic, behavior-based, goal oriented and reactive model, GOBT (Goal Oriented Behavior Tree) [3], in Gameme. GOBT is a data structure used in offline game design. It describes how to perform a set of behaviors in order to fulfill certain goals. The fundamental idea behind GOBT is that intelligent behavior can be created through a collection of simple behavior modules; a complex goal can be accomplished by a collection of simpler sub-goals. The GOBT collects these behavior or sub-goal modules into tree layers. Furthermore, the GOBT is reactive in nature, meaning that in the end, the tree architecture still simply maps inputs to goals without planning occurs. The basic premise of reactive is that we begin with a simple behavior or goal at lower levels of GOBTs, and once we have succeeded there, we extend with higher-level goals. Each GOBT contain an execution order which is a set of behaviors or goals, which is the execution sequence of achieving the highest level goal (the root of the GOBT).

2   The Procedural Planning System

The concept of distinguishing between procedural and descriptive has been advocated for a long time, and it still has explanatory power. We consider that the process of generating a game is a process from the descriptive stage to the procedural stage. So, the PPS

Y. Gao and N. Japkowicz (Eds.): Canadian AI 2009, LNAI 5549, pp. 245–248, 2009.

is designed based on the Procedural Reasoning System(PRS) [4]. Design of the PPS should incorporate the knowledge arrangement and transformations. All knowledge in the PPS is stored as modules which are built from underlying C++ code and instantiated in working memory at run-time. These knowledge modules include rules in RBSs, GOBTs and priority queues for goals. Knowledge modules are developed as a way of separating concerns in different aspects of agents' goals, and demonstrate maintainability by enforcing logical boundaries between components in the same modules. Agents' goals can be approximately classified to several categories based on their trigger properties. In Figure 1, the agent Xman has two groups of goals ordered in the priority queue "Monster is invisible"(trigger property) and "Monster is visible"(trigger property) by priority factors of itself.

Also, in order to have dynamic planning results, knowledge defined offline is very important because it provides the foundation for real-time planning. Since the agent planning in games has certain time limit, we do not propose too much real-time searching for agents. In addition, we did not apply dynamic processing in every level of the goal planning. For certain low-level goals, such as detail about the goal "eat", our implementation is based on the static execution order. So, some knowledge modules in Gameme should be defined offline and retrieved and parameterized by the interpreter in real-time. On the other hand, each agent's knowledge is processed as the form of *external* and *internal* knowledge. The external knowledge is other agent's knowledge which affects the planning of the agent. The internal knowledge is the internal factor of the agent such as the current priority queue and goal(on the top of the priority queue).

The agent architecture of Gameme contains three towers which are perception, predication and action tower. The predication tower is a PPS customized for each agent control in games. The PPS consists of *knowledge, desires, plans* and *intentions* which are connected by an interpreter as shown in [5]. We designed seven planning activities for each planning cycle in the PPS [5]. In cycle, the interpreter exchanges information with these four components and drives the process of agent's planning. Furthermore, goal arbitration happens all the time during planning. Some hybrid architectures consider this problem the domain of "deliberation" or "introspection" — the highest level of a three-layered architecture [2]. However, the PPS treats this problem as a general problem of goal selection by adding *Controllers* to the interpreter of the PPS. It is a reactive module that provides functionality of goal selection. It has unified mechanism in making decision, but the output of controller is different for each agent based on their own goals. During each cycle of planning, the controller change *intention*(priority queue), and then choose the top goal in the *intention* based on knowledge it received. Here are steps that the controller executes in each planning cycle for an agent:

1. *Analyze the external knowledge;*
2. *If context has no change, keep current priority queue and go to step 4;*
3. *Else, decide which priority queue is selected and retrieve it from the predefined knowledge;*
4. *Analyze the internal knowledge;*
5. *While the priority queue is not empty, check the top goal in the queue;*
   - *If the top goal in the queue is not matching the internal knowledge, pop it;*
   - *Otherwise, return the top goal as the result;*
6. *If the queue is empty and can't find the matching goal, return the previous goal;*

Fig. 1. Priority queues of the game agent Xman

The PPS is a simplified and efficient game agent architecture. It is ideal for planning where actions can be defined by predetermined procedures(execution orders) that are possible in the game environment. The PPS selects plans based on the run-time environment and the highest priority goals instead of generating plans. While planning is more about selection than search or generation, the interpreter ensures that changes to the environment do not result in inconsistencies in the plan [6]. In addition, the PPS can efficiently avoid slow planning in dynamical game environment. All predetermined plans are early selections based on possible circumstances during the game design process. Early selection of goals can eliminate most of these branches, making goal-driven search more effective in pruning the space [7]. While the selection is the main planning operation in each planning cycle, the 3D engine can accept the planning result in time, and render the correct visual reaction for game agents.

2.1 Experimental Results

In this section, we provide examples of game agent(Xman) planning in the PPS. We treat distance between the monster and Xman as the external knowledge, and Xman's power value as the internal knowledge. Xman can gain power when it is eating and resting; keep power when it is idle; and other activities(goals) make the power value reduce. Here, we explain how the controller works when Xman meets a monster.

1. The controller decides to focus on the "Monster-Visible" queue. If Xman is not in the full power level, the top goal "Chase" is popped.

$$Heap[1..3] = [3(Chase)][2(Fight), 1(Evade)]$$
$$\xrightarrow{Pop} Heap[1..2] = [2(Fight)][1(Evade)]$$

2. The goal "Fight" becomes top goal in the queue. If Xman is in medium power level, the goal "Fight" becomes the current goal.

$$Heap[1..2] = [2(Fight)][1(Evade)].$$

3. If Xman is still in the medium power level, the controller changes nothing in the queue that it focuses. So Xman keeps fighting with the monster, and the goal "Fight" still active.
4. If Xman defeats the monster, the controller detects that the monster is not around. So it switches to the "Monster-Invisible" queue. After fighting, the power level of Xman is not full, so the top goal "Rest" is popped. And the goal "Eat" becomes the current goal.

$$H[1..3] = [3(Rest)][2(Eat), 1(Idle)]$$
$$\xrightarrow{Pop} H[1] = [2(Eat)][1(Idle)]$$

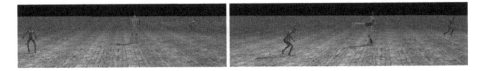

Fig. 2. The example of multi-agent planning in the PPS rendering in a game scene

We also implement the PPS in multi-agent planning. The results show that the agent planning is efficient, and there was no delay in 3D animation rendering with the Ogre 3D Graphics engine[1]. In Figure 2, when a Monster is visible, two Xmans have reactions based on their own power level. In the left picture, two Xmans are both in medium power; they both stagger. In the right picture, the Xman in the left side has medium power, so it still staggers; the Xman in the right side has less power, so it runs away.

3 Conclusion

The main contribution is the procedural knowledge approach used in game agent planning in real-time. It allows action selection to be reactive within a complex runtime game environment, rather than relying on memory of predefined plans. Also, the procedural planning benefits from the modularity design from the knowledge design to the whole system architecture. In addition, the idea of handling a light-weight agent architecture by aspects of balancing offline and real-time design and usage of the controller pattern is also useful in other game agent systems. We are planning to extend the functionality of the core AI module in Gameme in order to let game agents have more functionality such as reinforce learning and scheduling.

References

1. Soto, I., Garijo, M., Iglesias, C.A., Ramos, M.: An agent architecture to fulfill real-time requirements. Agents (2000)
2. Bryson, J.J.: The behavior-oriented design of modular agent intelligence. Agent Technologies, Infrastructures, Tools, and Applications for E-Services (2003)
3. She, Y., Grogono, P.: Goal oriented behavior trees: A new strategy for controlling agents in games. In: Proceedings of the 4th International North American Conference on Intelligent Games and Simulation (2008)
4. Georgeff, M.P., Lansky, A.L.: Reactive reasoning and planning. In: Proceedings of the 6th National Conference on Artificial Intelligence (1987)
5. She, Y., Grogono, P.: The procedural planning system used in the agent architecture of games. In: Proceedings of the Conference on Future Play: Research, Play, Share (2008)
6. Jones, M.T.: Artificial Intelligence: A System Approach. Infinity Science Press LLC (2008)
7. Luger, G.F.: Artificial Intelligence: Structures and Strategies for Complex Problem Solving. Addison-Wesley, Reading (2008)

---

[1] http://www.ogre3d.org/

An Empirical Study of Category Skew on Feature Selection for Text Categorization

Mondelle Simeon and Robert Hilderman

Department of Computer Science
University of Regina
Regina, Saskatchewan, Canada S4S 0A2
{simeon2m,hilder}@cs.uregina.ca

Abstract. In this paper, we present an empirical comparison of the effects of category skew on six feature selection methods. The methods were evaluated on 36 datasets generated from the 20 Newsgroups, OHSUMED, and Reuters-21578 text corpora. The datasets were generated to possess particular category skew characteristics (i.e., the number of documents assigned to each category). Our objective was to determine the best performance of the six feature selection methods, as measured by F-measure and Precision, regardless of the number of features needed to produce the best performance. We found the highest F-measure values were obtained by bi-normal separation and information gain and the highest Precision values were obtained by categorical proportional difference and chi-squared.

1 Introduction

Due to the consistent and rapid growth of unstructured textual data that is available online, text categorization is essential for handling and organizing this data. For a survey of well studied text categorization methods and other automated text categorization methods and applications, see [1] [2] [3] [4]. Feature selection methods [5] [6] are used to address the efficiency and accuracy of text categorization by extracting from a document a subset of the features that are considered most relevant.

In the literature, it is common for a feature selection method to be evaluated using a particular dataset or group of datasets. However, the performance of a method on different datasets where well-defined and well-understood characteristics are varied, such as the category skew (i.e., the number of documents assigned to each category), could prove to be useful in determining which feature selection method to use in certain situations. For example, given some text repository, where the number of documents assigned to each category possesses some level of skewness, which feature selection method would be most appropriate for a small training set? A large training set? A near uniform distribution of documents to categories? A highly skewed distribution? Some initial steps were made in addressing these important issues in [5]. However, this study has a fundamental problem in that it assumes the best performance regardless of the situation is obtained with no more than 2000 features.

Y. Gao and N. Japkowicz (Eds.): Canadian AI 2009, LNAI 5549, pp. 249–252, 2009.
© Springer-Verlag Berlin Heidelberg 2009

1.1 Our Contribution

In this paper, we present an extensive empirical comparison of six feature selection methods. The methods were evaluated on 36 multiple category datasets generated so that the selected documents possessed particular category skew characteristics based upon the standard deviation of the number of documents per category. In this study, we use an exhaustive search on each of the datasets to determine the set of features that produces the maximum F-measure and Precision values regardless of the number of features needed to achieve the maximum. We also contrast the best performance against the most "efficient" performance, the smallest set of features that produces F-measure and Precision values higher than those obtained when using all of the features. Finally, we study the performance of the feature selection methods using predetermined fixed percentages of the feature space.

2   Methodological Overview

In this work, we used the SVM classifier provided in the Weka collection of machine learning algorithms, stratified 10-fold cross-validation, and macro-averaged F-measure and Precision values. Six feature selection methods were used in conjunction with the SVM classifier. Five are the "best" performing methods reported in [5]: information gain(IG), chi-squared(χ^2), document frequency(DF), bi-normal separation(BNS), and odds ratio(OR). The sixth method, categorical proportional difference(CPD), is a method described in [7].

We utilize the OHSUMED [8], 20 Newsgroups [8], and Reuters-21578 [8] text corpora as repositories from which to randomly generate datasets which contain documents of uniform, low, medium, and high category skew. Clearly, in a dataset where documents are uniformly distributed across categories, the standard deviation is 0%. In datasets, where documents are not uniformly distributed across categories, the standard deviation will be non-zero. In this work, the low, medium, and high category skew datasets have a standard deviation of 16%, 64%, and 256%, respectively. Given these datasets and the set of feature selection methods, our approach to generating the experimental results consists of two phases, as described in [7].

3   Experimental Results

In this section, we present the results of our experimental evaluation of each of the feature selection methods on the 36 datasets.

3.1   Average Maximum Precision and F-Measure

In Table 1, the P and F columns for the *Base*, CPD, OR, BNS, DF, χ^2, and IG describe the Precision and F-measure values respectively, obtained when using the complete feature space (*Base*) and each of the respective methods. The highest values are highlighted. Table 1 shows that all the feature selection

Table 1. Average Maximum Precision and F-Measure

Skew	Base		CPD		OR		BNS		DF		χ^2		IG	
	P	F	P	F	P	F	P	F	P	F	P	F	P	F
Uniform	0.670	0.662	**0.901**	0.736	0.810	0.719	0.843	**0.774**	0.667	0.661	0.755	0.744	0.737	0.733
Low	0.669	0.646	**0.920**	0.727	0.850	0.698	0.845	**0.749**	0.678	0.666	0.762	0.735	0.756	0.726
Medium	0.670	0.576	**0.918**	0.652	0.829	0.610	0.865	0.660	0.672	0.623	0.786	0.692	0.765	**0.695**
High	0.379	0.273	0.565	0.322	0.697	0.476	0.731	0.524	0.565	0.408	**0.734**	0.522	0.722	**0.565**

methods obtained higher average maximum Precision values than the *Base* on all the datasets. The highest average maximum Precision values on the uniform, low, and medium category skew datasets is obtained by CPD, representing an increase of 34.4%, 37.5%, and 37.0%, respectively, over the *Base* values. On the high category skew datasets, although χ^2 obtained the highest maximum Precision value, the values obtained by χ^2, BNS, IG, and OR are not significantly different. The highest average maximum F-measure values on the uniform and low category skew datasets is obtained by BNS, representing an increase of 16.9% and 15.9% over the *Base* values, respectively. On the medium and high category skew datasets, the highest average maximum F-measure value is obtained by IG, representing an increase of 20.7% and 107% over the *Base* values, respectively.

3.2 Efficiency

Efficiency is defined as the smallest percentage of the feature space required to obtain F-measure or Precision values that are higher the *Base*. To obtain the efficiency values for each of CPD, OR, BNS, DF, χ^2, and IG, the smallest percentages of the feature space (satisfying the definition) from each of the uniform, low, medium, and high category skew datasets for each text corpora were averaged for

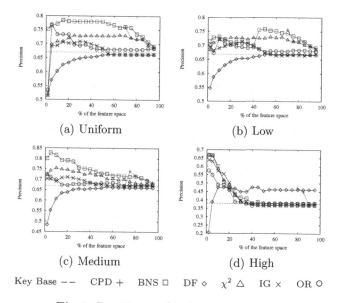

(a) Uniform   (b) Low

(c) Medium   (d) High

Key Base --   CPD +   BNS □   DF ◇ χ^2 △   IG ×   OR ○

Fig. 1. Precision vs. % of the feature space

each respective method. Efficiency values for CPD, OR, BNS, DF, χ^2, and IG were 71.5%, 13.9%, 7.0%, 13.3%, 4.1%, and 2.8% respectively for F-measure, and 63.5%, 2.8%, 2.0%, 17.3%, 3.8%, and 5.9%, respectively, for Precision.

3.3 Pre-determined Threshold

Fig. 1 shows the average Precision values (along the vertical axis) obtained for each feature selection method using a fixed percentage of the feature space, varying according to the sequence 2%, 5%, 10%, 15%, ..., 95% (along the horizontal axis).

Figs. 1(a), 1(b), 1(c) shows that BNS obtains the highest Precision values at most of the sampled cutpoints. In Fig. 1(d), at 2% of the feature space, the Precision values obtained by BNS, χ^2, OR, and IG represent a 76%, 76%, 51% and 67% improvement, respectively, over the *Base* values. Above 20%, the Precision values obtained by DF are higher than those obtained by any of the other methods, except at 80%, where CPD is highest. Similar results were also obtained for F-measure (not shown due to space constraints).

4 Conclusion

This paper presented a comparative study of six feature selection methods for text categorization using SVM on multiple category datasets having uniform, low, medium, and high category skew. We found that the highest F-measure values were obtained by bi-normal separation (uniform and low) and information gain (medium and high), while the highest Precision values were obtained by categorical proportional difference (uniform, low, and medium) and χ^2 (high). We also found the most efficient methods were bi-normal separation (F-measure) and information gain (Precision).

References

1. Sebastiani, F.: Machine learning in automated text categorization. ACM Comput. Surv. 34(1), 1–47 (2002)
2. Kim, S.B., Han, K.S., Rim, H.C., Myaeng, S.H.: Some effective techniques for naive bayes text classification. IEEE Trans. on Knowl. and Data Eng. 18(11), 1457–1466 (2006)
3. Joachims, T.: Text categorization with support vector machines: Learning with many relevant features. In: Nédellec, C., Rouveirol, C. (eds.) ECML 1998. LNCS, vol. 1398, pp. 137–142. Springer, Heidelberg (1998)
4. Han, E.-H., Karypis, G., Kumar, V.: Text categorization using weight adjusted k-nearest neighbor classification. In: Cheung, D., Williams, G.J., Li, Q. (eds.) PAKDD 2001. LNCS, vol. 2035, pp. 53–65. Springer, Heidelberg (2001)
5. Forman, G.: Feature selection for text classification. In: Liu, H., Motoda, H. (eds.) Computational Methods of Feature Selection, pp. 257–276. Chapman and Hall/CRC, Boca Raton (2008)
6. Yang, Y., Pedersen, J.O.: A comparative study on feature selection in text categorization, pp. 412–420. Morgan Kaufmann Publishers, San Francisco (1997)
7. Simeon, M., Hilderman, R.J.: Categorical proportional difference: A feature selection method for text categorization. In: AusDM, pp. 201–208 (2008)
8. Asuncion, A., Newman, D.: UCI machine learning repository (2007)

Opinion Learning without Emotional Words[*]

Marina Sokolova[1] and Guy Lapalme[2]

[1] Children's Hospital of Eastern Ontario Research Institute
msokolova@ehealthinformation.ca
[2] RALI-DIRO, Université de Montréal
lapalme@iro.umontreal.ca

Abstract. This paper shows that a detailed, although non-emotional, description of event or an action can be a reliable source for learning opinions. Empirical results illustrate the practical utility of our approach and its competitiveness in comparison with previously used methods.

1 Motivation

Humans can infer opinion from details of event or action description: I saw it several times can signal a positive opinion in a movie review; there was a long wait may indicate a negative opinion about health care services or a positive opinion about entertainment. We show that, under certain conditions, quantitative (few, small) and stance (probably) indicators extracted from texts provide for successful machine learning. Opinion learning has mainly studied *polarity* of texts (I enjoyed this movie expresses positive polarity, I hated the film – negative polarity). With an increased supply of free-form, unstructured or loosely structured texts, learning opinions can benefit from an assessment of the parameters of the language, other than polarity.

We conjecture that descriptive words which emphasize quantitative properties (high, some), time (old, yesterday) and confidence in happening (can, necessary, probably) can be used in learning opinions. We organize descriptive words in a three level hierarchy: words (the lexical level), word categories (semantics) and the text as a whole (pragmatics). The three levels are built upon quantitative and stance indicators found in text, e.g. stance/degree/time adverbs (probably,again), size/quantity/extent adjectives (long,many), degree pronouns (some). The hierarchy avoids the use of topic and emotionally-charged words. Whereas most sentiment analysis studies are stated as binary classification problems, we consider them as regression problems using numerical opinion tags and as classification problems according to opinion categories.

In [5], we built a hierarchy of verbs which are not explicit opinion bearers such as sit or eat. We showed that it worked well for texts in which negative opinions were expressed indirectly. The empirical results have confirmed that such lexical features can provide reliable and, sometimes, better opinion classification than statistically selected features. This paper is another exploration into language parameters other than polarity.

[*] Parts of this research were supported by NSERC funds available to both authors.

Y. Gao and N. Japkowicz (Eds.): Canadian AI 2009, LNAI 5549, pp. 253–256, 2009.
© Springer-Verlag Berlin Heidelberg 2009

2   Hierarchical Text Representation

Previous studies of sentiment and opinion mining did not explicitly investigate conversational[1], often spontaneously written, texts such as forum posts. Based on [2], we suggest that some lexical characteristics of these types of texts are worth exploring : (i) a larger proportion of descriptive degree pronouns (some, few, none); (ii) frequently used adverbs of time (today), degree (roughly), stance (maybe), frequency (often); (iii) a larger proportion of descriptive adjectives, e.g., size/quantity/extent (big), time (new), relational (same); their gradable derivatives (biggest); (iv) frequent use of order words, e.g. ordinal and cardinal numbers (first, one); (v) frequent use of stance verbs, i.e. modal verbs (can) and mental verbs (think).

We use those word categories to build a first level of a hierarchical text representation. We find seed words in [2] and add their synonyms from an electronic version of Roget's Interactive Thesaurus[2]. To accommodate negative comments, we added negations. We ignore derived, topical, affiliative, foreign words which are less frequent in conversational text. By choosing to omit evaluative and emotive adjectives, we emphasize the role of factual, easily quantified description in text. Starting from the bottom, the hierarchy defines the word categories used in detailed descriptions, then groups the categories into four types of comments, and finally combines the types into direct and indirect detailed categories:

direct primary clues of quantitative evaluation and attributes of the issues:
> **estimation** physical parameters, relative and absolute time (adverbs and adjectives).
>
> **quantification** the broadness of the discussed reference (adverbs, adjectives, pronouns of degree, cardinal numbers, frequency adverbs);

indirect secondary clues of the issue evaluation:
> **comparison** comparative evaluation of the issues, their qualities and relations (gradable and relation adjectives and ordinal numbers).
>
> **confidence** the certainty about the happening of events (stance adverbs and verbs from the lower level).

Our assumption is the following: descriptive words emphasize important characteristics of the discussed issues. In sentences, such words usually precede their references, especially in conversational text [2]. Hence, the extraction of words which follow the descriptors results in the set of words most emphasized in the text. The idea behind the extraction procedure is the following: two-word sequences - bigrams - which have descriptors on their left side capture the modified and intensified words. After extracting such bigrams, we find modified and intensified words. The probability of the word occurrence after a descriptor reveals frequently modified and intensified words. Concentrating on one-side bigrams prevents the multiple extraction of the same word. In [7], two-word patterns were used for classification reviews into *recommended* and *not recommended*. However, the patterns were not organized in a semantic hierarchy.

---

[1] A conversational style, phrase etc is informal and commonly used in conversation. http://www.ldoceonline.com/dictionary/

[2] http://thesaurus.reference.com/

Table 1. SVM's results in learning of opinions. The features sets are defined in Section 3. The best results are in **bold**. For regression, the mean overall value baseline: *Relative Absolute Error (rae)* – 100.2, *Root Relative Squared Error(rrse)* – 101.1. For classification, the majority class baseline: *Accuracy(Acc)* and *Recall(R)* – 52.6%.

	Regression Text Features						Classification (%) Text Features						
	I		II		III		I		II		III		
											numeric		binary	
Opinion	*rae*	*rrse*	*rae*	*rrse*	*rae*	*rrse*	*Acc*	*R*	*Acc*	*R*	*Acc*	*R*	*Acc*	*R*
summed	95.1	97.6	91.4	94.4	82.0	86.7	67.3	72.9	68.4	75.8	70.1	82.4	73.6	78.2
positive	75.8	82.6	77.7	84.4	**63.7**	70.6	73.3	79.4	75.0	81.2	76.1	83.7	**79.0**	80.3
negative	89.4	88.1	87.2	85.8	75.17	80.2	63.7	67.3	66.6	69.7	67.8	73.9	71.3	76.6

3 Empirical Results

We ran regression and classification experiments on consumer-written reviews which represent conversational text. Hu and Liu [3] describes 314 free-form comments marked with positive and negative opinion tags. The extensively labeled texts are a more realistic, albeit difficult, environment for machine learning than those which are tagged with positive/negative tags.

For regression problems, we computed three numerical labels for each text: the number of positive tags, the number of negative tags and a signed sum of the two numbers. For classification problems, we applied an unsupervised equal-frequency discretization to each of the three label sets [4]. The resulting classification problems are referred to as *stronger* vs *weaker* classes.

We construct three feature sets for text representation: **I**, direct descriptors enhanced by the most frequent extracted words; cut-off was determined by frequencies of personal pronouns; **II**, all descriptors enhanced by the most frequent extracted words; **III**, all the extracted words with frequency > 5, with numerical and binary attribute versions for the classification problems. Attribute normalization with respect to the number of words in the text eliminates the length bias. We applied SUPPORT VECTOR MACHINE, K-NEAREST NEIGHBOR and decision-based M5P TREES (regression) and REP TREES (classification) and their bagged versions. [3] For all the learning problems, SVM was considerably more accurate than the other algorithms. Table 1 reports SVM's best results (ten-fold cross-validation).

In regression problems, for both measures, the best results reported in this work are *statistically significant* better the previous best results (*rae*: paired t-test, $P = 0.0408$; *rrse*: paired t-test, $P = 0.0418$)[5]. In classification problems, current best results for *Recall* provide *very statistically significant* improvement (paired t-test, P=0.0071), whereas *Accuracy* improvement is not statistically significant (paired t-test, P=0.6292).

---

[3] http://www.cs.waikato.ac.nz/ml/weka/

4   Discussion and Future Work

Machine learning algorithms were successfully used in opinion classification [6]. Some of this work relied on characteristics of reviewed products [3]. In contrast, we opted for a method which does not involve the use of the domain's content words. Syntactic and semantic features that express the intensity of terms were used to classify opinion intensity [8]. We, instead, focus on the formalization and utilization of non-emotional lexical features. Whereas [1] compared the use of adverbs and adjectives with adjectives only, we concentrated on descriptive adjectives and adverbs.

We studied the relevance of detailed, specific comments to the learning of positive, negative, and summed opinions in both *regression* and *classification* settings. Learning results of positive opinion turned out to be more accurate. Hence relations between features representing descriptive details and positive opinion are easier to detect than those for other problems. The obtained empirical results show improvement over baselines and previous research. The improvement is specially significant for regression problems.

Our approach can be applied to analyze other types of texts, e.g. blogs and reviews published in traditional media. The first step would be to determine the distinctive word categories of such texts, and then find the semantic parameters of opinion disclosure which use them. Next, we can use the formalized patterns for information extraction. Finally, we would use the extracted information to represent texts in machine learning experiments. The application of our method could result in finding similarities imposed by shared communication characteristics, e.g. the one-to-many speaker-audience interaction, and differences that can be attributed to dissimilar characteristics, e.g. Internet and published media.

References

1. Benamara, F., Cesarano, C., Picariello, A., Reforgiato, D., Subrahmanian, V.: Sentiment analysis: Adjectives and adverbs are better than the adjectives alone. In: Proceedings of ICWSM 2007, pp. 43–50 (2007)
2. Biber, D., Johansson, S., Leech, G., Conrad, S., Finegan, E.: Longman Grammar of Spoken and Written English. Longman (1999)
3. Hu, M., Liu, B.: Mining and summarizing customer reviews. In: Proceedings of the KDD 2004, pp. 168–177 (2004)
4. Reinratz, T.: Focusing Solutions for Data Mining. Springer, Heidelberg (1999)
5. Sokolova, M., Lapalme, G.: Verbs Speak Loud: Verb Categories in Learning Polarity and Strength of Opinions. In: Bergler, S. (ed.) Canadian AI 2009. LNCS, vol. 5032, pp. 320–331. Springer, Heidelberg (2008)
6. Sokolova, M., Szpakowicz, S.: Machine Learning Application in Mega-Text Processing. In: Soria, E., Martin, J., Magdalena, R., Martinez, M., Serrano, A. (eds.) Handbook of Research on Machine Learning Applications, IGI Global (2009)
7. Turney, P.: Thumbs Up or Thumbs Down? Semantic Orientation Applied to Unsupervised Classification of Reviews. In: Proceedings of ACL 2002, pp. 417-424 (2002)
8. Wilson, T., Wiebe, J., Hwa, R.: Recognizing strong and weak opinion clauses. Computational Intelligence 22(2), 73–99 (2006)

Belief Rough Set Classifier

Salsabil Trabelsi[1], Zied Elouedi[1], and Pawan Lingras[2]

[1] Larodec, Institut Supérieur de Gestion de Tunis, Tunisia
[2] Saint Mary's University Halifax, Canada

Abstract. In this paper, we propose a new rough set classifier induced from partially uncertain decision system. The proposed classifier aims at simplifying the uncertain decision system and generating more significant belief decision rules for classification process. The uncertainty is repereseented by the belief functions and exists only in the decision attribute and not in condition attribute values.

Keywords: rough sets, belief functions, uncertainty, classification.

1 Introduction

The standard rough set classifiers do not perform well in the face of uncertainty or incomplete data [3]. In order to overcome this limitation, researchers have adapted rough sets to the uncertain and incomplete data [2,5]. These extensions do not deal with partially uncertain decision attribute values in a decision system. In this paper, we propose a new rough set classifier that is able to learn decision rules from partially uncertain data. We assume that this uncertainty exists in decision attribute and not in condition attributes (we handle only symbolic attributes). We choose the belief function theory introduced by Shafer [4] to represent uncertainty which enables flexible representation of partial or total ignorance. The belief function theory used in this paper is based on the transferable belief model (TBM). The paper also provides experimental evaluation of our classifier on different databases.

2 Rough Sets

This section presents the basic concepts of rough sets proposed by Pawlak [3]. $A = (U, C \cup \{d\})$ is a decision system, where U is a finite set of objects $U = \{o_1, o_2, \ldots, o_n\}$ and C is finite set of *condition* attributes, $C = \{c_1, c_2, \ldots, c_s\}$. In supervised learning, $d \notin C$ is a distinguished attribute called *decision*. The value set of d, called $\Theta = \{d_1, d_2, \ldots, d_t\}$. For every object $o_j \in U$, we will use $c_i(o_j)$ to denote value of a condition attribute c_i for object o_j. Similarly, $d(o_j)$ is the value of the decision attribute for an object o_j. We further extend these notations for a set of attributes $B \subseteq C$, by defining $B(o_j)$ to be value tuple of attributes in B for an object o_j. The rough sets adopt the concepts of indiscernibility relation to

Y. Gao and N. Japkowicz (Eds.): Canadian AI 2009, LNAI 5549, pp. 257–261, 2009.

partition the object set U into disjoint subsets, denoted by U/B, and the class that includes o_j is denoted $[o_j]_B$.

$$IND_B = U/B = \{[o_j]_B \mid o_j \in U\}, \text{ with } [o_j]_B = \{o_i \mid B(o_i) = B(o_j)\} \quad (1)$$

The equivalence classes based on the decision attribute is denoted by $U/\{d\}$

$$IND_{\{d\}} = U/\{d\} = \{[o_j]_{\{d\}} \mid o_j \in U\} \quad (2)$$

The lower and upper approximations for B on X, denoted $\underline{B}X$ and $\bar{B}X$ respectively where: $\underline{B}X = \{o_j \mid [o_j]_B \subseteq X\}$, and $\bar{B}X = \{o_j \mid [o_j]_B \cap X \neq \emptyset\}$
$Pos_C(\{d\})$, called a positive region of the partition $U/\{d\}$ with respect to C.

$$Pos_C(\{d\}) = \bigcup_{X \in U/\{d\}} \underline{C}X \quad (3)$$

A reduct of C is a minimal set of attributes $B \subseteq C$ where:

$$Pos_B(\{d\}) = Pos_C(\{d\}) \quad (4)$$

3   Belief Function Theory

In this section, we briefly review the main concepts underlying the belief function theory as interpreted in the TBM [6]. Let Θ be a finite set of elementary events to a given problem, called the frame of discernment. All the subsets of Θ belong to the power set of Θ, denoted by 2^Θ. The impact of a piece of evidence on the different subsets of the Θ is represented by a basic belief assignment (BBA). The BBA is a function $m : 2^\Theta \to [0,1]$ such that: $\sum_{E \subseteq \Theta} m(E) = 1$.

In the TBM, the BBA induced from distinct pieces of evidence are combined by the rule of combination [6].

$$(m_1 \bigcirc m_2)(E) = \sum_{F, G \subseteq \Theta : F \cap G = E} m_1(F) \times m_2(G) \quad (5)$$

In the TBM, beliefs to make decisions can be represented by probability functions called the pignistic probabilities denoted $BetP$ and is defined as [6]:

$$BetP(\{a\}) = \sum_{\emptyset \subset F \subseteq \Theta} \frac{|\{a\} \cap F|}{|F|} \frac{m(F)}{(1 - m(\emptyset))}, \text{ for all } a \in \Theta \quad (6)$$

4   Belief Rough Set Classifier

In this section, we will begin by describing the modified definitions of the basic concepts of rough sets under uncertainty. These adaptations were proposed originally in [7]. Second, we simplify the uncertain decision system to generate the more signification decision rules to create the belief rough sets classifier.

The basic concepts of rough sets under uncertainty are descibed as follows:

1. **Decision system under uncertainty:** Defined as follows: $A = (U, C \cup \{ud\})$, where U is a finite set of objects and C is finite set of *certain condition* attributes. $ud \notin C$ is an *uncertain decision* attribute. We propose to represent the uncertainty of each object by a BBA m_j expressing belief on decision defined on Θ representing the possible values of ud.

2. **Indiscernibility relation:** The indiscernibility relation for the decision attribute is not the same as in the certain case. The decision value is represented by a BBA. Therefore, we will use the $U//\{ud\}$ to denote the uncertain indiscernibility relation. We need to assign each object to the right decision classes. The idea is to use the pignistic transformation. For each object o_j in the decision system U, compute the pignistic probability, denoted $BetP_j$ to correspond to m_j. For every ud_i, a decision value, we define decision classes:

$$IND_{\{ud\}} = U//\{ud\} = \{X_i \mid ud_i \in \Theta\}, \text{ with } X_i = \{o_j \mid BetP_j(\{ud_i\}) > 0\} \tag{7}$$

3. **Set approximation:** To compute the new lower and upper approximations for our uncertain decision table, we follow two steps:

   (a) For each equivalence class based on condition attributes C, combine their BBA using the operator mean. The operator mean is more suitable than the rule of combination [6] to combine these BBAs, because they are beliefs on decision for different objects and not different beliefs on one object. After combination, check which decision classes have certain BBA.

   (b) For each decision class X_i based on uncertain decision attribute, we compute the new lower and upper approximations, as follows: $\underline{C}X_i = \{o_j \mid [o_j]_C \subseteq X_i \text{ and } m_j(\{ud_i\}) = 1\}$ and $\bar{C}X_i = \{o_j \mid [o_j]_C \cap X_i \neq \emptyset\}$

4. **Positive region:** We define the new positive region denoted $UPos_C(\{ud\})$.

$$UPos_C(\{ud\}) = \bigcup_{X_i \in U/ud} \underline{C}X_i, \tag{8}$$

5. **Reduct:** is a minimal set of attributes $B \subseteq C$ such that:

$$UPos_B(\{ud\}) = UPos_C(\{ud\}). \tag{9}$$

The main steps to simplify the uncertain decision system are as follows:

1. **Eliminate the superfluous condition attributes:** We remove the superfluous condition attributes that are not in reduct.

2. **Eliminate the redundant objects:** After removing the superfluous condition attributes, we will find redundant objects. They may not have the same BBA on decision attribute. So, we use their combined BBAs using the operator mean.

3. **Eliminate the superfluous condition attribute values:** In this step, we compute the reduct value for each belief decision rule R_j of the form: **If** $C(o_j)$ **then** m_j. For all $B \subset C$, let $X = \{o_k \mid B(o_j) = B(o_k)\}$ **If** $Max(dist(m_j, m_k)) \leq$ threshold **then** B is a reduct value of R_j. Where $dist$ is a distance measure between two BBAs. We will choose the distance measure described in [1] which satisfies properties such as non-degeneracy and symmetry.

5   Experimental Results

We have performed several tests on real databases obtained from the U.C.I. repository [1]. These databases are modified in order to include uncertainty in the decision attribute. We use three degrees of uncertainty: Low, Middle and High. The percent of correct classification (PCC) for the test set is used as a criterion to judge the performance of our classifier. The results summarized in Table 1 show that the belief rough set classifier works very well in certain and uncertain case for all databases with the different degrees of uncertainty. However, when the degree of uncertainty increases there is a slight decline in accuracy.

Table 1. Experimental results

Database	PCC certain case	PCC Low Unc	PCC Middle Unc	PCC High Unc
Wisconsin breast cancer database	82.61%	81.2%	79.53%	74.93%
Balance scale database	73.7%	72.3%	70.9%	59.34%
Congressional voting records database	94.11%	93.74%	92.53%	82.53%

6   Conclusion

In this paper, we have developed a new rough set classifier to handle uncertainty in decision attributes of databases. The belief rough set classifier is shown to generate significant belief decision rules for standard classification datasets. The proposed algorithm is computationally expensive. So, our future work will develop heuristics for simplification of the decision system for efficient generation of significant decision rules without costly calculation.

References

1. Bosse, E., Jousseleme, A.L., Grenier, D.: A new distance between two bodies of evidence. Information Fusion 2, 91–101 (2001)
2. Grzymala-Busse, J.: Rough set strategies to data with missing attribute values. In: Workshop Notes, Foundations and New Directions of Data Mining, the 3rd International Conference on Data Mining, Melbourne FL, USA, pp. 56–63 (2003)

[1] http://www.ics.uci.edu/ mlearn/MLRepository.html

3. Pawlak, Z., Zdzislaw, A.: Rough Sets: Theoretical Aspects of Reasoning About Data. Kluwer Academic Publishing, Dordrecht (1991)
4. Shafer, G.: A mathematical theory of evidence. Princeton University Press, Princeton (1976)
5. Slowinski, R., Stefanowski, J.: Rough classification in incomplete information systems. Mathematical and Computer Modeling 12, 1347–1357 (1989)
6. Smets, P., Kennes, R.: The transferable belief model. Artificial Intelligence 66, 191–236 (1994)
7. Trabelsi, S., Elouedi, Z.: Attribute selection from partially uncertain data using rough sets. In: The Third International Conference on Modeling Simulation, and Applied Optimization, Sahrjah, U.A.E., January 20-22 (2009)

Automatic Extraction of Lexical Relations from Analytical Definitions Using a Constraint Grammar*

Olga Acosta

National Autonomous University of Mexico
oacostal@iingen.unam.mx

Abstract. In this paper, we present the preliminary results of a rule-based method for identifying and tagging elements in a hyponymy-hypernymy lexical relation for Spanish. As a starting point, we take into account the role that verbal patterns have in analytical definitions. Such patterns connect each term with its possible genus. We built a constraint grammar based on the results of a syntactic parser. The results demonstrate that constraint grammars provide an efficient mechanism for doing more precise analysis of syntactic structures following a regular pattern.

1 Introduction

The automatic extraction of lexical relations is a core task in Natural Language Processing because these relations are used in building lexical knowledge bases, question-answering systems, semantic web analysis, text mining, automatic building of ontologies, and other applications [1], [2] and [3].

One widely recognized lexical relation is the hyponymy-hypernymy relation, which has been found in analytical definitions [4]. It is estimated that about 90% of the analytical definitions in dictionaries have a genus which is the head of the first noun phrase (NP) [5].

The main goal of this work is to sketch a methodology for automatic extraction of hyponymy-hypernymy relations from analytical definitions. In order to achieve our goal, we took into account those discursive fragments containing useful relevant information for defining a term. These fragments are mainly recognizable in specialized texts and are called definitional contexts [4], which are a rich source of lexical relations and have a pattern of the type:

Term + Verbal pattern + Definition.

Verbal patterns play an important role because they serve as link between a term and its corresponding definition. Our approach is to take advantage of the regularity underlying verbal patterns to identify terms involved in a hyponymy-hypernymy relation applying a formal grammar.

* This paper is supported by the National Council for Science and Technology (CONA-CYT), number of RENIECyT: 2249.

Y. Gao and N. Japkowicz (Eds.): Canadian AI 2009, LNAI 5549, pp. 262–265, 2009.
© Springer-Verlag Berlin Heidelberg 2009

2   Definitional Contexts and Definitional Verbal Patterns

Based on [4], we consider a definitional context (DC) as the discursive context where relevant information is found to define a term. The minimal constitutive elements of a DC are: a term, a definition, and usually linguistic or metalinguistic forms such as verbal phrases, typographical markers and/or pragmatic patterns.

At this point, it is convenient to recognize that there are different kinds of definitions. In a previous work, a typology of definitions was established based on the two basic constituents of the analytical model: genus and differentia. The typology depends on whether these constituents are present or absent in the definition. The analytical definition occurs when both constituents, genus and differentia, are present [4].

DCs show a set of regular verbal patterns in Spanish whose function is to introduce the definitions and to link them with their terms. These patterns are called definitional verbal patterns (DVPs). Based on [6], the DVP predicts the type of definition. In our case, DVPs associated to analytical definitions in English are: to be, to represent, to refer to, to characterize + as/for, to know + as, to understand + as, to define + as, to consider + as, to describe + as.

Finally, the regularity observed in the DVPs and the verbs involved in their construction allowed us to capture their behavior in rules. These rules consider the contexts where they are located in order to identify, on the one hand, the type of definition and, on the other hand, lexical relations occurring in DCs.

3   Constraint Grammar

Constraint Grammar (CG) is a parsing framework intended to solve morphological and syntactic ambiguities. Constraints are formulated on the basis of extensive corpus studies. They may reflect absolute, rule-like facts, or probabilistic tendencies where a certain risk is judged to be proper to take [7].

The rules of a CG consider local or global contexts of a particular element to assign or add tags to specific tokens [8]. Furthermore, it is possible to remove wrong readings or to select more appropriate readings depending upon the context. Here we used the tools proposed by [9].

4   Analysis

The analysis carried out in this work started with collecting a corpus of DCs extracted from specialized texts (books, journals and Internet). The corpus was divided into two parts. One part was used for building rules and another part as a test corpus.

One of the advantages of the CG formalism is that it allows us to preprocess the input before applying the constraint grammar. We performed a set of replacements to adjust the inputs of the syntactic parser and to prepare our corpus for identifying and tagging terms in a hyponymy-hypernymy relation. For example, the following replacement rule assigns one tag, @GEN, to nouns whose syntactic function is subject and they are within parentheses:

SUBSTITUTE (N @SUBJ) (@GEN) TARGET (N @SUBJ) IF (*-1 ("par-l" PU)) (*1 ("par-r" PU));

Finally, we applied mapping rules for identifying and tagging terms in the DCs to the second part of the corpus. Those new patterns that were not taken into account in the original mapping rules were added to the constraint grammar. For example, the following mapping rule identifies and tags terms in a DC:

MAP (§TERM) TARGET NOM IF (-1C (*)) (*1 V-PP BARRIER LIM-SENT LINK 1 (DET) OR (ADV) OR (PRP));

The previous rule assigns the tag §TERM to the elements of the set NOM (proper noun or noun phrase) if the context of the element satisfies two conditions: first, it has any word class before it; and second, it has a verb from the set V-PP after it (e.g., to be, to signify, to represent) before any element whose function is a sentence boundary. Also, the V-PP must be linked to a determinant, an adverb, or a preposition.

5   Results

The preliminary results show us the utility of the rule-based approaches for abstracting the syntactic regularity underlying specialized texts with the goal of extracting relevant information. Undoubtedly, constraint grammars allow us to perform more precise analysis of syntactic chains that potentially operate as conditional patterns of a type of definition.

We evaluated our results by determining what proportion of DCs were covered with the rules considered. Our CG rules identified both the term and the genus with roughly 98% precision.

6   Related Work

Most of the research in automatic extraction of lexical relations has focused on is-a and part-of relations extracted from Machine Readable Dictionaries, text corpora or web resources. Hearst [2] proposed an approach relied on lexical syntactic patterns for the acquisition of hyponyms. This author used a bootstrapping algorithm which has served as model for most subsequent pattern-based algorithms. However, it was not successful when applied to meronymy, because the common patterns were also indicative of other semantic constraints.

Other works identified lexical patterns of part-whole relations where statistical measures were used for evaluating quality of extracted instances [10]. There are approaches implementing machine learning techniques for automatically learning semantic constraints. Such approaches are aimed to improve the result of applying lexical syntactic patterns [1].

Furthermore, there are proposals for automatically discovering specific lexical relations using dependency paths generated for a syntactic parser [3]. Finally, our approach is similar to [2], because it relies on lexical syntactic patterns derived of extensive corpus studies for Spanish, although it considers new verbs as core of DVPs for analytical definitions. In future work, we do not discard the use of machine learning techniques for automatically discovering new syntactical patterns that indicate the hyponymy-hypernymy relation from analytical definitions.

7   Future Work

This work presents the first steps towards the extraction of lexical relations from DCs. The current results allow us to extract the terms involved in a hyponymy-hypernymy relation by considering the regularity of verbal patterns in analytical definitions. However, it is important to point out that we are interested in exploring synonymy and meronymy relations also.

By using the CG rules we accurately identify the term and the genus, but we still need an additional process to distinguish one from the other.

References

1. Badulescu, A., Girju, R., Moldovan, D.: Automatic discovery of part-whole relations. Computational Linguistics 32(1), 83–135 (2006)
2. Hearst, M.: Automatic acquisition of hyponyms from large text corpora. In: Proceedings of COLING 1992 (1992)
3. Jurafsky, D., Snow, R., Ng, A.Y.: Learning syntactic patterns for automatic hypernym discovery. In: Advances in Neural Information Processing Systems, vol. 17 (2005)
4. Sierra, G., Alarcon, R., Aguilar, C., Bach, C.: Definitional verbal patterns for semantic relation extraction. Terminology 14(1), 74–98 (2008)
5. Wilks, Y., Guthrie, L., Slator, B.: Electric Words. Dictionaries, Computers and Meaning. MIT Press, Cambridge (1996)
6. Aguilar, C.: Linguistic Analysis of Defintions in Definitional Contexts. PhD dissertation, National Autonomous University of Mexico, Departament of Linguistics (2009)
7. Karlsson, F., Voutilainen, A., Heikkila, J., Anttila, A.: Constraint Grammar: A Language-Independent System for Parsing Unrestricted Text. Mouton de Gruyter, Berlin (1995)
8. Agirre, E., et al.: Extraction of semantic relations from a basque monolingual dictionary using constraint grammar. In: Proceedings of 9th Euralex 2000 (2000)
9. Bick, E.: A constraint grammar parser for spanish. In: Proceedings of TIL 2006 (2006)
10. Berland, M., Charniak, E.: Finding parts in very large corpora. In: Proceedings of the 37th annual meeting of the Association for Computational Linguistics on Computational Linguistics (1999)

Grid-Enabled Adaptive Metamodeling and Active Learning for Computer Based Design

Dirk Gorissen

Ghent University - IBBT, Department of Information Technology (INTEC), Gaston Crommenlaan 8, Bus 201, 9050 Ghent, Belgium

Abstract. Many complex, real world phenomena are difficult to study directly using controlled experiments. Instead, the use of computer simulations has become commonplace as a feasible alternative. However, due to the computational cost of these high fidelity simulations, the use of neural networks, kernel methods, and other surrogate modeling techniques have become indispensable. Surrogate models are compact and cheap to evaluate, and have proven very useful for tasks such as optimization, design space exploration, prototyping, and sensitivity analysis. Consequently, in many scientific fields there is great interest in techniques that facilitate the construction of such regression models, while minimizing the computational cost and maximizing model accuracy. This paper presents a fully automated machine learning toolkit for regression modeling and active learning to tackle these issues. A strong focus is placed on adaptivity, self-tuning and robustness in order to maximize efficiency and make the algorithms and tools easily accessible to other scientists in computational science and engineering.

1   Introduction

For many problems from science, and engineering it is impractical to perform experiments on the physical world directly (e.g., airfoil design, earthquake propagation). Instead, complex, physics-based simulation codes are used to run experiments on computer hardware. While allowing scientists more flexibility to study phenomena under controlled conditions, computer experiments require a substantial investment of computation time. This is especially evident for routine tasks such as optimization, sensitivity analysis and design space exploration [1].

As a result researchers have turned to various approximation methods that mimic the behavior of the simulation model as closely as possible while being computationally cheap(er) to evaluate. This work concentrates on the use of data-driven, global approximations using compact surrogate models (also known as metamodels, behavioral models, or response surface models (RSM)). Examples of metamodels include: rational functions, Kriging models, Artificial Neural Networks (ANN), and Support Vector Machines (SVM).

However, in order to come to an acceptable model, numerous problems and design choices need to be overcome (what data collection strategy to use, what model type is most applicable, how should model parameters be tuned, etc.). This work describes the design of a state-of-the-art research platform that provides a fully automatic, flexible

Y. Gao and N. Japkowicz (Eds.): Canadian AI 2009, LNAI 5549, pp. 266–269, 2009.
Springer-Verlag Berlin Heidelberg 2009

and rigorous means to tackle such problems. This work lies at the intersection of Machine Learning/AI, Modeling and Simulation, and Distributed Computing. The methods developed are applicable to any domain where a cheap, accurate, approximation is needed to replace some expensive reference model.

2 Motivation

The scientific goal of global surrogate modeling is the generation of a surrogate that is as accurate as possible (according to some, possibly physics-based, measure), using as few simulation evaluations as possible, and with as little overhead as possible.

As stated in the introduction, the principal reason driving surrogate model use is that the simulator is too costly to run for a large number of simulations. One model evaluation may take many minutes, hours, days or even weeks [1]. A simpler approximation of the simulator is needed to make optimization, design space exploration, visualization, sensitivity analysis, etc. feasible. A second reason is when simulating large scale systems. A classic example is the full-wave simulation of an electronic circuit board. Electro-magnetic modeling of the whole board in one run is almost intractable. Instead the board is modeled as a collection of small, compact, accurate surrogates that represent the different functional components (capacitors, resistors, ...) on the board. These types of problems can be found in virtually every scientific discipline and field.

An important aspect of surrogate modeling is data collection. Since data is computationally expensive to obtain, it is impossible to use traditional, one-shot, space filling designs. Data points must be selected iteratively, there where the information gain will be the greatest. A sampling function is needed that minimizes the number of sample points selected in each iteration, yet maximizes the information gain of each iteration step. This process is called active learning, but is also known as adaptive sampling, Optimal Experimental Design, and sequential design. Together this makes that there are an overwhelming number of options available to the designer: different model types, different experimental designs, different model selection criteria, different hyperparameter optimization strategies, etc. However, in practice it turns out that the designer rarely tries out more than one subset of options and that readily available algorithms and tools to tackle this problem are scarce and limited (particularly in industry). There is room for a more extensible, flexible, and automated approach to surrogate modeling, that does not mandate assumptions (but does not preclude them either) about the problem, model type, sampling algorithm, etc.

The motivation stems from our observation that the primary concern of an application scientist is obtaining an accurate replacement metamodel for their problem as fast as possible and with minimal user interaction. Model (type) selection, model parameter optimization, active learning, etc. are of lesser or no interest to them. Thus the contribution of this work is in the general methodology, the various extensions [2,3,4,5] and the available software implementation (see [6]). This framework will be useful in any domain (biology, economics, electronics, ...) where a cheap, accurate approximation is needed for an expensive reference model. In addition the work makes advances in computational intelligence more easily available to application scientists. Thus lowering the barrier of entry and bridging the gap between theory and application.

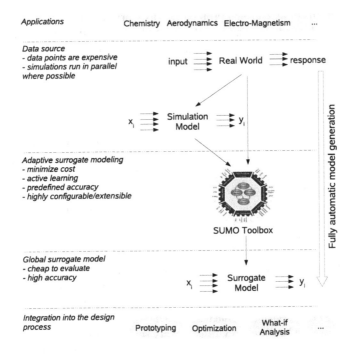

Fig. 1. Automatic Adaptive Surrogate Modeling

3 SUMO Toolbox

Most of the thesis work revolves around a software tool that forms the platform on which all now algorithms are implemented and benchmarked: The SUrrogate MOdeling Toolbox. The SUMO Toolbox [5] is an adaptive tool that integrates different modeling approaches and implements a fully automated, adaptive global surrogate model construction algorithm. Given a simulation engine, the toolbox automatically generates a surrogate model within the predefined accuracy and time limits set by the user (see figure 1). However, at the same time keeping in mind that there is no such thing as a 'one-size-fits-all', different problems need to be modeled differently. Therefore the toolbox was designed to be modular and extensible but not be too cumbersome to use or configure. Different plugins are supported: model types (neural networks, SVMs, splines, ...), hyperparameter optimization algorithms (Pattern Search, Genetic Algorithm (GA), Particle Swarm Optimization (PSO), ...), active learning (density based, error based, gradient based, ...), and sample evaluation methods (local, on a cluster or grid). The behavior of each component is configurable through a central XML configuration file and components can easily be added, removed or replaced by custom, problem-specific, implementations.

The difference with existing machine learning toolkits such as Rapidminer (formerly Yale), Spider, Shogun, Weka, and Plearn is that they are heavily biased towards classification and data mining (vs. regression), they assume data is freely available and cheap (no active learning), and they lack advanced algorithms for the automatic selection of the model type and model complexity.

4   Conclusion

The current status of the thesis work is that the main software engineering efforts to setup the infrastructure and interfacing code has been completed. A wide range of approximation methods, hyperparameter optimization strategies (both single and multi-objective), model (type) selection methods, and active learning techniques have been implemented and the software and algorithms can be considered stable for production use. As is, the toolbox is able to model a given data source with selection of the model type (ANN, SVM, ..), model complexity, and active learning all done fully automatically. The full implementation is available for download at http://www.sumo.intec.ugent.be.

The approach has already been applied successfully to a very wide range of applications, including RF circuit block modeling [6], hydrological modeling [7], aerodynamic modeling [5], and automotive data modeling [4]. Its success is primarily due to its flexibility, self tuning implementation, and its ease of integration into the larger computational science and engineering pipeline.

Future work will focus on further fine tuning the existing algorithms and tools and increase the number and type of applications in order to complete the thesis with a fully functional, well tested, methodology and software implementation that can easily be applied to any approximation problem.

References

1. Wang, G.G., Shan, S.: Review of metamodeling techniques in support of engineering design optimization. Journal of Mechanical Design 129(4), 370–380 (2007)
2. Gorissen, D., Hendrickx, W., Crombecq, K., Dhaene, T.: Integrating gridcomputing and meta-modelling. In: Proceedings of 6th IEEE/ACM International Symposium on Cluster Computing and the Grid (CCGrid 2006), Singapore, pp. 185–192 (2006)
3. Gorissen, D., Tommasi, L.D., Croon, J., Dhaene, T.: Automatic model type selection with heterogeneous evolution: An application to rf circuit block modeling. In: Proceedings of the IEEE Congress on Evolutionary Computation, WCCI 2008, Hong Kong (2008)
4. Gorissen, D., Couckuyt, I., Dhaene, T.: Multiobjective global surrogate modeling. Technical Report TR-08-08, University of Antwerp, Middelheimlaan 1, 2020. Antwerp, Belgium (2008)
5. Gorissen, D., Crombecq, K., Couckuyt, I., Dhaene, T.: Automatic Approximation of Expensive Functions with Active Learning. In: Foundation on Computational Intelligence, Learning and Approximation. Series Studies in Computational Intelligence. Springer, Heidelberg (accepted) (2009)
6. Gorissen, D., Tommasi, L.D., Crombecq, K., Dhaene, T.: Sequential modeling of a low noise amplifier with neural networks and active learning. In: Neural Computing and Applications (accepted) (2008)
7. Couckuyt, I., Gorissen, D., Rouhani, H., Laermans, E., Dhaene, T.: Evolutionary regression modeling with active learning: An application to rainfall runoff modeling. In: International Conference on Adaptive and Natural Computing Algorithms (2009)

Reasoning about Movement in Two-Dimensions

Joshua Gross

Ryerson University, Toronto, ON M5B 2K3, Canada

Abstract. Reasoning about movement in two-dimensions can be broken down to depth and spatial relations between objects. This paper covers previous work based on the formal logic descriptions in situational calculus. Also covered is a simulator that produces data for a logical reasoner to process and potentially make decisions about motion in two-dimensions.

1 Introduction

Motion of two-dimensional objects can be inherently related to depth profiles which result from the dynamic spatial relations between each object. This paper presents a description of depth profiles, their representation in a computer system, and how reasoning software can be used to describe the interaction of objects based on dynamic depth profiles. We build on previous work by[1] with the inclusion of plan recognition, a solution to the progression problem, and dealing with noisy sensor data.

This paper is an extension of [1], where the authors provided a sound and complete description of reasoning about depth profiles in situational calculus (SC). They provide a detailed description of depth profiles as a graphical representation between objects, as well as how changes in the profiles represent movement of bodies in relation to an observer over time. Work has been done on an implementation of [1] and continues with the work mentioned in the conclusion of this paper.

2 Background

2.1 Depth Profiles

A depth profile is a representation of a set of objects from the view of a single observer. Each depth profile corresponds to one slice of time in the scenario. The depth profile is made of depth peaks, each representing the relations between the object and the observer. It is important to note that objects in our problem are approximated as cylinders, and consequently, their 2D slices are circles. This approximation seems reasonable in a traffic domain, where vehicles can be approximated by elliptical cylinders. These encompassing cylinders can be constructed in real time using either a laser range finder or video data, [5][6]. The size of the depth peak is calculated by the amount that the object obstructs the observer's field of view (angular value) and the objects distance from the

Y. Gao and N. Japkowicz (Eds.): Canadian AI 2009, LNAI 5549, pp. 270–273, 2009.

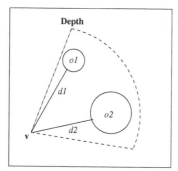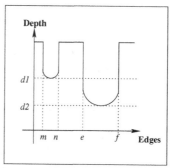

Fig. 1. Mapping of a observers view point to corresponding depth profile graph. The x-axis is the distance from the observer's left boundary. The y-axis is the distance from the observer to the closest point on the circle.

observer. These values are mapped onto a graph, as seen in Figure 1. As can be seen in Figure 1, there is a direct correlation between the amount an object covers of an observer's field of view and the size of its peak in the depth profile. Changes in the depth peaks over time can be used to interpret how objects are interacting with each other or with the observer. Three assumptions are made on the domain. **(1)** Object persistence refers to the assumption that objects can not spontaneously appear or disappear. **(2)** Smoothness of motion is the assumption that objects cannot jump sporadically around the domain. Finally, **(3)** non interpenetration is the assumption that objects cannot pass through each other. Depth peaks, however, can approach, recede, coalesce and split. An individual peak can extend, shrink, appear or vanish. More detail on the properties of these peaks can be found in [1].

2.2   Representation of Depth Profiles

At the initial stage of this project it seemed reasonable to start with simulated data to test our ideas. This lead to the development of a simulator to produce experimental data that could be used in an intelligent reasoner. From this point we had two algorithmic hurdles. First, from the viewpoint of the observer, what portion of the objects in the domain are visible. Second, how to represent this data, a depth profile, in a usable format for the reasoning software.

The solution for determining visible objects is focused around a rotational sweep line algorithm that can run in $O(n * log(n))$. Circles are reduced to being represented as a chord between the two tangent points calculated from the view of the observer. A chord calculated from these tangent points spans the visible portion of any circle. This is a much simpler problem and is easier related to the algorithm discussed by [2], but differs in its ability to calculate the partially visible portion of objects, in the case that an objects is occluded(partially).

Knowing the visible tangent points, from the view of the observer, allows us to produce the data in a usable format for the simulator. The set of visible tangent points is processed to produce the following information that describes the depth peak of an object: distance to an object from observer(depth), angular size of an object(size), and distance from the left boundary of the observer's field of view(left boundary distance). These three pieces of information, for each object, together, produce a depth profile of the current domain.

2.3 Reasoning Using the Situation Calculus (SC)

Depth Peak Attributes in SC: The SC is a predicate logic language for axiomatising dynamic worlds. It is based on actions, situations and fluents as described by [4]. Actions are first-order terms consisting of an action function symbol and its arguments. In this paper we consider three actions that affect a depth profile, *endMove, startMove, sense*. Sense relates to calulation of depth profiles, the rest of the actions are self-explanatory. All actions are considered to be instantaneous.

A situation is a first-order term denoting a sequence of actions. Such sequences are represented using a binary function symbol $do(\alpha, s)$, where α is an action and s is a previous sequence of actions.

Relations or functions whose values vary from situation to situation are called fluents, and are denoted by predicate or function symbols whose last argument is a situation term. We use the following fluents in our reasoner: *facing, location, moving* and *rotating*, which are characterized by their names. Also included are *depth, size* and *lbd* which have been discussed earlier. The final fluent is *dist* which is the angular distance between two peaks in the domain. Further explanation about the SC can be obtained from [4] and about the functions described in this section can be found in [1]. Formal definitions in SC of transitions between peaks can be found in [1] and are not discussed in the scope of this paper.

3 Conclusion

We have implemented a simulator that can produce relevant depth profiles to be used by a reasoning system. The simulator as well as the underlying visibility software have been implemented in C++. The reasoning system has been implemented in ECLiPSe PROLOG and is interfaced with the simulator. The potential for this software is great but relies on a couple additions to its functionality. First, is some form of plan recognition software for recognizing vehicle actions based on an observers own motion and perceptions. A plan recognition framework would allow more complex decisions to be made as we would be able to look forward based on the predicted movement of vehicles around the observer. It would allow an observer to act intelligently when faced with a variety of situations. As of right now, there is only a proof of concept implementation that proves the correctness of the reasoner. Using plan recognition as defined by [7] and [4] more complex scenarios in all areas of the traffic domain are within the range of possible applications for the software. Secondly, to be able to use this software on complex scenarios a solution to the progression problem would be needed. Large amounts of data need to be processed as scenarios become increasingly complex. To accommodate this in a suitable amount of decision making time it is necessary to be able to ignore past situations that are no longer relevant. This problem is currently being solved by [11] and has been described in detail by [4]. Other relevant work in the traffic domain [13,12] provide direction for plan recognition. Also, it is known that sensor data can be incomplete or noisy. Work by [14,15], outlines a solution that can work in situational calculus.

References

1. Soutchanski, M., Santos, P.: Reasoning about Dynamic Depth Profiles. In: 18th biennial European Conference on Artificial Intelligence, pp. 30–34 (2008)
2. de Berg, M., van Kreveld, M., Overmars, M., Schwarzkopf, O.: Computational Geometry, Algorithms and Applications Second Revised Edition, 2nd edn. Springer, Heidelberg (2000)
3. Lee, D.T.: Proximity and Reachability in the Plane (1978)
4. Reiter, R.: Knowledge in Action. In: Logical Foundations for Specifying and Implementing Dynamical Systems. MIT Press, Cambridge (2001)
5. Nagel, H.: Steps Towards a Cognitive Vision System. Artificial Intelligence Magazine 25(2), 31–50 (2004)
6. Howarth, R.J., Buxton, H.: Conceptual Descriptions From Monitoring and Watching Image Sequences. Image and Vision Computing Journal 18, 105–135 (2000)
7. Goultiaeva, A., Lesperance, Y.: Incremental Plan Recognition in an Agent Programming Framework. In: Proceedings of the 2006 AAAI Workshop on Cognitive Robotics, AAAI Technical Report WS-06-03 (2006)
8. Miene, A., Lattner, A., Visser, U., Herzog, O.: Dynamic-Preserving Qualitative Motion Description for Intelligent Vehicles. In: IEEE Intelligent Vehicles Symposium, pp. 642–646 (2004)
9. Lattner, A., Timm, I.J., Lorenz, M., Herzog, O.: Representation of occurrences for road vehicle traffic. In: IEEE International Conference on Integration of Knowledge Intensive Multi-Agent Systems, pp. 191–196 (2005)
10. Gehrke, J., Lattner, A., Herzog, O.: Qualitative Mapping of Sensory Data for Intelligent Vehicles. In: International Joint Conference on Artificial Intelligence, pp. 51–60 (2005)
11. Vassos, S., Levesque, H.: On the Progression of Situation Calculus Basic Action Theories: Resolving a 10-year-old Conjecture. In: Proceedings of the Twenty-Third AAAI Conference on Artificial Intelligence (2008)
12. Pynadath, D.V., Wellman, M.P.: Accounting for context in plan recognition, with application to traffic monitoring. In: Proceedings of the Eleventh Conference on Uncertainty in Artificial Intelligence (1995)
13. Lin, D., Goebel, R.: A Message Passing Algorithm for Plan Recognition. In: Twelfth International Joint Conference on Artificial Intelligence (1991)
14. Bacchus, F., Halpern, J.Y., Levesque, H.J.: Reasoning about Noisy Sensors and Effectors in the Situation Calculus. Artificial Intelligence 111, 171–208 (1999)
15. Gabaldon, A., Lakemeyer, G.: ESP: A Logic of Only-Knowing. In: Noisy Sensing and Acting Twenty-Second Conference on Artificial Intelligence, AAAI 2007 (2007)

Executable Specifications of Fully General Attribute Grammars with Ambiguity and Left-Recursion

Rahmatullah Hafiz

School of Computer Science, University of Windsor, Canada

Abstract. A top-down parsing algorithm has been constructed to accommodate any form of ambiguous context-free grammar, augmented with semantic rules with arbitrary attribute dependencies. A memoization technique is used with this non-strict method for efficiently processing ambiguous input. This one-pass approach allows Natural Language (NL) processors to be constructed as executable, modular and declarative specifications of Attribute Grammars.

1 Introduction

In an Attribute Grammar (AG) [6], syntax rules of a context-free grammar (CFG) are augmented with semantic rules to describe the meaning of a context-free language. AG systems have been constructed primarily as compilable parser-generators for formal languages. Little evidence could be found where fully-general AGs have been used for declaratively specifying directly-executable specifications of NLs. None of the existing approaches support arbitrary attribute dependencies (including dependencies from right) in semantics and general context-free syntax (including ambiguity and left-recursion) altogether in one-pass within a single top-down system. Basic top-down parsers do not terminate while processing left-recursion, and are normally inefficient while parsing exponentially ambiguous grammars (e.g., $S ::= SSa|\epsilon$, on input "$aaa....$" [1]).

A top-down approach that supports executable and declarative specifications of AGs with unconstrained syntax and semantics, offers following advantages:

a. Language descriptions can be specified and executed directly without worrying about syntax-semantics evaluation method and order.

b. Individual parts of AGs can be efficiently tested separately. Modularity assists with systematic and incremental development.

c. Accommodating ambiguity is vital for some domain e.g. NLP.

d. Transforming a left-recursive CFG to a weakly equivalent non-left-recursive form may introduce loss of parses and lack of completeness in semantic [8,4].

e. Arbitrary attribute dependencies provide unrestricted and declarative construction of complex semantic (e.g., set-theoretic Montague semantics with CFG).

f. If implemented using a modern functional language, then benefits such as purely-functional, declarative and non-strict environment are readily available.

Y. Gao and N. Japkowicz (Eds.): Canadian AI 2009, LNAI 5549, pp. 274–278, 2009.

2   Our Contributions

This paper describes an extension of our previous work by accommodating the executable specifications of arbitrary semantic rules coupled with general syntactic rules. In [4,5], we have shown how top-down parsers can be constructed as executable specifications of general CFGs defining the language to be parsed:

- To accommodate direct left-recursion, a *left-recursive context* is used, which keeps track of the number of times a parser has been applied to an input position j. For a left-recursive parser, this count is increased on recursive descent, and the parser is curtailed whenever the condition "$left - recursive\ count$ of parser at $j > \#input - j + 1$" succeeds.
- To accommodate indirect left-recursion, a parser's result is paired with a set of curtailed non-terminals at j within its current parse path, which is used to determine the context at which the result has been constructed at j.

Our extended parser-application is a mapping from an input's start position to a set of end positions with tree structures. We also thread attribute values along with the start and end positions so that they are available for any dependencies that are specified in semantic rules. These attribute values are defined in terms of *expressions* (as our method is referentially transparent and non-strict) that represent arbitrary operations on syntax symbols' attributes, which are computed from the surrounding *environment* when required.

The structured and lazily computed *result* of parser-applications allow us to establish full call-by-need based dependencies between attributes - including inherited dependencies from the right and top. For example, when a parser p_i with syntax $p_i ::= p_m\ p_n$, is applied to input position 1 and successfully ends at position 5, we represent one of p_i's input/output relations (as a result) as:

$$SubNode\ p_i\ inh_{p_i 1}\ =\ syn_{p_i 5}\ [..Branch\ [SubNode\ p_m\ (inh_{p_m 1}, syn_{p_m 3})$$
$$, SubNode\ p_n\ (inh_{p_n 3}, syn_{p_n 5})]..]$$

where, p_m starts at 1 and ends at 3, p_n starts at 3 and ends at 5, inh_{xy} and syn_{xy} represent inherited and synthesized attributes of parser x at position y respectively. Now arbitrary semantics can be formed with non-strict expressions e.g., $inh_{p_m i} \leftarrow f(inh_{p_n j}, syn_{p_i k})$, where f is a desired operation on attributes.

We have achieved our objective by defining a set of combinators for modular, declarative and executable language processors, which are similar to textbook AG notation. Our combinators (e.g., <|> and *> correspond to `orelse` and `then`, `rule_s` and `rule_i` for synthesized and inherited semantic rules, `parser` and `nt` for complete AG rules and non-terminals etc.) are pure, higher-order and lazy functions that ensure fully-declarative executable specifications. We have implemented our method in a lazy functional language Haskell. We execute syntax and semantics in one-pass in polynomial time using a memoization technique to ensure that results for a particular parser at a particular position are computed at most once, and are reused when required. We represent potentially exponential results for ambiguous input in a compact and shared tree structure to achieve polynomial space. The *memo-table* is systematically threaded through parser executions using state-monad [7].

3 An Example

We illustrate our approach with the `repmax` example [2,3], which we have extended to accommodate ambiguity, left-recursion and arbitrary attribute dependencies. Our goal is to parse inputs such as "1 5 2 3 2" with an ambiguous left-recursive CFG $tree ::= tree\ tree\ num\ |\ num$, $num ::= 1|2|3|4|5|...$, and to extract all possible trees (for syntactic correctness) with all terminals replaced by maximum value of the sequence. The AG that specifies this problem is as follows (where $\uparrow$ and $\downarrow$ represent synthesized and inherited attributes respectively):

$$start(S_0) \quad ::= tree(T_0)\ \{RepVal.T_0 \downarrow = MaxVal.T_0 \uparrow\}$$
$$tree(T_0) \quad ::= tree(T_1)\ tree(T_2)\ num(N_1)$$
$$\{\ MaxVal.T_0\uparrow = Max(MaxVal.T_1 \uparrow, MaxVal.T_2 \uparrow, MaxVal.N_1 \uparrow),$$
$$RepVal.T_1 \downarrow = RepVal.T_0 \downarrow, RepVal.T_2 \downarrow = RepVal.T_0 \downarrow,$$
$$RepVal.N_1 \downarrow = RepVal.T_0 \downarrow\}$$
$$|num(N_2)\ \{MaxVal.T_0 \uparrow = MaxVal.N_2 \uparrow, RepVal.N_2 \downarrow = RepVal.T_0 \downarrow\}$$
$$num(N_0) \quad ::= 1\ \{MaxVal.N_0 \uparrow = 1\}\ |\ 2\ \{MaxVal.N_0 \uparrow = 2\}\ ....$$

According to this AG, there are two outputs for the input sequence "1 5 2 3 2":

Using our combinators and notation, an executable specification of the general AG above can be constructed declaratively in Haskell as follows:

```
start = memoize Start parser (nt tree T0)[rule_i RepVal OF T0 ISEQUALTO findRep
                                          [synthesized MaxVal OF T0]]
tree  = memoize Tree  parser (nt tree T1 *> nt tree T2 *> nt num T3)
        [rule_s MaxVal OF LHS ISEQUALTO
         findMax [synthesized MaxVal OF T1,synthesized MaxVal OF T2,synthesized MaxVal OF T3]
         ,rule_i RepVal OF T1 ISEQUALTO findRep [inherited RepVal OF LHS]....]
 <|>   parser (nt num N1)
        [rule_i RepVal OF N1  ISEQUALTO findRep [inherited RepVal OF LHS]
         ,rule_s MaxVal OF LHS ISEQUALTO findMax [synthesized MaxVal OF N1]]
num   = memoize Num terminal (term "1") [MaxVal 1] <|> ... <|> terminal (term "5") [MaxVal 5]
```

When `start` is applied to "1 5 2 3 2", a compact representation of ambiguous parse trees is generated with appropriate semantic values for respective grammar symbols. For example, `tree` parses the whole input (starting at position 1 and ending at position 6) in two ambiguous ways. The `tree`'s inherited and synthesized attributes (represented with I and S) are with its start and end positions respectively. The attributes are of the form '`attribute_type value`' e.g. `RepVal` 5. The compact results have pointing sub-nodes (as parser-name, unique-id pairs e.g. `(Tree,T1)`) with inherited and synthesized attributes:

```
Tree START at: 1 ; Inherited atts: T0 RepVal 5 T0 RepVal 1
     END at : 6 ; Synthesized atts: T0 MaxVal 5
       Branch [SubNode (Tree,T1) ((1,[((I,T1),[RepVal 5])]), (4,[((S,T1),[MaxVal 5])]))
             ,SubNode (Tree,T2) ((4,[((I,T2),[RepVal 5])]), (5,[((S,T2),[MaxVal 3])]))
             ,SubNode (Num, T3) ((5,[((I,T3),[RepVal 5])]), (6,[((S,T3),[MaxVal 2])]))]
     END at : 6; Synthesized atts: T0 MaxVal 5
       Branch [SubNode (Tree,T1) ((1,[((I,T1),[RepVal 5])]), (2,[((S,T1),[MaxVal 1])]))

             ,SubNode (Tree,T2) ((2,[((I,T2),[RepVal 5])]), (5,[((S,T2),[MaxVal 5])]))
             ,SubNode (Num, T3) ((5,[((I,T3),[RepVal 5])]), (6,[((S,T3),[MaxVal 2])]))] ...
Num START at: 1 ; Inherited atts: N1 RepVal 5
    END at : 2 ; Synthesized atts: N1 MaxVal 1
    Leaf (ALeaf "1",(S,N1))...
```

Our semantic rules declaratively define arbitrary actions on the syntax symbols' attributes. For example, the second semantic of `tree` is an inherited rule for the second parser T2, which depends on its ancestor T0's inherited attribute `RepVal`. The T0's `RepVal` is dependent on its own synthesized attribute `MaxVal`, and eventually this `RepVal` is threaded down as every `num`'s inherited attribute.

4 Concluding Comments

The goal of our work is to provide a framework where general CFGs (including ambiguous left-recursion) can be integrated with semantic rules with arbitrary attribute dependencies as directly-executable and modular specifications. Our underlying top-down parsing method is constructed with non-strict combinators for declarative specifications, and uses a memoization technique for polynomial time and space complexities. In the future we aim to utilize grammatical and number agreements, and conditional restrictions for disambiguation. By taking the advantages of general attribute dependencies, we plan to efficiently model NL features that can be characterized by other grammar formalisms like unification grammars, combinatory categorical grammars and type-theoretic grammars. We have implemented our AG system in Haskell, and have constructed a Natural Language Interface based on a set-theoretic version of Montague semantic (more can be found at: `cs.uwindsor.ca/~hafiz/proHome.html`). We are constructing formal correctness proofs, developing techniques for detecting circularity and better error-recognition, and optimizing the implementation for practical uses. We believe that our work will help domain specific language developers (e.g. computational linguists) to build and test their theories and specifications without worrying about the underlying computational methods.

References

1. Aho, A.V., Ullman, J.D.: The Theory of Parsing, vol. I. Prentice-Hall, Englewood Cliffs (1972)
2. Bird, R.S.: Intro. to Functional Programming in Haskell. Prentice-Hall, Englewood Cliffs (1998)

3. De Moor, O., Backhouse, K., Swierstra, D.: First-class attribute grammars. In: WAGA (2000)
4. Frost, R.A., Hafiz, R., Callaghan, P.: Modular and efficient top-down parsing for ambiguous left-recursive grammars. In: 10th IWPT, ACL, pp. 109–120 (2007)
5. Frost, R.A., Hafiz, R., Callaghan, P.: Parser combinators for ambiguous left-recursive grammars. In: 10th PADL, vol. 4902, pp. 167–181. ACM, New York (2008)
6. Knuth, D.E.: Semantics of context-free languages. Theory of Comp. Systems 2(2), 127–145 (1968)
7. Wadler, P.: Monads for functional programming. First International Spring School on Advanced Functional Programming Techniques 925, 24–52 (1995)
8. Warren, D.S.: Programming the ptq grammar in xsb. In: Workshop on Programming with Logic Databases, pp. 217–234 (1993)

$\mathcal{K}$-$\mathcal{MORPH}$: A Semantic Web Based Knowledge Representation and Context-Driven Morphing Framework

Sajjad Hussain

NICHE Research Group, Faculty of Computer Science, Dahousie University, Canada

Abstract. A knowledge-intensive problem is often not solved by an individual knowledge artifact; rather the solution needs to draw upon multiple, and even heterogeneous, knowledge artifacts. The synthesis of multiple knowledge artifacts to derive a 'comprehensive' knowledge artifact is a non-trivial problem. We discuss the need of knowledge morphing, and propose a Semantic Web based framework $\mathcal{K}$-$\mathcal{MORPH}$ for deriving a context-driven integration of multiple knowledge artifacts.

1 Introduction

Knowledge morphing is a specialized knowledge modeling and execution approach that aims to formulate a comprehensive knowledge object, specific to a given context, through "the intelligent and autonomous fusion/integration of contextually, conceptually and functionally related knowledge objects that may exist in different representation modalities and formalisms, in order to establish a comprehensive, multi-faceted and networked view of all knowledge pertaining to a domain-specific problem"–Abidi 2005 [1]. The knowledge morphing approach extends the traditional notion of knowledge integration by providing the ability to 'reason' over the morphed knowledge to (a) infer new knowledge, (b) test hypotheses, (c) suggest recommendations and actions, and (d) query rules to prove problem-specific assertions or theorems.

The concept of knowledge morphing is grounded in the belief that typically knowledge artifacts entail knowledge that is broader than a specific problem's scope. For instance, knowledge artifacts in healthcare such as clinical guidelines may contain knowledge about the diagnosis, treatment, prognosis and follow-up care for a particular disease [2]. Therefore, the integration of entire knowledge artifacts unnecessarily exacerbates the complexity of establishing interoperability between multiple artifacts for no meaningful purpose. Therefore, in our approach we attempt a context-driven reconciliation of ontology-encoded knowledge artifacts by: (i) identifying contextualized sub-ontologies from selected ontology-encoded knowledge artifacts that are validated for conceptual/contextual consistency and completeness; and (ii) aligning and then merging contextualized sub-ontologies, under the *context-specific axioms*, to generate a new sub-ontology that represents the 'morphed' knowledge artifact as shown in Figure 1. In this way, our knowledge morphing approach pursues highly-specific ontology alignment guided by the

Y. Gao and N. Japkowicz (Eds.): Canadian AI 2009, LNAI 5549, pp. 279–282, 2009.

problem-context–i.e. a single knowledge morphing context (akin to a query) forms the basis of the process. This also means that as the problem-context changes a new morphed knowledge artifact will be developed.

By modeling knowledge artifacts as ontologies, semantic interoperability among knowledge artifacts can be achieved via *ontology reconciliation*. However, ontology reconciliation under different contexts is still an undertaking challenge [3]. The literature suggests other approaches for knowledge integration problem from different perspectives. For instance, the ECOIN framework performs semantic reconciliation of independent data sources, under a defined context, by defining *conversion functions* between contexts as a network. ECOIN takes the *single ontology, multiple views* approach [4], and introduces the notion of *modifiers* to explicitly describe the multiple specializations/views of the concepts used in different data sources. It exploits the modifiers and conversion functions, to enable context mediation between data sources, and reconcile and integrate source schemas with respect to their conceptual specializations. Another recent initiative towards knowledge integration is the OpenKnowledge project [5] that supports the knowledge sharing among different knowledge artifacts, not by sharing their asserted statements, instead by sharing their *interaction models*. An interaction model provides a context in which knowledge can be transmitted between two (or more) knowledge sources (peers).

2 $\mathcal{K}$-$\mathcal{MORPH}$ Architecture

We adopt a Semantic Web (SW) architecture [6] to address the problem of knowledge morphing. Given that at the core of knowledge morphing is the need to semantically model the different knowledge artifacts, we believe that the SW offers a logic-based framework to (a) semantically model the knowledge of various knowledge artifacts found in different modalities as ontologies; (b) semantically annotate the heterogeneous knowledge artifacts based on their respective ontologies; (c) capture and represent the underlying domain concepts, and the semantic relationships that are inherent within a problem-context, in terms of a domain ontology; (d) ensure interoperability between multiple ontologically defined knowledge artifacts; and (e) maintaining changes, evolution and management of ontologies.

In this paper, we present our approach to pursue knowledge morphing. We propose a Semantic Web based Knowledge Morphing framework $\mathcal{K}$-$\mathcal{MORPH}$, which is comprised of the following steps: (a) modeling the domain knowledge as ontologies–each type of knowledge artifact is represented as a distinct ontology, termed as Knowledge Artifact Ontology (KAO), where the knowledge is captured as an instantiation of the respective KAO; (b) defining the problem-context in terms of morphing constructs that serve as the knowledge morphing specification; (c) selecting the knowledge artifacts that are to be used to create the morphed knowledge object; (d) selecting from the candidate knowledge artifact the specific knowledge objects that are pertinent to the problem-context. Since the knowledge artifacts are modeled as KAO the selected problem-specific knowledge objects appear as contextualized sub-ontologies; (e) applying reasoning techniques to synthesize the contextualized sub-ontologies to yield the morphed knowledge. This

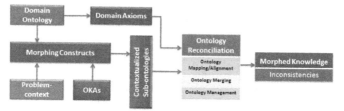

Fig. 1. $\mathcal{K}$-$\mathcal{MORPH}$: Knowledge Morphing Process

is achieved via a context-driven ontology reconciliation method that infers the inherent interactions between different contextualized sub-ontologies using domain-specific axioms; and (f) validation and verification of the morphed knowledge. Figure 1 illustrates the $\mathcal{K}$-$\mathcal{MORPH}$ morphing process with more details available in see [7].

3 Application of $\mathcal{K}$-$\mathcal{MORPH}$ in Clinical Decision-Making

Clinical decision making involves an active interplay between various medical knowledge artifacts to derive pragmatic solutions for a clinical problem [8]. We are currently developing a prototype *Healthcare Knowledge Morpher* (as shown in Figure 2) that deals with the three different medical knowledge artifacts, namely, (i) Electronic Patient Records (EPR), (ii) Clinical Practice Guidelines (CPG), and (iii) Clinical Pathways (CP). Each knowledge artifact, despite targeting the same domain knowledge, has a different purpose. For instance, CPGs are systematically developed disease-specific recommendations to assist clinical decision-making in accordance with the best evidence [2]. CP serve as institution-specific workflows that guide the care process in line with the evidence-based medical knowledge found in CPG. EPR are containers of patient's longitudinal medical information.

For clinical decision making we need an active interplay between these three distinct artifacts as follows: The EPR determines the clinical context that in turn determines which CPG need to be referred to make the 'right' clinical decisions. Based on the context, the morphing construct will determine the clinical intention, the knowledge needs for the given intention and the knowledge resources to be utilized. The two knowledge sources in this case–i.e. the CPG and CP –both now need to be integrated to optimally apply the knowledge for clinical decision making. The CPG will provide the declarative knowledge and it needs to be aligned with the procedural knowledge contained by CP. Knowledge morphing is therefore needed at two levels: (a) morphing the different knowledge components from multiple knowledge artifacts of the same type–i.e. recommendations from multiple CPGs; and (b) morphing different knowledge artifact types–i.e. synthesizing CPG and CP. The morphed knowledge artifact will consist of operational relations between EPR, CPG, and CP knowledge artifacts and serve

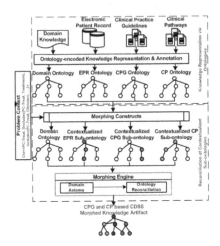

Fig. 2. $\mathcal{K}$-$\mathcal{MORPH}$: Healthcare Knowledge Morphing

as a holistic knowledge artifact to support clinical decision making in terms of (a) evidence-based recommendations based CPG-based knowledge, based on the patient scenario recorded in EPR, and also (b) institution-specific workflow knowledge to pragmatically execute the recommendations.

This research is funded by a grant from Agfa Healthcare (Canada).

References

1. Abidi, S.S.R.: Medical knowledge morphing: Towards the integration of medical knowledge resources. In: 18th IEEE International Symposium on Computer-Based Medical Systems, Dublin, June 23-24 (2005)
2. Abidi, S.S.R., Shayegani, S.: Modeling the form and function of clinical practice guidelines: An ontological model to computerize clinical practice guidelines. In: Knowledge Management for Healthcare Processes Workshop at ECAI 2008, Patras (2008)
3. Shvaiko, P., Euzenat, J.: Ten challenges for ontology matching. In: Meersman, R., Tari, Z. (eds.) OTM 2008. LNCS, vol. 5332, pp. 1164–1182. Springer, Heidelberg (2008)
4. Firat, A., Madnick, S., Grosof, B.: Contextual alignment of ontologies in the eCOIN semantic interoperability framework. Inf. Tech. and Management 8(1), 47–63 (2007)
5. Robertson, D., et al.: Open knowledge: Semantic webs through peer-to peer interaction. Technical Report DIT-06-034, University of Trento, Povo, Italy (May 2006)
6. Berners-Lee, T., Hendler, J., Lassila, O.: The semantic web. Scientific American 284(5), 34–43 (2001)
7. Hussain, S., Abidi, S.S.R.: K-MORPH: A semantic web based knowledge representation and context-driven morphing framework. In: Workshop on Context and Ontologies, at ECAI 2008, Patras, Greece, July 21-25 (2008)
8. Patel, V., Arocha, J., Zhang, J.: Thinking and reasoning in medicine. In: Cambridge Handbook of Thinking and Reasoning. Cambridge University Press, Cambridge (2004)

Background Knowledge Enriched Data Mining for Interactome Analysis

Mikhail Jiline

University of Ottawa
mjiline@site.uottawa.ca

Abstract. In recent years amount of new information generated by biological experiments keeps growing. High-throughput techniques have been developed and now are widely used to screen biological systems at genome wide level. Extracting structured knowledge from amounts of experimental information is a major challenge to bioinformatics. In this work we propose a novel approach to analyze protein interactome data. The main goal of our research is to provide a biologically meaningful explanation for the phenomena captured by high-throughput screens. We propose to reformulate several interactome analysis problems as classification problems. Consequently, we develop a transparent classification model which while perhaps sacrificing some accuracy, minimizes the amount of routine, trivial and inconsequential reasoning that must be done by a human expert. The key to designing a transparent classification model that can be easily understood by a human expert is the use of the Inductive Logic Programming approach coupled with significant involvement of background knowledge into the classification process.

1 Introduction

In recent years the amount of available information in biological datasets keeps growing. New high-throughput experimental techniques have been developed and now are widely accepted and used. We should not forget as well about already accumulated datasets such as genome sequences, protein sequences, protein structural and 3D information, protein motifs and literature which represents curated but not well structured knowledge.

At the same time it is becoming increasingly clear that no significant biological function can be attributed to one molecule. Each function is a result of the complex interaction between all types of biological molecules such as DNA, RNA, proteins, enzymes and signal molecules.

Given all these circumstances, the reductionist approach, which served very well the scientific community for ages, may not be used so productively in modern molecular biology. While it was used successfully to identify and describe the molecular machinery, there is simply no way the reductionism can help us understand how the system properties emerge from the interactions of myriad of relatively basic elements.

The Systems Biology approach is quite opposite to the reductionism, of the traditional science. Instead of moving from complex behavior/system to the simpler ones, Systems Biology tries to reconstruct and understand the complex behavior by observing

Y. Gao and N. Japkowicz (Eds.): Canadian AI 2009, LNAI 5549, pp. 283–286, 2009.

how the elementary pieces of the system operate and interact with each other. Speaking in terms of molecular biology, Systems Biology observes and studies individual genes, proteins, enzymes and metabolites trying to infer metabolic pathways, and then in turn find more complex interactions between pathways, regulatory mechanisms, mechanisms of the cell specialization and interaction.

Therefore, one of the main challenges of Systems Biology is the development of methods and techniques that can be used to infer knowledge from the accumulated datasets.

Due to numerous technical difficulties such as sheer volume of data (dozens of thousands of genes, hundreds of thousands of interactions), large amount of noise, heterogeneity and incompleteness of data, a systematic study of biological dataset is not an easy task.

2 Explanatory Models

A wide range of approaches has been proposed to address the data mining needs of Systems Biology: from pure statistical methods to graph-theoretical methods, to classical data mining algorithms. Different studies concentrated on a range of different goals: from increasing the reliability of experimental data to prediction of new interactions, to prediction of protein functions.

However, there seems to be lack of studies trying to develop logical explanatory models of the observed data. While conventional classification models are good at providing additional information about the data (assigning labels to objects of interest, clustering objects, etc.), they are not designed to provide insights into the essence and the structure of the observed data.

Explanatory models, on the other hand, may be more difficult to build. They may be less precise than dedicated classification models. However, they are capable of uncovering and presenting to a human domain expert information about the fundamental nature of observed data.

As an illustration of the difference between traditional 'result'-oriented and explanatory models we may consider a clustering problem. A result-oriented model provides a human researcher with the best set of clusters according to given criteria (some sort of a compactness measure). After that, it's up to the expert to find the underlying reason for the clustering. Meanwhile, with an explanatory model, such reasoning will be an integral part of each cluster definition.

Given the growing amount and the diversity of data and knowledge in Systems Biology, we think explanatory models will be an indispensable tool. They have potential to improve the efficiency of human experts allowing to speed up the data processing, to increase the amount of processed data, and eventually to keep the pace of research in Systems Biology.

3 Inductive Logic Programming

The computational foundation of Background Knowledge Enriched Data Mining lays in the Inductive Logic Programming approach. Inductive Logic Programming (ILP) is a subfield of Machine Learning which works with data represented as first order logic statements. Moreover, not only are the positive examples and negative examples logic formulas, but the background knowledge, and hypotheses are first order logic formulas as well.

Inductive Logic Programming has a history dating back to 1970 when Plotkin defined such fundamental to ILP notions as θ-subsumption and Least General Generalization (LGG). LGG allowed him to formulate an algorithm performing induction on a set of first order statements representing the data.

Since then, a lot of researchers have improved ILP algorithms[1], especially from the efficiency point of view. New directions such as stochastic search, incorporation of explicit probabilities, parallel execution, special purpose reasoners, and human-computer collaboration systems are under active research.

4 Background Knowledge Enriched Data Mining

We propose to apply a novel approach – Background Knowledge Enriched Data Mining - to help biologists and other live science specialists analyze and understand accumulated experimental data together with the existing structured domain specific knowledge. Automatic reasoning minimizes the amount of routine, trivial and inconsequential reasoning that must be done by a human expert.

Inductive logic programming algorithms may be used to perform an automatic dataset analysis as well as to facilitate a semi-automatic knowledge discovery process [2,3].

We see as significant advantages of the inductive logic programming approach

- the ease of integration of accumulated domain specific knowledge,
- the ease of representation of heterogeneous data sources,
- the high expressiveness of the logic,
- the ease of interpretation for obtained results and hypotheses.

A combination of such advantages allows us to try to attack the problem of finding an explanation for the observed phenomena. Contrarily to most approaches that value the final accuracy of the built classification system, we value the structured knowledge discovered by the algorithm as well.

Background Knowledge Enriched Data Mining can be applied to such problems as

- a protein functional module analysis;
- the discovery of biologically meaningful labels for clusters;
- a verification of the experiment integrity.

It should be stressed that the main goal of our approach is not the production of new data (such as protein function prediction), at least in the bare form. Our approach is geared towards producing hypotheses and links between existing data, hopefully, helping to explain the observed phenomena, pointing to subtle inconsistencies and trouble spots in the data, showing ways to generalize the data, etc.

From a methodological point of view, our approach follows typical data mining steps outlined below:

1. Formulate a biological problem as a classification task.
2. Collect data to be analyzed.

3. Collect relevant background knowledge.
4. Extract features from the original dataset and the background knowledge.
5. Learn an ILP classifier.
6. Interpret the classifier structure.

While the sequence of the steps seems straightforward, all of them indeed contain uncertainties and require further research. During the initial research project we identified the areas critical for the success of our approach: a background knowledge selection; a background knowledge representation; an integration of specialized algorithms; a search optimization, and the development of the evaluation metrics for explanatory models.

We are planning to address the background knowledge selection and representation problems by selecting/developing a feature selection algorithm best suited for the types of data specific for the domain of Systems Biology.

The integration of specialized algorithms problem will be handled via a generalization of the lazy evaluation approach to the ILP. The lazy evaluation approach was originally proposed to handle numerical reasoning in an ILP system. We extended this approach to allow any specialized algorithm to work inside the ILP system. By further developing this approach we can even think of the ILP as the universal glue allowing many specialized algorithms to work cooperatively in one learning system.

The search optimization is one of the most difficult tasks to handle when one is working with ILP algorithms. We would like to develop search heuristics involving metrics specific for the data of Systems Biology. We will introduce heuristics sensitive to the heterogeneous noise in data (a data-type specific probability of error in the background knowledge). We believe that such heuristics are extremely helpful when one has to combine data from very unreliable experimental datasets with highly accurate curated datasets and analyze them in one learning system.

Another contribution of this work is the development of evaluation metrics for explanatory models based on the idea of mapping ILP models to the natural language then mining biological literature to find the correlation between discovered ILP models and the human's natural way of describing the phenomena.

References

1. Muggleton, S.: Inverse Entailment and Progol. New Gen. Comput. 13, 245–286 (1995)
2. Clare, A., King, R.D.: Data Mining the Yeast Genome in a Lazy Functional Language. In: Practical Aspects of Declarative Languages: 5th International Symposium, pp. 19–36 (2003)
3. Sternberg, M.J.E., King, R.D., Lewis, R.A., Muggleton, S.: Application of Machine Learning to Structural Molecular Biology. Biological Sciences 344(1310), 365–371 (1994)

Modeling and Inference with Relational Dynamic Bayesian Networks

Cristina Manfredotti

DISCo Comp. Science Dept, University of Milano-Bicocca,
Viale Sarca, 336, 20126 Milano, Italy
Dept. Computer Science, University of Toronto,
10 King's College rd. M5S3G4 Toronto, ON, Canada
cristina.manfredotti@gmail.com

Abstract. The explicit recognition of the relationships between interacting objects can improve the understanding of their dynamic domain. In this work, we investigate the use of Relational Dynamic Bayesian Networks to represent the dependencies between the agents' behaviors in the context of multi-agents tracking. We propose a new formulation of the transition model that accommodates for relations and we extend the Particle Filter algorithm in order to directly track relations between the agents.

Many applications can benefit from this work, including terrorist activities recognition, traffic monitoring, strategic analysis and sports.

1 Introduction, Motivations

Understanding the dynamic behaviors of many objects interacting with each other is a particularly challenging problem. The explicit recognition of relationships between targets can improve our understanding of the whole system. This is the case in *Multi-target tracking*, the problem of finding the tracks of an unknown number of moving objects from noisy observations.

Representing and reasoning about relations can be useful in two ways:

- *Relations can improve the efficiency of the tracking.* The information contained in the relationships can improve the prediction, resulting in a better final estimation of the behavior. Also the *data association* task can become more precise, reducing the possibilities of errors.
- *Monitoring relations can be a goal in itself.* This is the case in many activities like traffic prediction or consumer monitoring, anomaly detection or terrorist activity recognition.

With these two aims in mind we considered *Relational Dynamic Bayesian Networks* (RDBNs) (that can be seen as an extension of *Probabilistic Relational Model* [2] to dynamic domain) as a formalism to monitor relations between moving objects in the domain.[1]

---

[1] A literary review cannot be included because of the limited space. We just mentioned that our approach is different from the work of Milch ans Russel [3], where the concept of class is used to develop an inference system without an explicit representation of relationships.

Y. Gao and N. Japkowicz (Eds.): Canadian AI 2009, LNAI 5549, pp. 287–290, 2009.

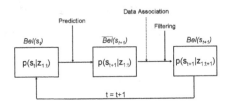

Fig. 1. Graphical sketch of the Bayes filter iteration

In our RDBN-based model, relationships are considered as variables whose values may change over time, and therefore, while tracking the objects in the domain, we can also track the evolution of their relationships. For this purpose, in the next section we propose a formalization of a new dynamic model and a version of Particle Filter that has been adapted to this setting. We conclude with some final remarks and considerations on future works.

2   Modeling and Inference

In order to make inference in a multi-agents setting, we extend the algorithms traditionally used in tracking to represent relations. As in classic tracking, the aim is to estimate the current posterior distribution of the state space $p(s_t|z_{1:t})$ conditioned on the sequence of observations $z_{1:t}$ up to time t. This distribution is often called the *target's belief.*

The tracker *predicts* the probability distribution of the future state s_{t+1}, given the knowledge about the current state s_t, by means of a *state transition model* $p(s_{t+1}|s_t)$ (Figure (1) gives a sketch of the filter iteration). Once measurements about the state at time $t+1$ (z_{t+1}) are acquired, each of them is associated to the most probable prediction (this is the non-trivial task of *data association* step) and *filtered* using the *measurement model* $p(z_{t+1}|s_{t+1})$ that relates (potentially noisy) measurements to the state.

In order to define our RDBN model we introduce the following components:

The state of the system s is divided in two parts: the *state of the objects* s^o and the *state of the relations* s^r; so we can write: $s = [s^o, s^r]$. We assume the relations to be *unobservable*[2].

The relational transitional model $p(s_t|s_{t-1}) = p(s^o_t, s^r_t|s^o_{t-1}, s^r_{t-1})$ is a joint probability of the state of all instances and relations (see Figure (2)). We assume that the state of relations are not directly affected by the state of the objects at the previous time step. This assumption simplifies the transition model without loosing in generality. Indeed, information about the previous state of the objects are included in the respective state of relations which

---

[2] Assume that we want to decide whether two vehicles are traveling together; while we can observe their individual position at a given time step (part of s^o), to discriminate whether the two objects are accidentally close to each other we need to make inference about the whole objects histories.

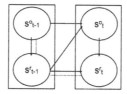

Fig. 2. Relational Transition Model

influences relations at the current time step (see the dashed path in Figure (2)). Therefore the transition model can be rewritten as:

$$p(s_t^o, s_t^r | s_{t-1}^o) = p(s_t^o | s_{t-1}^o, s_{t-1}^r) p(s_t^r | s_{t-1}^r, s_t^o). \tag{1}$$

The measurement model $p(z_t | s_t)$ gives the probability of the measurements obtained at time t given the state at the same time. Since the part of the state relative to relations, s^r, is not directly measurable, the observation z_t is independent of the relations between objects: $p(z_t | s_t^o, s_t^r) = p(z_t | s_t^o)$.

Under the Markov assumption, the most general algorithm for computing the belief of the combined state is given by the *Bayesian filter algorithm*:

$$Bel(s_t) = \alpha \, p(z_t | s_t^o) \int p(s_t^o, s_t^r | s_{t-1}^o, s_{t-1}^r) Bel(s_{t-1}) ds_{t-1} \tag{2}$$

where α is a normalization constant. The prediction $\overline{Bel}(s_t)$, according to the state transition model, can be written as:

$$\overline{Bel}(s_t) = p(s_t^o, s_t^r | z_{1:t-1}) = \int p(s_t^o | s_{t-1}^o, s_{t-1}^r) p(s_t^r | s_{t-1}^r, s_t^o) Bel(s_{t-1}) ds_{t-1}. \tag{3}$$

In the most general case we can represent the two partial transition models of Equation (1) by a First Order Probabilistic Tree (FOPT). A FOPT is a Probabilistic Tree whose nodes are First Order Logic formulas. There exist particular situations in which the exponential complexity of FOPTs might be a burden; in those situations other formalism, as rule-based systems, might be a better choice.

2.1 Relational Particle Filter

The complex nature of the setting makes impossible to use filters that require a probabilistic function in closed form, as Kalman filter. To solve this issue we developed a Particle Filter (PF) algorithm [1], whose properties are appealing for our case. In a traditional PF the state of different targets would evolve separately, therefore a traditional PF would not account for correlated behaviors. To deal with the complexity of our state we integrate the relational transition model introduced in Equation (1) with a new *relational PF* (Algorithm (1)) that is able to do inference on both the state of the objects and the state of the relations of our domain.

Algorithm 1. Pseudo Code for the Relational Particle Filter algorithm

$Bel(s_t) = RPF(Bel(s_{t-1}), z_t)$

for all $m = 1 : M$ **do**

$x^o_{t,(m)} \sim p(s^o_t | s^o_{t-1}, s^r_{t-1})$; hypothesis for the state of the objects

$x^r_{t,(m)} \sim p(s^r_t | s^r_{t-1}, s^o_t = x^o_{t,(m)})$; hypothesis for the state of relations

$\omega_{(m)} = p(z_t | x^o_{t,(m)})$; weights computation

for all $m = 1 : M$ **do**

$\widetilde{\omega}_{(m)} = \frac{\omega_{(m)}}{\sum_{m=1}^{M} \omega_{(m)}}$; weights normalization

Resample $Bel(s_t)$ from $\{[x^o_{t,(m)}, x^r_{t,(m)}]\}$ according to weights $\{\widetilde{\omega}_{(m)}\}$ with repetition.

3   Applications and Future Works

There are a number of possible applications of this approach that involve the need for inference from sensed data. A first assessment of our relational tracker was performed on a simulation environment representing cars on a highway. As the behavior of a driver influences the driving of other cars nearby, our relational tracker is able to propagate information and generate more correct tracks than a traditional tracker.

The approach needs to be validated on real situations and on different domains. Particular attention will be given to the activity recognition domain. An interesting research question is the evaluation of different representation models (FOPTs, rules, etc.) for the transition probability, and their practical efficiency. Also the development of efficient techniques for the automatic learning of priors in relational domains is a challenging research topic.

References

1. Arulampalam, S., Maskell, S., Gordon, N.: A tutorial on particle filters for online nonlinear/non-gaussian bayesian tracking. IEEE Transactions on Signal Processing 50, 174–188 (2002)
2. Friedman, N., Getoor, L., Koller, D., Pfeffer, A.: Learning probabilistic relational models. In: IJCAI, pp. 1300–1309 (1999)
3. Milch, B., Russell, S.J.: General-purpose mcmc inference over relational structures. In: UAI (2006)

A Semi-supervised Approach to Bengali-English Phrase-Based Statistical Machine Translation

Maxim Roy

Computing Science, Simon Fraser University,
Burnaby, BC, V5A1S6, Canada
maximr@cs.sfu.ca

Abstract. Large amounts of bilingual data and monolingual data in the target language are usually used to train statistical machine translation systems. In this paper we propose several semi-supervised techniques within a Bengali English Phrase-based Statistical Machine Translation (SMT) System in order to improve translation quality. We conduct experiments on a Bengali-English dataset and our initial experimental results show improvement in translation quality.

1 Introduction

We explore different semi-supervised approaches within a Bengali English Phrase-based SMT System. Semi-supervised learning refers to the use of both labeled and unlabeled data for training in SMT. Semi-supervised learning techniques can be applied to SMT when a large amount of bilingual parallel data is not available for language pairs. Sarkar et al. [1] explore the use of semi-supervised model adaptation methods for the effective use of monolingual data from the source language in order to improve translation accuracy.

Self-training is a commonly used technique for semi-supervised learning. In self-training a classifier is first trained with a small amount of labeled data. The classifier is then used to classify the unlabeled data. Typically the most confident unlabeled points, together with their predicted labels, are added to the training set. The classifier is re-trained and the procedure repeated. Note the classifier uses its own predictions to teach itself. The procedure is also called self-teaching or bootstrapping.

Since a sufficient amount of bilingual parallel data between Bengali and English for SMT is not publicly available, we are exploring the use of semi-supervised techniques like self-training in SMT. We have access to approximately eleven thousand parallel sentences between Bengali and English provided by the Linguistic Data Consortium (LDC) Corpus Catalog[1]. The LDC corpus contains newswire text from the BBC Asian Network website and other South Asian language websites (eg. Bengalnet). We also have a large monolingual Bengali dataset which contains more than one million sentences. The monolingual corpus was provided by the Center for Research on Bangla Language Processing, BRAC University,

---

[1] LDC Catalog No.: LDC2008E29.

Y. Gao and N. Japkowicz (Eds.): Canadian AI 2009, LNAI 5549, pp. 291–294, 2009.

Bangladesh. The corpus was built by collecting text from the Prothom Alo newspaper website and contains all the news available for the year of 2005 (from 1st January to 31st December) - including magazines and periodicals. There are 18,067,470 word tokens and 386,639 distinct word types in this corpus.

2 Our Approach

We are proposing several self-training techniques to effectively use this large monolingual corpus (from the source language in our experiments) in order to improve translation accuracy. We propose several sentence selection strategies to select sentences from a large monolingual Bengali corpus, which are briefly discussed below along with the baseline system where sentences are chosen randomly. In our baseline system the initial MT system is trained on the bilingual corpus L and we randomly select k sentences from a large monolingual corpus U. We translate these randomly selected sentences with our initial MT system $M_{B \to E}$ and denote these sentences along with their translation as U^+. Then we retrain the SMT system on $L \cup U^+$ and use the resulting model to decode the test set. We also remove these k randomly selected sentences from U. This process is continued iteratively until a certain level of translation quality, which in our case is measured by the BLEU score, is met. Below in algorithm-1, we describe the baseline algorithm.

Algorithm 1. Baseline Algorithm Semi-supervised SMT
1: Given bilingual corpus L, and monolingual corpus U.
2: $M_{B \to E} = \textbf{train}(L, \emptyset)$
3: **for** $t = 1, 2, ...$ **do**
4:     Randomly select k sentence pairs from U
5:     $U^+ = \textbf{translate}(k, M_{B \to E})$
6:     $M_{B \to E} = \textbf{train}(L, U^+)$
7:     Remove the k sentences from U
8:     Monitor the performance on the test set T
9: **end for**

Our first sentence selection approach uses the reverse translation model to rank all sentences in the monolingual corpus U based on their BLEU score and only select sentences above a certain BLEU score. Mainly we want to select those sentences from the monolingual corpus U for which our MT system can generate good translations. In order to obtain a BLEU score for sentences in the monolingual corpus U we used the reverse translation model. While a translation system $M_{B \to E}$ is built from language B to language E, we also build a translation system in the reverse direction $M_{E \to B}$. To measure the BLEU score of all monolingual sentences B from monolingual corpus U, we translate them to English sentences E by $M_{B \to E}$ and then project the translation back to Bengali using $M_{E \to B}$. We denote this reconstructed version of the original Bengali sentences by $\hat{B}$. We then use B as the reference translation to obtain the BLEU score for sentences $\acute{B}$. Below in algorithm-2, we describe the reverse model sentence selection algorithm.

Algorithm 2. Reverse Model sentence selection Algorithm

1: Given bilingual corpus L, and monolingual corpus U.
2: $M_{B \to E} = \textbf{train}(L, \emptyset)$
3: $M_{E \to B} = \textbf{train}(L, \emptyset)$
4: **for** $t = 1, 2, \ldots$ **do**
5:   $U^+ : (B, E) = \textbf{translate}(U, M_{B \to E})$
6:   $U^* : (\acute{B}, E) = \textbf{translate}(E, M_{E \to B})$
7:   Use B and $\acute{B}$ to rank all sentences in U^+ based on the BLEU score
8:   Select k sentence and their translation $\acute{k}$ from ranked U^+
9:   $M_{B \to E} = \textbf{train}(L, k \cup \acute{k})$
10:   $M_{E \to B} = \textbf{train}(L, k \cup \acute{k})$
11:   Remove the k sentences from U
12:   Monitor the performance on the test set T
13: **end for**

Our next sentence selection approach uses statistics from the training corpus L for sentence selection from the monolingual corpus U. In this approach we first find most frequent words in the training corpus L. We call them seed words. Then we filter the seed words based on their confidence score which is how confidently we can predict their translation. Seed words with confidence scores lower than certain threshold values are removed. Then we use these remaining seed words to select sentences from monolingual corpus U and remove selected sentences from monolingual corpus U. Next we look for the most frequent words other than seed words in the selected sentences to be used for the next iteration as new seed words. We translate these selected sentences and add them to the training corpus L. After that we re-train the system with the new training data. In the next iteration we select new sentences from the monolingual corpus U using the new seed words and repeat the steps. We keep on repeating the steps until no more seed words are available. In each iteration we monitor the performance on the test set T. Below in algorithm-3, we describe the frequent word sentence selection algorithm.

3   Experiments

The SMT system we used in our experiments is PORTAGE [2]. The models (or features) which are employed by the decoder are: (a) one or several phrase table(s), which model the translation direction $P(f|e)$, (b) one or several n-gram language model(s) trained with the SRILM toolkit [3]; (c) a distortion model which assigns a penalty based on the number of source words which are skipped when generating a new target phrase, and (d) a word penalty. These different models are combined log linearly. Their weights are optimized with respect to BLEU score using the algorithm described in [4].

Algorithm 3. Frequent word sentence selection Algorithm
1: Given bilingual corpus L, and monolingual corpus U.
2: $M_{B \to E} = \textbf{train}(L, \emptyset)$
3: $S = \textbf{select seed}(L)$
4: $S^+ = \textbf{compute confidence score and filter}(S)$
5: **for** $t = 1, 2, \dots$ till convergences **do**
6:    Select k sentences from U based on S^+
7:    $U^+ = \textbf{translate}(k, M_{B \to E})$
8:    $S = \textbf{select seed}(U^+)$
9:    $S^+ = \textbf{compute confidence score and filter}(S)$
10:   Remove the k sentences from U
11:   $M_{B \to E} = \textbf{train}(L, U^+)$
12:   Monitor the performance on the test set T
13: **end for**

Our initial experimental results look promising and all sentence selection approaches perform better than the baseline random sentence selection strategy. Currently we are conducting full-scale experiments and looking into other sentence selection strategies (e.g. combining the reverse model and frequent word sentence selection model) and other approaches that can be applied in a semi-supervised setting. In addition to the Bengali and English bilingual corpus, we are in the process of testing our approaches on English-French langauge pair for which there are several large parallel corpora available.

4   Conclusion

Since there are not enough bilingual corpora available between Bengali and English, it is hard to achieve higher accuracy using fully SMT systems. We are currently in the process of building a large Bengali-English parallel corpus, which will help improve translation accuracy. Also we only use one reference translation to evaluate the performance of our approaches. Since unavailability of multiple reference translations can have impact on the BLEU score we are in the process of creating a better test set with multiple reference translations.

References

1. Sarkar, A., Haffari, G., Ueffing, N.: Transductive learning for statistical machine translation. In: Proc. ACL (2007)
2. Ueffing, N., Simard, M., Larkin, S., Johnson, J.H.: NRC's Portage system for WMT 2007. In: Proc. ACL Workshop on SMT (2007)
3. Stolcke, A.: SRILM - an extensible language modeling toolkit. In: Proc. ICSLP (2002)
4. Och, F.J.: Minimum error rate training in statistical machine translation. In: Proc. ACL (2003)

Author Index